Geographical Information Systems (GIS)
and Computer Cartography

Geographical Information Systems and Computer Cartography

by
CHRISTOPHER B. JONES

LONGMAN

Addison Wesley Longman Limited
Edinburgh Gate, Harlow
Essex CM20 2JE
England

and Associated Companies throughout the World

First published 1997

ISBN 0 582 04439 1

British Library Cataloguing-in-Publication Data
A catalogue record for this book is available from the British Library.

Library of Congress Cataloging-in-Publication Data
A catalog entry for this title is
available from the Library of Congress.

Set by 30 in 9/11 Times

Produced by Longman Singapore Publishers (Pte) Ltd.
Printed in Singapore

To my parents David and Nora

Contents

Preface

The last few years have seen a proliferation of geographical information systems (GIS), employed by a wide range of users. The growing interest in GIS reflects the fact that so much of the information needed for management and decision making in government and commerce is spatially referenced, yet conventional information technology developed for commercial purposes is poorly suited to answering apparently simple queries framed in space and in time. GIS technology is still very much in flux with many of its users finding their systems harder to operate and often less well adapted to their needs than anticipated. A characteristic of the technology is the fact that, in order to make effective use of it, there is usually a need to become familiar with concepts that are peculiar to spatial data processing. This book was written with the intention of demystifying these concepts and providing a sound appreciation of the technology on which GIS is based.

The emphasis here is on understanding principles and techniques for representing, processing and visualising spatial data in GIS. The book provides a brief summary of applications of GIS and describes many of the techniques for data analysis, retrieval and graphic display that are to be found in commercial GIS. In focusing on the technical foundations of spatial data processing in GIS and cartography, the book may be regarded as complementing existing texts in which the focus is on interpretation and statistical analysis of data in the environmental and social sciences.

The intended audience for the book is primarily that of students on undergraduate and masters-level programmes that include GIS or cartography. It should also however be of considerable value to professionals using and implementing GIS in government, industry and commerce. No prior knowledge of GIS is assumed and there is no assumption of any specialised computing or mathematical background. There is a progression from the introductory material to more advanced topics, and as such the book should be of interest to those with some experience of GIS or computer science who wish to gain more specialised knowledge of techniques in representation and visualisation of spatial data.

It is anticipated that for teaching purposes the book will be used in combination with practical classes which provide experience of using a GIS package to apply the methods described here. In the UK, support for such practical work can be found in the practical exercises produced at the Midlands Regional Research Laboratory, at the University of Leicester under the project ASSIST. These are largely based on the ARC/INFO and IDRISI packages. In the USA, practical exercises are provided with the NCGIA Core Curriculum in GIS.

The book consists of sixteen chapters that are grouped into five parts as follows:

1. Introduction
2. Acquisition and Preprocessing of Geo-referenced Data
3. Data Storage and Retrieval
4. Spatial Data Modelling and Analysis
5. Graphics and Cartography

The Introduction consists of three chapters that together give an overview of the subject matter. The first chapter reviews the relationships of GIS and cartography to each other and to other disciplines, and reviews applications of the technology in a representative set of fields. Chapter 2 introduces concepts of spatial information and the spatial models used in GIS. The third chapter summarises functional capability of typical GIS with the stress on using spatial data for a range of types of geographically-referenced analyses.

Part 2 focuses on the characteristics of geo-referenced data and the processes required to obtain data and make it usable in a GIS. The term geo-referenced refers here to the linking of data to locations on the Earth's surface, typically by means of geographical and map coordinates. Chapter 4 explains the basics of coordinate systems used for referencing spatial data, and the transformations that are applied for map making. Chapter 5 describes techniques for digitising existing maps, while Chapter 6 reviews primary spatial data acquisition techniques with particular attention to the subject of remote sensing. This subject is reviewed here, but it is a major topic in its own right which is covered very much more thoroughly in a text such as that of Lillesand and Kiefer (1994). This is followed in Chapter 7 with a discussion of the issue of data quality and of standards for exchanging data between systems and organisations. Part 2 concludes in Chapter 8 with a more advanced coverage of techniques for transforming between vector and raster data, which are the two most commonly employed representations of spatial data (introduced in Chapter 2).

Part 3 deals with the subject of storage and retrieval of spatial data. Chapter 9 introduces some basic concepts for computer data storage of spatial data, including techniques for data compression. The second half of the chapter describes some conventional data structures and methods for file indexing, and relates them to spatial data. Chapter 10 addresses the main principles underlying database technology and indicates how existing DBMS can be used for spatial data, as well as highlighting some of their limitations. Chapter 11 builds upon the topics introduced in Chapter 9 and is of interest particularly to those concerned with implementation and design of spatial databases. It describes spatial indexing techniques and the computational geometry procedures that are essential to processing queries on geometric data.

Part 4 develops the themes of spatial data modelling and analysis that were introduced in Chapter 3. Chapter 12 deals with a range of widely used methods for modelling surfaces from point, line and area samples. In particular, the problem of interpolating between observations is addressed, with an introduction to the statistical concepts of autocorrelation. Chapter 13 describes a variety of techniques for using spatial data in decision support. The techniques are concerned particularly with identifying optimal locations taking account of multiple criteria, and with finding paths through networks and studying interactions in space.

Part 5, on Graphics and Cartography, is composed of three chapters. Chapter 14 summarises computer graphics functionality for plotting map symbols and text and gives an overview of graphics hardware and software that may be used in implementing GIS. Chapter 15 introduces principles of cartographic design and communication and reviews methods for display of two-dimensional and three-dimensional data. Chapter 16 focuses on the subject of map generalisation, which concerns the derivation of small-scale maps from more detailed representations. This is a major challenge to automation and the chapter is intended to help raise awareness of important issues and provide some directions for future work.

Regarding the use of this text in teaching, coverage of all material presented here would be expected to extend over at least two semesters. The organisation of the book is intended to encourage selective use of material. Part 1 has been designed with a view to providing a general overview of GIS and might be presented in half of a one semester module, perhaps in combination with some practical exercises. A single introductory GIS module might be based on Chapters 1 to 7 with selected material drawn from subsequent chapters such as 9 and 10, 13, 14 and 15. More specialised teaching modules may be expected to make more use of Chapters 8 to 16. Thus a module focusing on visualisation and cartography might use all of Chapters 14, 15 and 16. A module with emphasis on spatial analytical methods could be expected to use most of Chapters 12 and 13 and possibly Chapter 15, while a module with a bias towards computational methods could include Chapters 8, 11, 14 and 16.

The author has received help in producing this book from many quarters. Mike Goodchild provided very useful comments on early versions of many of the chapters. Several individuals have helped in providing illustrative material or in reading and commenting on parts of the book. In particular thanks are due to David Kidner, Paul Beynon-Davies, Mark Ware, Geraint Bundy, John McBride, Tjark van Heuvel (Directorate General for Public Works and Water Management, Netherlands), Mick Lee (British Geological Survey), Brigitte Husen (University of Hannover), Michael Barnsley, David Martin, and Daniel Dorling. Several organisations have assisted by providing illustrative material. These include Laser-Scan, ESRI, ERDAS, Sokkia, Numonics, National Remote Sensing Centre, Leica, Zeiss, Trimble, PAFEC, Welsh Water (Hyder), Brecon Beacons National Park, Grintec and Calcomp.

Thanks are also due to the University of Glamorgan for use of their facilities in the course of writing the book.

Finally, I would like to thank Anne for her tolerance of the intolerable.

Chris Jones
Abergavenny

Acknowledgements

We are grateful to the following for permission to reproduce copyright material:

The Directoraat-Generaal Rijkswaterstaat for plates 4, 5 and 13; Sokkia for figure 6.3; PAFEC Ltd. for plate 3; Laserscan for plate 6, figure 5.1, 5.8 and 14.17; Dr B. Husen and The Institute of Cartography at the University of Hannover for figure 16.8; ERDAS (UK) Ltd for figure 15.11; Carl Zeiss Ltd for figure 6.1; The Natural Environment Research Council (NERC) for figures 6.5 and 6.23; Ordnance Survey for figures 1.3, 1.5, 1.6, 1.7, 1.8, 12.2, 12.4, 12.5, 12.6, 12.7, 16.1, 16.11, 16.14 and 16.15; Brecon Beacons National Park for figure 1.8; ESRI (UK) Ltd for figure 3.9 and plate 1; Reed Books for figure 4.18; ACSM for plate 15, figure 15.5, 15.15; The Director, British Geological Survey for plate 16 ©NERC. All rights reserved; The Academic Press, Michael F Goodchild and Yang Shiren for figure 4.20; John Wiley & Sons Ltd for figures 6.6, 6.14, 6.15, 6.19, 6.20, 6.22, 6.24, 6.25, 6.26 and 6.27 from **Lillesand** *Remote Sensing and Image Interpretation (3rd edn)* © 1994 John Wiley & Sons Ltd; Lars Schylberg for figure 16.13; the Office for Official Publications of the EC for Table 6.1; The Cartographic Department of the Automobile Association for figures 1.4, 15.3, 15.10 and plate 14; the University of Wisconsin Cartography Lab for figure 4.14; CalComp Ltd for figure 5.8; the National Remote Sensing Centre (NRSC) for figures 6.16, 6.17, plates 9, 10, 11 and 12; Numonics for figure 14.6 and Keith Clarke for figure 16.12.

Whilst every effort has been made to trace owners of copyright material, in a few cases this has proved impossible and so we would like to offer our apologies to any copyright holders whose rights we may have unwittingly infringed.

PART 1 Introduction

CHAPTER 1

Origins and applications

Introduction

The need to place information in a geographical context pervades many aspects of human activity. In public and commercial organisations, many of these activities are concerned with the recording and planning of the human-made environment, with monitoring and managing the natural environment, with transport and navigation, and with understanding social structures. It is an inevitable consequence of the revolution in information technology that we should attempt to build computing systems to handle this geographical information. The results of these technological efforts are reflected in the fields of geographical information systems (GIS) and computer cartography which are the subject of this book.

When compared with the development of computing systems for maintaining commercial and financial information, progress in the field of geographical information systems has been remarkably slow. One of the earliest clearly identifiable geographical information systems is the Canada Geographic Information System (CGIS), which was developed for planning purposes (Tomlinson, 1985). Although the system can be regarded as having laid the foundations, in the mid 1960s, for many subsequent GIS, it was not in fact followed by a proliferation of similar systems. It was only in the late 1980s that we saw the introduction of proprietary GIS which could claim to meet a significant proportion of the data-handling requirements of organisations concerned with geographical information. Examples of organisations in which these

requirements arose include environmental mapping agencies, local and regional government administrations, marketing companies, mineral exploration companies, the military, and utility companies supplying water, electricity, gas and telecommunications.

The relatively late introduction of commercially marketed GIS technology may be explained, to some extent, by the fact that the type of information to be stored in these systems is more complex, and more difficult to process at a basic level, than that found in conventional business information systems. The reasons why geographical data processing is more complex than commercial data processing relate both to the nature of geographical information itself and to the type of retrieval and analysis operations performed upon it. Geographical information is typically concerned with spatially referenced and interconnected phenomena, such as towns, roads and administrative areas, as well as less precisely defined regions with environmental attributes, such as woodlands and marshes. Physical structures and locations are defined by geometric data consisting of combinations of points, lines, areas, surfaces and volumes, in association with classifications and statistical data that attach real-world meaning. These collections of data must be treated in a manner which retains the integrity of the whole objects to which they refer, at different levels of abstraction, rather than as isolated pieces of data. Enquiries on geographical information frequently require some form of spatial search or analysis to be performed on individual regions or on combinations of particular phenomena. Such procedures often require quite sophisticated geometric procedures for manipulation and transformation.

In contrast, commercial data processing can, in general, be reduced to sets of comparative operations on the names or identifiers of, for example, personnel or goods, and to arithmetic operations on attribute values, such as salary and price, which are associated with them. These operations involve less complex algorithms than those required for spatial data. It could also be argued that the development of information technology was initially in response to non-spatial data-processing problems and as such it has been adapted to those requirements. GIS may be seen as one of a number of classes of information processing that require additional layers of special-purpose procedures.

Computer systems for storing and retrieving geographical data are now at a relatively advanced stage of development, but it is still a rapidly developing field and many problems remain to be solved if these systems are to meet all the requirements of spatial analysis and decision-making. Because many organisations need to access a mix of data relating to technical, commercial and human resource issues, a measure of the effectiveness of GIS technology in the future may be the extent to which it becomes absorbed within the information infrastructure and hence disappears as an information processing system in its own right!

In the remainder of this chapter we examine the relationship between GIS and computer cartography, before placing GIS in the context of a range of allied disciplines that have contributed to its development. We then go on to give an overview of the main application areas of GIS and computer cartography. This chapter is intended to provide an appreciation of the uses of GIS before introducing the generic characteristics of geographical information and spatial data models in Chapter 2 and the typical functional capabilities of GIS in Chapter 3.

Cartography and GIS

The fact that geographical information is spatially referenced means that it is associated, at least conceptually, with the field of cartography, as the traditional method of recording the location of spatial phenomena and the relationships between them. The application of computing technology to geographical information handling impacts therefore upon the discipline of cartography. Historically, the development of GIS may be seen to have paralleled efforts to automate cartographic production methods. The growth in the application of GIS technology is now so great however that, to some, cartography appears to be becoming subsumed within the field of GIS. This viewpoint may be understood if we see that the traditional role of cartography has combined the function of helping us understand spatial relationships with that of providing a database recording the form of the earth's surface and the objects located upon it. The introduction of GIS does not necessarily eclipse the role of cartography in the visualisation of spatial knowledge but, as a means of storing, managing and analysing that knowledge, a GIS provides immense benefits when compared to the analogue technology of conventional maps.

Geographical information systems may then be seen to be taking over and greatly extending the role of spatial data storage which was previously played by maps. Once spatial data have been represented in digital form it becomes very much easier to carry out measurements on the data, to perform analyses in various ways, and to make changes to it. Some of the operations can be applied without recourse to a graphic map of any sort. For example, one could enquire about the distance between two named places or, say, the area of coniferous forest within a named county, without referring to a map.

It remains the case however that, for many applications of geographical information, a map may serve for communication purposes as the ideal means of identifying objects of interest and for understanding their spatial properties. In the context of a GIS, an individual map may be seen as one of an infinite number of possible visual symbolisations of the stored data. Thus computer technology opens up opportunities for visualising and exploring spatial data in new ways or in ways that in the past were too expensive or time-consuming to contemplate. This sense of exploration of spatial data through cartography is enhanced when the possibilities of interaction with the map are considered (MacDougall, 1992; MacEachren, 1994). Interactive graphics enable the map to be linked to (non-map) graphical displays of statistics directly associated with the mapped data. Maps may also be regarded as a powerful form of user interface to a wide range of information sources. Thus we can expect to see interactive maps becoming more commonplace in public information systems and networked information resources such as the World Wide Web.

The relationship of GIS to other fields

The purpose of the early emphasis here on the close relationship between GIS and cartography is to highlight their common function in storing and communicating geographical information. However, it should be stressed that in general a GIS can be expected to provide tools for data manipulation and analysis, and often for visualisation, which are far more versatile than any associated with traditional cartography. These strengths of GIS for spatial information handling derive from several other disciplines, notably from the fields of computer science, surveying and quantitative geography (Figure 1.1).

From a computing viewpoint, the early development of GIS may be seen to be built largely upon a combination of *database technology*, for storing information, and of *computer graphics* for digitising and displaying spatially referenced information. Some of the early GIS included map-overlay facilities which automated procedures used in *landscape architecture* to evaluate spatial coincidence of environmental and socio-economic factors. These procedures employ

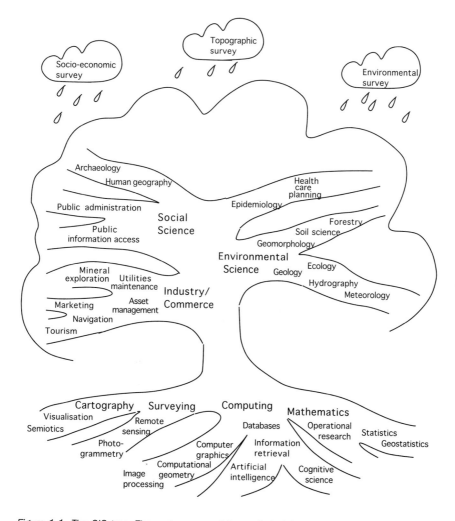

Figure 1.1 The GIS tree. The roots represent the technical foundations of GIS. The branches represent applications of GIS, the results and requirements of which feed back to the roots. Raining down on the tree are the data sources on which the individual applications depend.

techniques from the field of *computational geometry*, which concerns the development of algorithms to manipulate geometric data. More recently we can see the importance of *image processing* in interpreting remotely sensed data and of *artificial intelligence* applied to several aspects of GIS, including image understanding, database design and query languages, planning and automated cartography.

Analytical uses of GIS for understanding spatial patterns and correlations build upon foundations for spatial data analysis and *spatial statistics* laid down in the 1960s and 1970s by quantitative geographers (Taylor and Johnston, 1995). Despite the subsequent reaction among many geographers and social scientists against the use of quantitative techniques in spatial analysis, the recent improvements in GIS technology and the wider availability of spatial data have resulted in a resurgence of interest in quantitative spatial analysis, particularly in some specific application areas such as epidemiology (deLepper *et al.*, 1995) and in marketing and retailing (Longley and Clarke, 1995).

As GIS become more widely used for decision support in applications such as urban and regional planning, agriculture and the utilities, so the analytical tools of *operational research* will increasingly become integrated within existing GIS software. In this respect GIS may be seen as an adjunct to *management science*, and for some organisations the GIS-related functionality may become entirely integrated with a corporate or management information system.

Much of the data used in GIS are derived from topographic, socio-economic and environmental surveys. Because there are large quantities of such data recorded in paper documents, there has been an enormous demand for digitising such documents, which we refer to as *secondary digital data acquisition*. The largest primary source of digital data for use in GIS is undoubtedly that created by *remote-sensing technology* on board satellites and other aircraft. The discipline of remote sensing, which deals with the acquisition and analysis of remotely sensed data, is as an important relative of GIS, and from some points of view is regarded as a subdiscipline of GIS. In the future, a progressively higher proportion of acquired digital data will be from primary, ground-based or remote surveys. It is therefore essential for some applications that remote sensing and GIS be closely linked.

Uses of geographical information systems

As the range of information which can be placed in a geographical context is almost infinite, there are in principle few limits to the variety of possible applications of GIS. This is reflected in a continuing growth in GIS usage across many disciplines. It was indicated earlier however that the spur to development of commercial GIS packages has come from a much narrower range of interests among organisations that stand to benefit from the introduction of GIS technology. Major corporate users are found in the utility organisations, and in Western Europe and America there appear to be few local or regional government organisations that do not either currently exploit GIS or express a strong intention of doing so.

The reason why organisations such as these have become interested in GIS technology is that so much of the information that they need to conduct their business is spatially referenced. Utility companies and government agencies often have very large collections of paper maps that had been used to record the locations of their assets and to record maintenance work and transactions in property. Consulting and updating these large and often unwieldy collections of maps can be a very laborious process. GIS technology has provided relatively instant access to the information via computer terminals which enable locational data to be displayed and, if necessary, updated directly on screen. Having acquired geographical data in digital form, it becomes possible to display and analyse the data in ways that are often much quicker and more effective than was possible using manual techniques. Figure 1.2 provides an overview of typical procedures involved in creating and using GIS. The importance of particular types of procedure varies considerably between organisations. The main categories employed in the figure form the basis for review in Chapter 3 of the functions of GIS.

Most of the users of GIS technology build information systems by combining their own data with other data obtained from state or private organisations. Particularly important sources of such acquired data are national topographic mapping agencies which provide a base of topography, infrastructure, and property and administrative boundaries to which other data may be referenced. Another commonly used source of data is that provided by the census and other demographic and

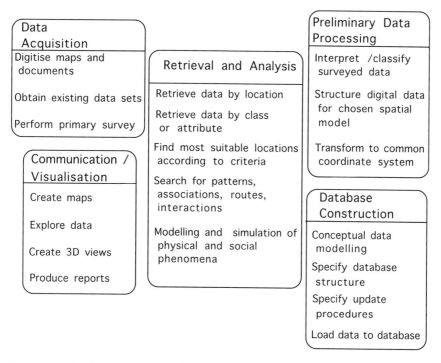

Figure 1.2 Procedures in GIS usage. The boxes represent the main categories of GIS-related activity.

socio-economic surveys. In the environmental sciences there are several types of survey organisations that supply data to others. These are concerned particularly with geology, soils, hydrology, ecology, oceanography and the atmospheric sciences. Many survey organisations engage in research and contract work that involves analysing their data using GIS technology. In the following sections we review the various types of data provided by survey organisations and then consider some of the applications of the data, both in the context of the provider organisations and of the end users.

Surveying and monitoring

Topographic and land survey

The type of data with the most widespread use in GIS is obtained from topographic surveys. The nature of the data collected varies somewhat between

different countries. The British national organisation for topographic survey, the Ordnance Survey (OS), is notable for the wide range of scales of mapping, extending between the most detailed urban plans at 1:1250 to small-scale national maps at 1:625 000. The classes of feature appearing on these maps, depending upon the scale, include buildings, land parcel boundaries, administrative boundaries, hydrography, transport (roads, canals, railways), forests, mountain peaks, ground elevation contours and tourist information. In applying computing technology to mapping, the OS concentrated initially on digitising its large-scale 1:1250 and 1:2500 map products, as well as the smallest-scale 1:625 000 and 1:250 000 products. Figure 1.3 illustrates some 1:1250 scale OS data in which individual buildings and land parcels are represented, while Figure 1.4 illustrates the more generalised nature of 1:200 000 scale data from the Automobile Association in which buildings have been aggregated within urban areas, which are shaded. The large-scale digitising provides valuable data for use in GIS operated by utility organisations and local government. Since every building and land

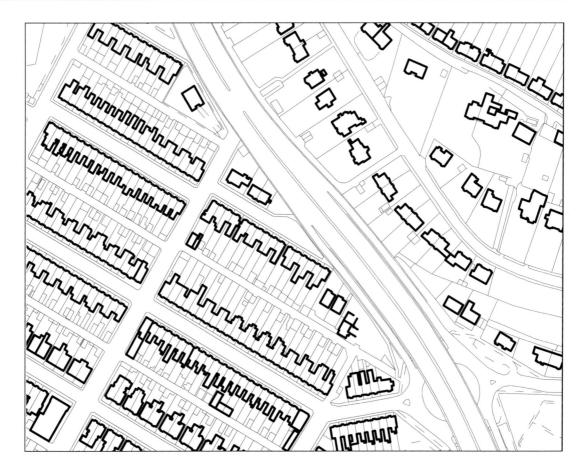

Figure 1.3 An example of a map created from Ordnance Survey 1:1250 scale topographic data. The main features displayed are buildings, land parcels and roads. Ordnance Survey, ©Crown copyright.

parcel is recorded on these maps it is possible to use them in combination with information systems that record property-based data for individual households and landowners. It is also possible to combine the data with records of facilities such as electricity, water and telephones linked to the properties. As an interim measure, digital data at the intermediate scale of 1:50 000 has been provided in a digital image format that allows it to be used as a backdrop to other digital data.

The smaller-scale data are particularly useful for applications such as navigational routeing, tourism and regional planning. The presence of administrative boundaries facilitates the representation of socio-economic data recorded for the various areal units at corresponding levels of aggregation. In addition to general topographic maps, the OS have been marketing digital elevation models representing land surface elevations on a regular grid interpolated from the contours on the 1:50 000 map series. Such data can be used to create three-dimensional (3D) landscape views in which topographic and environmental data may be draped on the 3D surface. They can also be used for intervisibility analysis and as the basis for an analysis of terrain features such as drainage basins. In the longer term the Ordnance Survey may be able to derive its main small-scale maps (including 1:10 000, 1:25 000 and 1:50 000) from the large-scale data. However, this will depend upon significant advances in the present state of software for performing the automated generalisation process that this would require.

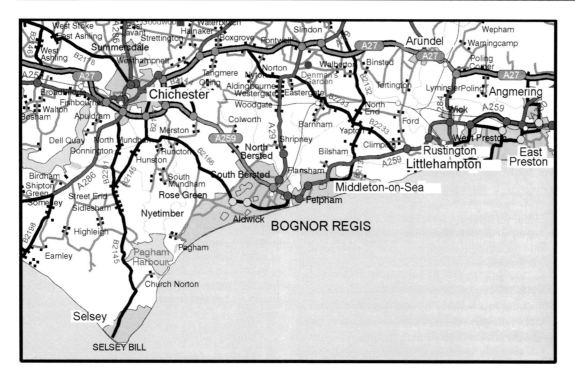

Figure 1.4 An example of a map created from 1:200 000 scale Automobile Association topographic data.

In the USA, the topographic surveying function is carried out by the United States Geological Survey (USGS). The largest-scale digital map product is at a scale of 1:24 000. Unlike the Ordnance Survey, the USGS has concentrated on digitising particular thematic elements of their map series in response to user demand (Starr and Anderson, 1991). The OS digitises the entire contents of a map sheet, since the digitising programme has been geared towards internal map production, as well as external user demand. Another very important source in the USA of digital map data, which includes details of properties and some urban infrastructure, is the TIGER system of the US Bureau of the Census (Marx, 1990). Many other national topographic mapping and survey organisations around the world have been investing in the creation of digital databases, either by digitising existing map products or by carrying out primary surveys from which digital data are generated.

Information systems that store general-purpose topographic data combined with administrative and land ownership boundaries are often referred to as Land Information Systems (LIS). There is some debate about the distinction between GIS and LIS. Dale (1991) suggests that an LIS combines institutional and human resources with the technology characterised by GIS, and that the term LIS is in any case in more general usage than GIS in most parts of the world, except the USA and the UK. There is little doubt, however, that much of the functionality associated with an LIS is equivalent to that found in systems described as GIS.

Hydrology, marine survey and glaciology

Surveys of the location of rivers and lakes are carried out in basic form by topographic mapping agencies. However, detailed surveying of the hydrological characteristics is usually carried out by specialist organisations, such as the Institute of Hydrology in the UK. Such organisations may use digital topographic base maps to provide a context for their more detailed observations. Studies of ground surface water depend upon the use of terrain models which, when created to sufficient accuracy, can be used to model surface water flow and hence to determine drainage basins and predict runoff and flooding under various conditions of weather and climate (Hogg *et al.*, 1993; Romanowicz *et al.*, 1993). The terrain models provided by national mapping agencies cannot always be guaranteed to be of adequate quality to model water flow in a realistic manner, though it is sometimes possible to improve their quality by imposing constraints of height and slope based upon additional survey data and on knowledge of the behaviour of existing drainage networks (Band, 1986).

The use of digital terrain models for coastal zone management in the Netherlands is illustrated in Plates 4 and 5, in which different methods of representing the ground relief are used. The terrain model is used to predict areas that would be affected by a 3.5-m flood. Plate 1.13 shows a related image in which a GIS is being used to identify suitable locations for development of slufter environments (wet dune areas that are subject to occasional tidal flooding).

Marine surveys are usually dependent upon ship-based depth sounding and water-sampling instrumentation and on remote sensing of sea surface character and sea temperatures. Surveys of the sea floor result in surface models comparable to digital terrain models. Oceanographic studies of currents, nutrients, salinity and temperature variation, however, require complex modelling facilities to represent variations in three-dimensionally referenced phenomena over time (see Mason *et al.* (1994) for an account of such four-dimensional GIS).

The link between ice formation and climate change has long been appreciated and we are now seeing the use of remote sensing and the 3D modelling facilities of GIS in monitoring the development of glaciers on global and local scales (Diament *et al.*, 1993). The presence of terrestrial and sea ice can be detected using satellite imagery, while terrain modelling techniques can be used to represent the form of ice surfaces and of fluvial channels associated with the melt waters.

Soil surveys

The application of GIS technology to soil science has been particularly well documented, since it formed a major theme of one of the first widely read textbooks on GIS (Burrough, 1986). The representation of soil information is in fact relatively challenging, in that soils are often subject to complex local variations. Soil maps frequently take the form of discretely bounded polygonal regions. However, boundaries between different soil types are not always so clear-cut on the ground. The polygons often represent idealised interpretations based on point samples and vertical profiles which actually record a variety of associated soil types. GIS provide a context in which to apply statistical techniques to modelling the natural spatial variation in soils as derived from discrete soil samples.

A GIS applied to soil data may be referred to as a soil information system and can be expected to provide facilities for digitising soil sample points and integrated soil boundaries from maps, for storing the data in a database, for interpolating and re-classifying soil attributes, and for overlaying and plotting these data with other types of data relating for example to terrain models and climatic zones.

Geological surveys

GIS for geological data have some commonality with those for soil information, in that the original data frequently consist of isolated sample points and interpreted boundaries. Data are derived from a combination of direct surface and subsurface samples, remotely sensed data from geophysical surveys, borehole sensors and airborne imaging devices, and from interpretations of geological phenomena derived from the directly and remotely sensed data.

The value of a GIS in helping to integrate geological data from a variety of sources, including maps, geophysical surveys and geochemical surveys, has been illustrated by Bonham-Carter (1991, 1994). However, the full potential of spatial information systems for geological data handling is only beginning to be realised with the introduction of flexible 3D tools for modelling, interpretation and visualisation. Computer systems have been used in oil exploration geology for a considerable time, though in the earlier systems they were based largely on a 2.5D model of the world in which geological boundaries could only be represented very effectively if they were relatively low lying and did not fold back on themselves. More fully 3D geological data models are now quite commonplace and are beginning to rival the capabilities of 3D modelling that have been developed in the medical sciences for manipulating data from body scanners (Jones, 1989a). It should be pointed out however that geoscientific exploration introduces challenges of computer modelling beyond those of the medical sciences in that data are, in general, sparsely sampled and inadequate to create an entirely reliable model (Raper 1989; Hamilton and Thomas 1992). There is therefore a greater dependence upon interpolation and interpretation.

The biosphere

Concerns about conservation of the natural environment and rehabilitation from the effects of industrial and agricultural development have led to increased use of GIS for environmental monitoring that focuses on the biosphere, both on a global scale and at the level of local ecological studies. GIS can be used to create ecological databases that integrate information derived from remote-sensing surveys, especially satellites such as Landsat and SPOT, and from ground surveys of flora and fauna and of physical and geo-chemical environmental parameters. The latter could relate for example to temperature, altitude, moisture content and to the presence of pollutants borne by the atmosphere, and by surface and subsurface water. Recognition of the importance of monitoring environmental data is reflected in databases such as the European Community CORINE project (Wyatt *et al.*, 1988) and the development of global databases such as the biodiversity database of the World Conservation Monitoring Centre.

There are numerous organisations worldwide that are concerned only with particular aspects of plant and animal life, concentrating on distinct types of habitat such as forests, moorland, rivers, lakes, open ocean and on coastal zones of particular sorts such as coral reefs, mud-flats and salt marshes.

Demographic and socio-economic surveys

Demographic surveys, in the form of a census and of population registers, provide the principal source of geographically referenced statistics on human population. The data vary between different countries but may include information such as age, sex, number of people per household, occupation, income, fuel use and access to amenities (Rhind, 1991). Uses of the data include planning for schools and roads, studies of the spatial variation of disease, and analyses of the location of customers for marketing purposes. Data

are usually aggregated to small areal units. In the UK these units are called enumeration districts (EDs) and they form the building blocks for further aggregation to higher level administrative units of electoral wards, administrative districts, counties, regions and countries. Figure 1.5 illustrates boundaries for wards, districts and counties respectively, within the country of Wales.

Spatial referencing for the EDs was originally provided digitally only by the coordinates of a centroid, though their digitised boundaries have now become generally available. Currently in the UK the postcode is also included in census returns and this can be translated to a representative geographical coordinate. The spatial referencing of postcodes makes them very valuable when they are linked to socio-economic data. In Scotland, EDs consist of aggregations of postcode areas, which makes it easier to study spatial variation at a finer resolution than the ED. In England and Wales there is no such simple hierarchical relationship, with the result that finer resolution studies involve errors due to the methods of re-sampling, whereby statistics for a postcode may be contributed by data from more than one ED.

In addition to relatively infrequent national censuses, government organisations such as the UK Office of Population and Census Surveys (OPCS) may carry out regular surveys of demographic and socio-economic data, such as that for employment, which can be used to study short-term trends in industrial and commercial growth and decline.

Archaeology and historical sites

Since archaeology is at least partially concerned with past geographies, it is natural that a wide range of the techniques of GIS have found practical application in the reconstruction of ancient landscapes (Allen *et al.*, 1990). While land survey techniques can be used to map visible signs of previous landscape features, particularly through observation of crop marks that show up in aerial photographs, it is also the case that structures in the shallow subsurface can be modelled using techniques similar to those employed in geoscientific surveys. These include seismic, magnetic and resistivity surveys. Three-dimensional visualisation techniques, including virtual-reality animations, have proven to be of value in simulating historic buildings and settlements.

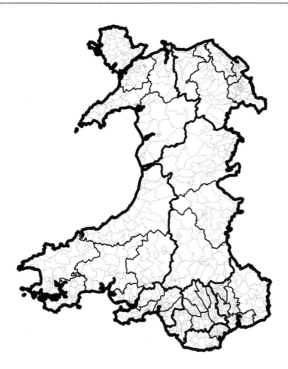

Figure 1.5 Administrative area units in Wales. The figure illustrates a hierarchical structure dividing Wales into counties, districts and wards. ©Crown Copyright.

Conventional GIS technology is now an essential tool in the management of data on historic monuments, enabling digital map-based surveys and photographic evidence to be integrated with aspatial databases recording the characteristics of artefacts and with building maintenance records.

Utilities maintenance

The rapid growth in the application of GIS technology in the late 1980s is clearly exemplified by utility companies providing public or private services for electricity, water, gas and telecommunications. These organisations are often responsible for maintaining plant, such as cabling and pipes that may have considerable capital value and may require regular maintenance (Plate 1). Procedures for maintenance and for servicing customers are heavily dependent upon map-referenced information to determine the location of existing facilities and to plan new installa-

tions. It is not surprising, therefore, that these organisations should have looked to GIS technology to provide an efficient means of storing, querying and analysing this information.

Since positional information must be correct to within a metre or less, and is frequently defined relative to other topographic features, the digital map base for a utilities GIS needs to be at a large scale, ideally showing the positions of individual houses. In the UK, large-scale digital data are generally available for urban areas, partly as a result of the demands of utility companies (see Figure 1.6 in which water supply services are displayed in combination with large-scale topographic map data). One of the benefits of storing utility data in digital form is that, provided they are structured appropriately, they can be used for computer-based network analyses to perform tasks such as estimating demand and reorganising supply routes in the event of the failure of part of a network. Another benefit of computerisation is in reducing the cost of repairs to damaged plant and roads, resulting from errors in locating pipes or cables needing servicing (Koop and Ormeling, 1990).

Sensitivity to the impact on the landscape of services associated with energy-generating utilities has resulted in the use of 3D visualisations to predict the visible impact of electricity pylons, wind farms and power stations. Figure 1.7 illustrates the use of a terrain model to visualise the location of wind turbines, along with the areas from which different numbers of the turbines are visible.

Another use of digital terrain models is in siting transmitters and receivers for purposes of radiocommunications (Plate 2). The terrain model can be used to study expected variation in the quality of radio reception due to landscape features.

Navigation

Whenever geographical information is employed for navigational purposes, GIS and computer cartography can be expected to provide significant benefits. By representing geographical information in a spatial database, a great deal of flexibility can be obtained to generate maps which are closely tailored to specific user requirements relating to routes, types of information displayed, scales of map and map projections. Not only can maps be produced, but as with the

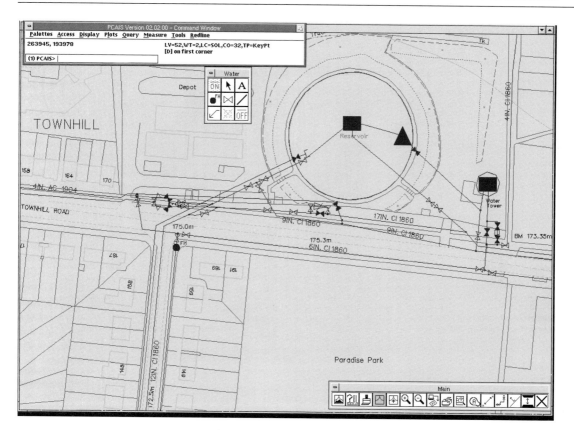

Figure 1.6 Water supply network data combined with Ordnance Survey large-scale topographic data. Courtesy of Welsh Water and Ordnance Survey. ©Crown Copyright.

utility supply information, appropriate computer representation of map data enables network analyses to be performed automatically. Thus routes can be generated which automatically meet the user's constraints relating for example to shortest path, fastest path, paths passing through specified locations and paths avoiding specified locations. This type of facility is of great value in optimising the use of vehicles operated for deliveries of commercial goods.

A growing application of map data for routeing purposes is that of in-vehicle navigation systems (White, 1991). Typically these products combine routeing software with cartographic data used to display a regularly updated map of the driver's immediate vicinity. Increasingly these systems use speech synthesisers to provide spoken instructions, thereby obviating the need for drivers to divert their attention from the road. An important issue for road navigation is being kept up to date on road-works or accidents and traffic jams which may temporarily alter the choice of an optimum route. These types of data need to be fed into the in-vehicle navigation systems. A cheaper alternative to in-car navigation is to use the services of automobile associations which employ a GIS to generate a route plan. There are also personal computer products which individuals may purchase to perform the same task.

Public administration

Spatially referenced information is an essential requirement for a large proportion of the tasks expected of local, regional and national government. Examples of these tasks are planning the location of new industrial sites, hospitals, domestic housing developments and leisure facilities such as sports centres and parks; highway building and maintenance;

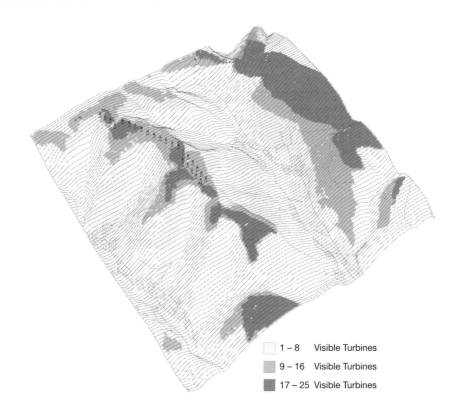

1 – 8   Visible Turbines
9 – 16   Visible Turbines
17 – 25 Visible Turbines

Figure 1.7 Digital terrain models can be used to help in the visibility aspects of the environmental impact assessment required when siting windfarms. Courtesy D.B. Kidner. ©Crown copyright.

emergency response planning, including locating fire hydrants (Plate 3), keeping track of hazardous materials, and evacuation routeing; districting for political representation and for school intake; maintenance of property owned and managed by an authority. Processing of planning applications is a prime example of an application in which much of the required information is spatially referenced (see Figure 1.8 in which a building subject to a planning application is highlighted on the map along with associated administrative data). Thus when planning applications are received it must be possible to find what legislation, such as preservation orders and planning consent, relates to the relevant land and buildings and to make spatial searches of the neighbouring land and properties to determine proximity to constraining phenomena such as natural habitats, other dwellings and sources of pollutants, that might be affected or have an effect upon the proposed development. Clearly GIS have an important role to play in the process of *environmental impact assessment* which is one component of development control.

The needs of planners are notable amongst GIS users for the range of spatial data types which may have to be combined in a single system. The capacity for information integration is, however, often seen as one of the major strengths of GIS. In addition to data integration, the usefulness of a GIS for planning will depend upon the extent to which it can serve as a decision support tool. Clearly access to information in itself is a tool in supporting decision-making, but the power of a GIS in the context of planning is also dependent upon the ability to view information at different levels of detail or generalisation, to generate solutions to planning problems, and to perform analyses of the relative merits of alternative solutions.

Automatic generation of solutions to planning problems can be achieved in some applications through the use of software for network analysis, automatic districting and for spatial overlay to

(a)

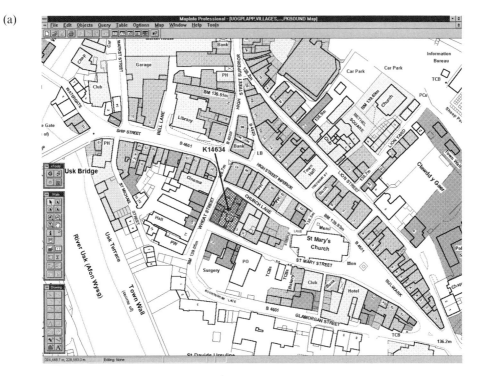

Figure 1.8 GIS can be used to assist in processing planning applications by displaying the site in context (**a**), determining automatically whether a particular location lies within the boundaries of zones subject to restrictions, and linking the map data to administrative details (here falsified) of the application (**b**). Topographic map base uses Ordnance Survey data, ©Crown Copyright.

(b)

determine the coincidence of a variety of attributes, such as land use and transportation facilities. There is considerable potential in planning to use artificial intelligence techniques to generate solutions to complex problems. Smith *et al.* (1988) demonstrated this by applying knowledge-based system technology to a planning GIS in order to generate possible land-use classifications, given a variety of environmental data sets and constraints dictated by planning policy. More recently, stress has been placed upon the need to integrate spatial analytical tools with interactive user interfaces that facilitate visualisation and co-operative human decision-making (Armstrong *et al.*, 1992).

Marketing and retailing

Decisions about where to locate new shops and who to target when distributing junk mail depend upon knowledge of the spatially variable characteristics of the population, including factors such as income, age and lifestyle. A considerable industry has built up around the need to service retailers with this *geodemographic data* about their potential customers, and GIS technology is well suited to providing facilities to analyse the data (Brown, 1991; Longley and Clarke, 1995). As previously indicated, census data provide some of the relevant information, and some geodemographic classification systems such as ACORN, from CACI, are based entirely on census data at the level of enumeration districts. Geodemographic data can however be updated and refined using a variety of other data sources, including electoral registration, credit activity, and questionnaires attached to retail products or carried out by telephone or on the street. In the UK, the CCN MOSAIC system uses data additional to the census and is able to distinguish groups of people at the level of unit postcodes, which occupy smaller regions than the enumeration districts (Birkin, 1995).

Targeting customers for mailshotting is relatively straightforward once the resident population have been classified in terms of their spending power and likely interests. The problem of where to locate new retail outlets is however potentially much more complicated as it must take account of the attractiveness of competing outlets and of the costs of travel to reach the proposed facility. Just as in some local and regional planning applications, the decision-making process may be supported by the use of tools for spatial interaction modelling (discussed in Chapter 13). Plate 6 illustrates the use of a GIS in market analysis. The pie charts are located at individual stores and show predicted demand for each of the nearby postal sectors.

Summary

In this introductory chapter we have attempted to relate GIS to the subject of computer cartography and to place it in the context of a set of contributory disciplines. We have also summarised the wide range of actual and potential applications of the technology. In providing a computer representation of spatially referenced information, GIS can be seen to have supplanted the traditional role of cartography in storing and managing spatial information, while the technology for visualisation has also been revolutionised. Equally importantly, the computer representation opens up an impressive array of options for modelling and simulating landscape and environmental processes in ways that depend almost entirely upon the computer's powers for numerical and symbolic manipulation.

In reviewing the applications of GIS, in a representative but by no means complete manner, we have drawn a distinction between uses of GIS in the context of surveying and monitoring of the physical and socio-economic environment; utilities maintenance; public administration; navigation; and marketing and retailing. The category of surveying and monitoring covers a considerable range of organisations and activities concerned with acquiring data for national archives, for subsequent redistribution and for monitoring purposes by environmental and social science agencies. The types of data that are dealt with include topographic and land survey, hydrology and oceanography, geology, soils, biosphere, demographic survey and archaeology.

The utilities are major users of GIS, the primary requirement being to record the locations of cables and pipelines that were once maintained on paper maps. Having digitised these records they can then also be exploited for routine maintenance and for network analysis. Perhaps the fastest growing area of application of GIS is public administration, where a very large proportion of the information maintained is spatially referenced. There are many specific applications of GIS including keeping track of land assets, highways maintenance, emergency planning, development control and decision support for planning domestic, industrial, educational, health and leisure developments.

This introduction to the origins and applications of GIS serves to lay the foundations for the following two chapters, which address issues of geographical information concepts and their representation and, in Chapter 3, the functional capabilities that can be expected of existing GIS.

Further reading

A set of articles reviewing applications of GIS is to be found in Volume 2 of Maguire, Goodchild and Rhind (1991). Special Issues of the journals *The American Cartographer* (Vol. **15**(3)) and *Cartography and Geographic Information Systems* (Vol **18**(3)) have been devoted to accounts of various aspects of the history of cartography, with many of the individual articles referring to automation and the link with GIS. For a sense of the early development of GIS the *Harvard Papers on Geographic Information Systems*, edited by Dutton (1978), make interesting reading. In the last few years several books have reviewed or collected together articles on specific application areas of GIS: see, for example, Scholten and Stillwell (1990), Worrall (1990, 1991) and Fischer and Nijkamp (1993) on GIS in administration and planning; Grimshaw (1994) and Longley and Clarke (1995) on business application; Openshaw (1995) for issues concerning the use of census data in the UK; Haines-Young and Green (1993) on GIS and ecology; Allen *et al.* (1990) on archaeological applications; deLepper *et al.* (1995) on medical and health-related uses of GIS; Goodchild *et al.* (1993) on environmental modelling and GIS. For some views on the social context and implications of GIS, see Pickles (1995).

To keep abreast of applications of GIS in a wide range of industrial, commercial and academic contexts, see the journals *GIS Europe*, *GIS World* and *Mapping Awareness*. The longest-established academic journal on GIS is the *International Journal of Geographical Information Systems*. More recent GIS-related journals are *Geographical Systems*, *Transactions on GIS* and *GeoInformatica*. Other journals that regularly publish articles on GIS include *Photogrammetric Engineering and Remote Sensing*, and *Environment and Planning A and B*. Journals that frequently publish articles on computer cartography include *Cartography and Geographic Information Systems*, *Cartographica* and *The Cartographic Journal*.

CHAPTER 2

Geographical information concepts and spatial models

Introduction

Geographical information systems are used for storing and analysing a very wide variety of subject matter ranging from the social sciences to the natural environmental sciences and from public administration to the management of the human-made environment. While it is difficult to generalise about the information content of social, environmental or physical phenomena, it is nevertheless possible to identify several types of information that are characteristically geographical in that they place information in a spatial context. In this chapter we provide a summary of the characteristics of geographically referenced information, making a distinction between the semantic and statistical data required to describe the non-spatial components of the information and the data required to describe the location, dimensions, shape, patterns, interactions, associations and relationships that are characteristically spatial in nature.

Having provided an overview of distinguishing concepts of geographical information, we continue in the second part of the chapter by introducing the main spatial models that are employed in GIS for the purpose of storing and manipulating spatially referenced information. Spatial models are important in that the way in which information is represented affects the type of analysis that can be performed and the type of graphic display that can be obtained. In GIS systems there is a major distinction between what are usually referred to as vector GIS and raster GIS. These two approaches to spa-

tial data processing, often to be found in the same GIS package, reflect two different methods of spatial modelling: the former focusing on discrete objects that are to be described, and the latter concerned primarily with recording what is to be found at a predetermined set of locations that may be grid cells or points. We will describe the important differences between these approaches and indicate their relative merits for different types of spatial data processing. This chapter serves as a precursor to Chapter 3, which introduces the various spatial data processing facilities that are provided by GIS and that manipulate and transform the spatial models which are introduced here.

Semantics

To be able to manipulate and retrieve data in an information system it is necessary to attach meaning that constitutes a description or an interpretation of the data for some particular purpose. For many purposes, meanings are standardised in the form of classifications consisting of a set of categories. Often categories are grouped or classified into successively higher levels that enable phenomena to be referred to at different levels of abstraction. Thus we talk of transportation systems consisting of roads, railways, bus routes, canals and air routes (Figure 2.1). Each of the subclasses can usually be further subdivided according to its purpose, often with respect to physical or socio-economic characteristics. Administrative regions provide another example of hierarchical clas-

Table 2.1 CORINE land-cover classification.

Level 1	Level 2		Level 3	
1 Artificial surfaces	1.1	Urban fabric	1.1.1	Continuous urban fabric
			1.1.2	Discontinuous urban fabric
	1.2	Industrial, commercial and transport units	1.2.1	Industrial or commercial units
			1.2.2	Road and rail networks and associated
			1.2.3	Port areas
			1.2.4	Airports
	1.3	Mine, dump and construction sites	1.3.1	Mineral extraction sites
			1.3.2	Dump sites
			1.3.3	Construction sites
	1.4	Artificial non-agricultural vegetated areas	1.4.1	Green urban areas
			1.4.2	Sport and leisure facilities
2 Agricultural areas	2.1	Arable land	2.1.1	Non-irrigated arable land
			2.1.2	Permanently irrigated land
			2.1.3	Rice fields
	2.2	Permanent crops	2.2.1	Vineyards
			2.2.2	Fruit trees and berry plantations
			2.2.3	Olive groves
	2.3	Pastures	2.3.1	Pastures
	2.4	Heterogeneous agricultural areas	2.4.1	Annual crops associated with permanent crops
			2.4.2	Complex cultivation patterns
			2.4.3	Land principally occupied by agriculture, with significant areas of natural vegetation
			2.4.4	Agro-forestry areas
3 Forests and semi-natural areas	3.1	Forests	3.1.1	Broad-leaved forest
			3.1.2	Coniferous forest
			3.1.3	Mixed forest
	3.2	Shrub and/or herbaceous vegetation associations	3.2.1	Natural grassland
			3.2.2	Moors and heathland
			3.2.3	Sclerophyllous vegetation
			3.2.4	Transitional woodland-shrub
	3.3	Open spaces with little or no vegetation	3.3.1	Beaches, dunes and sand plains
			3.3.2	Bare rock
			3.3.3	Sparsely vegetated areas
			3.3.4	Burnt areas
			3.3.5	Glaciers and perpetual snow
4 Wetlands	4.1	Inland wetlands	4.1.1	Inland marshes
			4.1.2	Peatbogs
	4.2	Coastal wetlands	4.2.1	Salt-marshes
			4.2.2	Salines
			4.2.3	Intertidal flats
5 Water bodies	5.1	Inland waters	5.1.1	Water courses
			5.1.2	Water bodies
	5.2	Marine waters	5.2.1	Coastal lagoons
			5.2.2	Estuaries
			5.2.3	Sea and ocean

sification, with which further classifications relating to, for example, employment, education or land use, may be associated. Semantics, or meanings, provide an essential means of distinguishing between information and hence enabling access to it for particular purposes. Some organisations devise their own classification system for the phenomena of interest to the organisation. Alternatively they may use a nationally or internationally recognised classification. An example of a land-cover classification scheme from the CORINE project, with an emphasis on natural and semi-natural vegetation, is illustrated in Table 2.1,

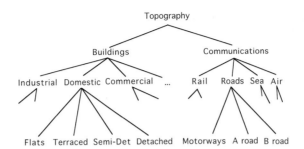

Figure 2.1 Classification systems attach meaning to data. They are often hierarchical.

while a commercial geodemographic classification system is illustrated in Table 2.2.

Statistics

Having attached meaning to particular phenomena, any attempt to understand the spatial characteristics must be accompanied by records of the presence or absence of the phenomena, which may in turn be quantified at varying levels of precision, according to the method of measurement. Measurement systems are often categorised as nominal, ordinal, interval and ratio (Table 2.3). A *nominal* measurement is simply the name of a category such as that of a clay soil type or a wheat crop type. It distinguishes one thing from another, but does not necessarily imply any relative order or priority. *Ordinal* observations use categories that have been ranked and therefore do imply order. Examples are provided by a set of descriptors such as coarse-grained, medium-grained and fine-grained, and more formal classifications such as the Richter scale of earthquakes and the Beaufort scale of wind strength. Agricultural land units could be classified as an ordinal scale from 1 to 5 where 1 means highly fertile soil, and 5 means poor-quality soil, not suited to most crops. Although each category is clearly greater or lesser in some sense than each other category, any associated numerical value does not correspond to an exact quantity and cannot therefore be used in conventional arithmetic to form ratios or differences.

Interval scales can be used to quantify the difference between particular observations, though ratios between values are not meaningful, because the zero point on an interval scale does not correspond to a zero quantity in the phenomenon being measured. Classic examples of such scales are the Centigrade and Fahrenheit temperature scales. *Ratio* measurements are distinguished by the fact that zero on the scale means a zero quantity in the observed phenomenon. The Kelvin scale of temperature is an example, as would be the count of population in a city. In the latter case, zero count means no persons and a count of 400 000 compared with 200 000 does mean that the former population is twice as large as the latter.

Non-spatial data in GIS are usually linked to spatial units to which the data refer, being for example a count of the number of occurrences of some phenomenon in the spatial unit. The data may also be statistically transformed, being for example a minimum, maximum, mean or median of a set of observations corresponding to the spatial unit. Data obtained from airborne remote-sensing instruments may refer to an imprecise spatial unit, represented by a pixel in the image, even though the actual observation may be derived from integrated measurements that are centred on the pixel locations.

Spatial information

Spatial characteristics of information can be broadly distinguished between those that describe *where things are*, using locations consisting of reference positions, spatial units and spatial relationships; those that describe the *form of phenomena*, using qualitative or quantitative descriptions of shape and structure; and those that describe *associations and interactions* between different phenomena.

Table 2.2 The CACI ACORN geodemographic classification scheme.

ACORN categories		ACORN groups		ACORN types
A thriving	1	Wealthy achievers, suburban areas	1.1	Wealthy suburbs, large detached houses
			1.2	Villages with healthy commuters
			1.3	Mature affluent home-owning areas
			1.4	Affluent suburbs, older families
			1.5	Mature, well-off suburbs
	2	Affluent, greys, rural communities	2.6	Agricultural villages, home-based workers
			2.7	Holiday retreats, older people, home-based workers
	3	Prosperous pensioners, retirement areas	3.8	Home-owning areas, well-off older residents
			3.9	Private flats, elderly people
B expanding	4	Affluent executives, family areas	4.10	Affluent working families with mortgages
			4.11	Affluent working couples with mortgages, new homes
			4.12	Transient workforces, living at their place of work
	5	Well-off workers, family areas	5.13	Home-owning family areas
			5.14	Home-owning family areas, older children
			5.15	Families with mortgages, younger children
C rising	6	Affluent urbanites, town & city areas	6.16	Well-off town & city areas
			6.17	Flats & mortgages, singles & young working couples
			6.18	Furnished flats & bedsits, younger single people
	7	Prosperous professionals, metropolitan areas	7.19	Apartments, young professional singles & couples
			7.20	Gentrified multi-ethnic areas
	8	Better-off executives, inner city areas	8.21	Prosperous enclaves, highly qualified executives
			8.22	Academic centres, students & young professionals
			8.23	Affluent city centres areas, tenements & flats
			8.24	Partially gentrified multi-ethnic areas, young families
			8.25	Converted flats & bedsits, single people
D settling	9	Comfortable middle agers, mature home-owning areas	9.26	Mature established home-owning areas
			9.27	Rural areas, mixed occupations
			9.28	Established home-owning areas
			9.29	Home-owning areas, council tenants, retired people
	10	Skilled workers, home-owning areas	10.30	Established home-owning areas, skilled workers
			10.31	Home owners in older properties, younger workers
			10.32	Home-owning areas with skilled workers
E aspiring	11	New home owners, mature communities	11.33	Council areas, some new home owners
			11.34	Mature home-owning areas, skilled workers
			11.35	Low-rise estates, older workers, new home owners
	12	White collar workers better-off multi-ethnic areas	12.36	Home-owning multi-ethnic areas, young families
			12.37	Multi-occupied town centres, mixed occupations
			12.38	Multi-ethnic areas, white collar workers
	13	Older people, less prosperous areas	13.39	Home owners, small council flats, single pensioners
			13.40	Council areas, older people, health problems
F striving	14	Council estate residents, better-off homes	14.41	Better-off council areas, new home owners
			14.42	Council areas, young families, some new home owners
			14.43	Council areas, young families, many lone parents
			14.44	Multi-occupied terraces, multi-ethnic areas
			14.45	Low-rise council housing, less well-off families
			14.46	Council areas, residents with health problems
	15	Council estate residents, high unemployment	15.47	Estates with high unemployment
			15.48	Council flats, elderly people, health problems
			15.49	Council flats very high unemployment, singles
	16	Council estate residents, greatest hardship	16.50	Council areas, high unemployment, lone parents
			16.51	Council flats, greatest hardship, many lone parents
	17	People in multi-ethnic, low-income areas	17.52	Multi-ethnic, large families, overcrowding
			17.53	Multi-ethnic, severe unemployment, lone parents
			17.54	Multi-ethnic, high unemployment, overcrowding

Table 2.3 Examples of statistical data attributes of different types.

Nominal	Ordinal	Interval	Ratio
Crop type	Sand grains	Temperature	Population
Wheat	Fine	20°	24 930
Rye	Medium	33°	153 208
Barley	Coarse	24°	7 965
Maize		10°	61 806

Descriptions of location may be specified nominally or metrically and in terms of spatial relationships. Nominal locations express position in terms of named topographic places, such as London or the Gobi Desert; named administrative units such as census enumeration districts; and addresses that may include both topographic and administrative names. Metric locations define position using coordinate systems, measuring distances or angles relative to a named place. Spatial relationships enable the position of one place to be specified relative to another referenced placed.

Named places

Named topographic places and landmarks

Place names provide what is probably the most fundamental method of specifying location in natural language. The named places will often serve as landmarks to help in navigating from one place to another. If in conversation we ask where something or somebody is, we are likely to receive an answer that contains a named place such as a town, a building, a street or a natural or human-made feature such as a hill, a valley or a river (Table 2.4a). The name may be a standardised, widely recognised name, as in the case of towns and cities, or it may be informal, being only locally or personally familiar, such as 'the back yard', 'the bog' and 'the wobbly stone'.

Addresses

Locations of properties are described nominally by addresses, which may refer either to a single building and the land it occupies, or a plot of land, or to a subdivision of a building, such as an apartment or a floor within a building (Table 2.4c). Addresses are an essential form of locational referencing, but because they can consist of a non-standardised combination of building names or numbers and of street and town names, there is often the possibility of imprecision or of duplication of the same address for different places. The possibility of confusion is reduced by the use of postcodes and zip codes. Postcodes typically correspond to a group of buildings or addresses and they may be associated with a single point which, in the UK postcode system, records the centroid of the first property in the list of properties belonging to the postcode unit. Increasingly postcodes are being used for spatial referencing of socio-economic data, since they provide a formalised link between a person's identity and a close approximation to their home location (Gatrell, 1989; Raper *et al.*, 1992). They are also easier to process in computers than are free text addresses. In the UK, it is now possible to purchase data that provide a link between individual addresses and their map coordinates.

Administrative units

Socio-economic data are usually linked to administrative areal units, such as electoral wards, counties and census enumeration districts (Table 2.4b). As census enumeration districts are quite small geographic areas, census data can be provided at varying levels of aggregation represented by the higher level administrative areas.

Table 2.4 Examples of named places to describe location.

(a)

Topographic names
Glasgow
River Taff
Wye Valley
Brecon Beacons

(b)

Administrative units
Avon County
Grampian Region
Borough of Lambeth
Lower Saxony

(c)

Addresses
42 Bilge Street, Grantham, Lincolnshire
Flat 3, Down House, Belford Avenue, Penge
Red Darren, Peterchurch, Radnorshire

Quantitative locational survey data

Coordinate systems

Whenever some precision is required in specifying locations of phenomena, for purposes of surveying and cartography, it is normal to supplement, and sometimes largely replace, the use of named places with measurements relative to an agreed frame of reference. The most global of these systems use latitude and longitude, but this is not convenient for detailed surveying since the units do not correspond to a single fixed distance on the ground. Consequently various other coordinate systems are used which provide consistent distance and angular measurement, at least in a local context. Of these, rectangular coordinate systems record locations as a pair (or triple) of distances from an agreed origin, which may be a place on the ground. Polar coordinate systems define location in terms of a combination of distance and orientation, as well as relative to a specified origin. Coordinate systems are of course fundamental to surveying and map-making and the subject is addressed in detail in Chapter 4.

Geometric primitives

The locations of surveyed spatial data are normally expressed in terms of one of a set of geometric primitives consisting of points, lines, areas, surfaces and volumes, or to sets of, or combinations of, these primitives that constitute more extensive or complex structures (Figure 2.2). The geometric primitives are, with the partial exception of points, intended to represent real-world phenomena that are continuous across some region of space. Sample data used to record the phenomena are, however, usually discrete, consisting of, for example, points used to define a line. The choice of a particular type of geometric primitive depends upon the level of detail or the degree of generalisation with which a phenomenon is being recorded.

Point-referenced observations are frequently used when sampling the value of physical phenomena such as temperature, noise levels or soil type. In practice such observations may relate to a volume of space, the location of which is typified by an individual point. In topographic surveys, single points may be used to record the location of phenomena such as radio masts and telephone kiosks which have an areal extent or volumetric extent that is small relative to the scale of mapping.

Similarly a line may be used to describe a boundary between physical phenomena that may in fact be a linear zone, with measurable area, or it may describe the path of a phenomenon such as a road or a river which, though linear in general form, in the real world is again clearly not linear in any geometrically precise sense of the word. On the basis of a ground survey, linear forms in the real world may be represented by an ordered sequence of point locations, described by coordinates.

Examples of phenomena treated, for sampling purposes, as areas include natural features, such as lakes and forests, and administrative regions, such as counties and districts which may be defined originally with regard to ground features, though their official representation may be as lines on maps. Areas are typically defined for sampling purposes by linear boundaries that form a closed area, or by sets of cells on a regular grid.

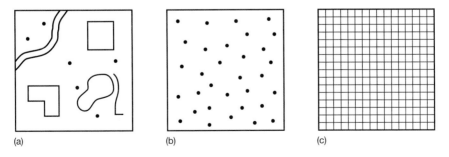

(a)   (b)   (c)

Figure 2.2 Examples of types of locational data used in surveys. (**a**) Points, lines and areas, representing, for example, trees, fences and buildings. (**b**) A set of irregular point-referenced observations, e.g. heights or temperature measurements. (**c**) Samples on a regular grid of cells, e.g. soil types, or dominant plant species.

Examples of regular grid-cell sampling (Figure 2.2c) occur in the context of some vegetation studies, soil studies and in satellite remote sensing, in which each sample of radiation is regarded as emanating from a rectangle on the earth's surface (though in fact it comes from a region centred on the cell). Regular grid sampling schemes are often appropriate for representing real-world phenomena, the locations of which are difficult to define precisely due to a gradual change of their characteristics in space and possibly in time. Vegetation, soil and sediments are examples of such phenomena, which are sometimes described as fuzzy. *Fuzzy modelling* schemes address the issue by categorising the phenomena in terms of a range of possible values that may be recorded within a given sampling unit. In some schemes each cell may be associated with the probabilities of occurrence of one or more categories of phenomena.

Area-based sampling schemes are common in the social sciences, since census data are usually aggregated within government administrative units such as census enumeration districts, electoral wards and counties, as indicated above. Unlike the boundaries of phenomena in the physical environment, the boundaries of social science data are often entirely independent of the phenomenon being studied. This can lead to fundamental difficulties in interpreting the data. The problem is that the social characteristics of interest for a given area may be homogeneous or heterogeneous according to the location of the boundary. Hence when statistics are derived from the lumped data of the areal units, spurious correlations between measured factors may result. For example, if a particular zone in a city included several streets dominated by young unemployed people and students, and several streets dominated by well-off middle-aged professionals, data consisting of average age and average income would blur the distinction between the two groups, giving numerical values that were representative of neither of the predominant groups. Redrawing the boundaries could result in quite different statistics for the individual zones. The problem of misleading correlations between statistics of areal units is sometimes called the *ecological fallacy*. The fact that areal units for social-science data vary from place to place and at the same place over time, due to administrative changes, results in what is called the *modifiable areal unit problem* (MAUP), which we return to in Chapter 12.

The most common example of a surface in geographic observations is that of the earth's surface. It is usually sampled as a set of points in three dimensions, or as contour lines (of equal value). Contours, however, have limited use for representing surfaces beyond that of visualisation. More practical sampling schemes used to describe surfaces for analytical purposes are regular grids of point observations or irregularly distributed point samples, which may form part of a triangulation surveying scheme.

Volumetric data are important in studies of several branches of the earth sciences, but sample data relating to volumes are often collected in the form of local point, linear and areal samples that are subsequently used to create volume models. For example, subsurface observations in geology are often point-referenced, such as the location down a borehole, or line-referenced, such as a boundary of a rock body as seen in an outcrop.

Size, shape and pattern descriptors

Data referenced to the various geometric primitives and higher level structures may include simple measurements of size, such as length, area or volume, and of orientation, such as the angular direction of a lineation or the slope of a surface. There may also be more subtle descriptions which may be qualitative, as in the description of a river as meandering, or of a surface as angular; or quantitative, based on analyses of sets of measurements using various statistical parameters (Figure 2.3). Thus, given a set of point-based observations, an analysis may indicate a random, regular or clustered pattern. Analyses of lines and surfaces, defined originally by point locations, may result in quantitative measures of sinuosity or roughness respectively. Information of this sort may be derived from survey descriptions (perhaps based on subjective field observations).

Spatial relationships

When a named place is given as a location, the implication is that the phenomenon of interest is *in*, *on* or *at* the named place, meaning that the relationship is that of equivalence with the named place. If a location of interest is not exactly that of a named place, or one needs to be more precise, then a variety of spatial relationships may be used (Freeman, 1975). Spatial relations may be categorised in several ways (Figure 2.4). One important distinction is between those that are orientation-independent,

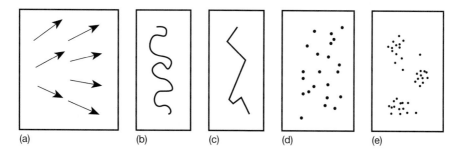

Figure 2.3 Shape and pattern descriptors: (a) set of quantitative orientation observations; (b) meandering shape; (c) angular shape; (d) random distribution; (e) clustered distribution.

called *topological relationships*, and those that are orientation-dependent, called *directional relationships*. The topological relationships differ in terminology, but can generally be reduced to those of *equivalence* (equal), *partial equivalence* (overlap, cross), *containment* (inside), *adjacency* (connected, or meets) and *separateness* (disjoint). These are illustrated in Figure 2.4. Directional relationships include those of *in front*, *behind*, *above* and *below*; the cardinal directions of *north*, *south*, *east* and *west*, and their combinations; and metric descriptions of angle of azimuth. A further category that is related to directional relationships is that of *between*, which implies a location somewhere on a bearing, or a route, from one place to another. *Proximity relationships* describe distance of separation either quantitatively as a measured distance, or qualitatively with terms such as *near*, *far* and *in the vicinity*. They are frequently used in combination with directional relationships.

Associations and correlations

Studies of combinations of phenomena originating in point and area-based surveys may result in the detection of the coincidence or association of particular phenomena (Cliff and Ord, 1981). Having identified a structural pattern such as clustering in point data, it could be of interest to attempt to establish a correlation between the observed clusters and some other phenomenon, in an effort to suggest causation. The search for such associations frequently occurs in studies of the distribution of socio-economic parameters such as health, income, education and employment. An attempt might be made, for example, to determine whether there was a correlation between an unusually high incidence of a particular disease and the occurrence of one or more known sources of pollution.

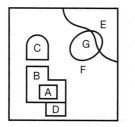

Topological	Proximal	Directional
A inside B	C near B	G east of C
D connected to B	D far from E	C north of D
C disjoint from B		
G overlaps E		

Figure 2.4 Spatial relationships may be categorised as topological, proximal and directional. Examples of each are provided. The examples of proximal and directional relations are expressed qualitatively. They could also be described quantitatively.

Temporal information

GIS are often used to maintain a record of changes over time and to predict future change. Thus it is important to be able to record when a real-world phenomenon or event relates to, called *real-world time*, as well as when it was recorded in the database, called *database time* or *transaction time* (Snodgrass, 1992). Time past is one-dimensional and gives rise to one-dimensional temporal relationships that are comparable to the spatial topological relationships (Allen, 1983). Temporal relationships include *before* (*precedes*) in some indeterminate way, *immediately before*, *equivalent*, *immediately after*, *after* (or *succeeds*) in an indeterminate sense, and *overlaps* (Figure 2.5). Note that the indeterminacy in before and after refers to the fact that they could include the possibility of being immediately before or after. When considering future events, it is possible, for example in a planning context, to envisage multiple possibilities. This gives rise to the concept of *branching time* or *parallel time*.

Although it is possible to identify discrete points in time, it is important to realise that in practice, the date attached to a particular phenomenon may relate to a period of time. Events may be recorded with widely differing granularity, ranging from less than a second, to a day, a month, a year or longer periods, such as the reign of a monarch or the period of a war. The date attached to a human-made phenomenon such as a building might be that of latest completion, that when building commenced, or a pair of dates encompassing the period of construction.

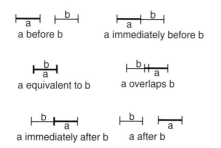

Figure 2.5 Six temporal relationships between the two intervals of time, labelled a and b.

Interactions in space and time

Dynamic interactions between phenomena arise due to changes in spatial distribution over time. Examples include that of population migration towards sources of employment; flows of water across the ground surface and through the subsurface; and flows of vehicles and aircraft along roads and flight paths respectively (Figure 2.6). Sample data, based on spatial units of measurement, are accumulated over time to create discrete spatio-temporal data that may be modelled by often highly complex mathematical and statistical functions that include time as a parameter.

It may be noted that time-dependent dynamic interactions may be distinguished from static interactions that occur primarily in space, such as the intervisibility between fixed points on the earth's surface in which the erection of a building or a bridge would create a perceived visual interaction between the structure and the surrounding locations. Clearly such interactions may be given a time dimension when change is considered.

Conceptual models of spatial information

Efforts to interpret and analyse spatial data are generally accompanied by the use of several abstract conceptual models of space, each of which emphasises a different aspect of spatial phenomena, depending upon the purpose of the analysis. Here we will focus on three such models which have influenced the way in which data are organised and processed within GIS. They are based respectively on *objects*, *networks* and *fields* (Figure 2.7)

Object-based models

Object-based spatial models emphasise individual phenomena that are to be studied in isolation or in terms of their relationships with other phenomena. Any phenomenon however big or small may be designated as an object, provided that it can be separated conceptually from neighbouring phenomena. Objects may be composed from other objects and they may have specific relationships with other separate objects

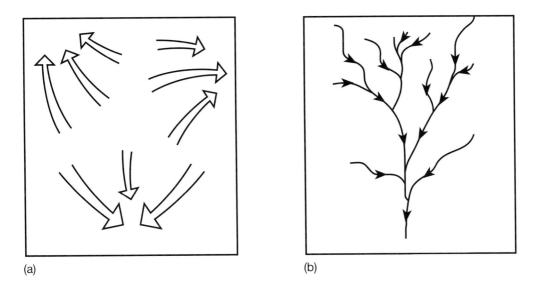

Figure 2.6 Dynamic interactions in space and time take place (**a**) across surfaces and (**b**) through discrete flow paths.

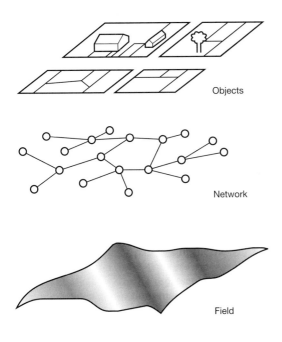

Figure 2.7 There are three widely used conceptual views of spatial phenomena. The object-based view focuses on discrete representations of phenomena. The network model represents paths between places. The field model represents the phenomena of interest as a continuous variable, which can be evaluated at any location in space.

(see Figure 2.8). In an application concerned with recording the ownership of land and property, an object-based view is likely to be taken, since each land parcel and each building must be distinguished and must be uniquely identifiable and individually measurable. Similarly, large-scale topographic surveys, which provide data for such applications, take an object-based view of the world, because a primary objective is to demarcate land, property and transport infrastructure, and to provide accurate locational descriptions of them.

An object-based view is appropriate, though not confined, to phenomena that have well-defined boundaries. Hence it is suited to human-made phenomena, such as buildings, roads, utilities and administrative regions. Some natural phenomena, such as lakes, rivers, islands and forests, are often represented in object-based models because they need to be treated as discrete phenomena for some purposes, but it should be remembered that the boundaries of such phenomena are rarely fixed over time, and hence there may be less precise definition of their actual locations at any one instant.

Networks

Network-based spatial models share some aspects of the object-based model in that they often deal with

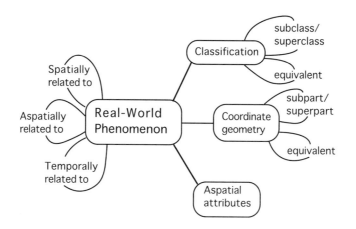

Figure 2.8 The object-based conceptual view identifies individual phenomena that may be described in terms of spatial, temporal and non-spatial attributes as well as in terms of spatial, non-spatial (semantic) and temporal relationships with other phenomena.

discrete phenomena, but the essential characteristic is the need to consider interactions between multiple objects, often along discrete paths or routes that connect them. The exact shape of the phenomena concerned may not be of great importance. What is important is some measure of the distance, or the impedance, between specified phenomena. Typical examples of applications for which a network model is appropriate are studies of traffic on road, sea and air routes; and analyses of the flow of water, gas and electricity through pipelines and cables. An organisation such as a gas supply company might well adopt both an object-based and a network-based view of their facilities, depending upon whether for example they were concerned with replacing a particular conduit, in which case an object-based view would be appropriate, or analysing flow for purposes of re-routeing, in which case the network model would be appropriate.

Fields

The field-based view is appropriate for modelling phenomena that are regarded as continuously variable across some region of space. Examples of phenomena that may be treated as fields include the concentration of pollutants in the atmosphere, the temperature of the ground surface, the moisture levels in soil, and the speed and direction of flow in bodies of air and water. Fields may represent either two or three spatial dimensions, depending on the application. A 2D field is one for which at any given location in two dimensions the field has a single value for the represented phenomenon. A 3D field is one that has a single value for any location defined in three dimensions. Some phenomena such as atmospheric pollution are naturally 3D in space, but may still be represented by a 2D field, if the nature of the available data is such as only to provide the possibility of estimating one value for each 2D location.

A field-based spatial model is often adopted when the data to be modelled are not known in sufficient detail to provide precise boundaries, even though at some resolution they could be said to exist. An example of this is the distribution of plant species, in which, in theory, it might be possible to represent every single specimen, but in practice it is neither practicable nor of interest to do so, particularly if the phenomena are continuously changing to some degree over short time-scales.

There are several types of data that are sometimes regarded as fields and sometimes as objects. One reason for choosing one model rather than another is the form in which the data were surveyed. Studies of crop type or forestry might adopt a field-based view if the data were obtained from satellite imagery, in which a value for the phenomenon is initially provided for every location within the region of interest; while an object-based view might be adopted if the data were provided in the form of surveyed linear boundaries of areas which were to be treated as internally homogeneous. If the application is concerned with categorising space into broad subdivisions, then a field-based model may be transformed to an object-based model, since the latter is more suited to measurement and analysis of discrete areal or linear features.

Representations of geographic information

Spatial data models for GIS

Geographical information systems provide methods for representing spatial data that allow the user to adopt conceptual models resembling to a large extent the three classes of model just described. Often, however, the spatial data concepts presented to the user are closely related to the source data from which a computer-based spatial model may be constructed. In this respect there tends to be an emphasis on the geometric primitives that characterise surveyed spatial data. There are two broad categories of spatial data model encountered in commercial GIS, often within the same system. These are the *vector data* model, which represents phenomena in terms of the spatial primitives, or components, consisting of points, lines, areas, surfaces and volumes; and the *raster data* model (sometimes referred to as the *grid model*) which represents phenomena as occupying the cells of a predefined, grid-shaped tessellation.

The vector approach emphasises the existence of discrete phenomena, delineated by their boundaries (points, lines and surfaces), and hence may be regarded as *object-based*. However, the facility to represent surfaces, in some vector-based GIS, brings with it the possibility of modelling 2D fields, with perhaps the most common example being the elevation of the earth's surface. Raster technology places emphasis on the contents of grid-cell locations in space and hence is sometimes described as *location-based* (Peuquet, 1984). The raster data model may appear to resemble the field view described above, but the stored spatial data model is not a description of a continuous variable, rather it is a set of grid-cell values that can certainly be regarded as sampling a field model, but can equally be regarded as sampling an object-based model.

Before elaborating on the characteristics of vector- and raster-based data models, we will consider the use of non-spatial attributes that describe the real-world phenomena represented in the models.

Non-spatial attributes

Both location-based and object-based models share a common purpose of representing the distribution in space of phenomena, whether physical or social. The way these phenomena are referred to in terms of the non-spatial attributes of *identification* and *classification* is not, in general, dependent upon the spatial model, although location-based models tend to highlight the distribution of classes of data as opposed to specific, named instances of the classes.

Representation of the identity of specific instances of spatial phenoma is achieved in computers by recording either text, such as the name of a town or county or the address of an individual building, or by means of unique numerical or alphanumeric (numbers and letters) codes. Unique codes may often be used in addition to textual names to avoid the possibility of ambiguity and to enable all discrete instances of phenomena of interest, such as parcels of land, to be given identifiers whether or not they have distinguishing names in the context of the application. When dealing with geometric data in the form of points and lines, it is not uncommon for the same geometric primitive to represent more than one real-world phenomenon. Thus a line might represent both the path of part of a river and part of a county boundary. Hence geometric primitives may have a single unique geometric object identifier and one or more real-world object identifiers. Equally, in some systems, the same location, again such as a river and county boundary, may be represented by more than one geometric primitive, each uniquely identified and associated with its respective feature class.

The way in which data are classified in a GIS is of paramount importance in enabling the user to access selectively information relevant to a particular purpose or query. Many organisations create their own classification systems designed to suit their own purposes. However, national survey organisations have a responsibility, in theory at least, to adopt and be consistent in the use of 'standard' classification schemes that will be recognised by other organisations that need to use their data. Since individual classifications may be described by relatively long-winded pieces of text, such as 'divided dual carriageway primary route', it is normal practice to create a set of short codes that designate such classifications. Each item of spatial data that has the classification may be accompanied by the relevant code, while textual explanation of the codes is stored only once in a code dictionary.

Many classification schemes are hierarchical in nature and this hierarchical structure may be reflected in the form of the codes. For example, transport system codes might all begin with a 5, while all road classes of a transport system begin with 51. Each subclass of road might then be indicated by a further digit, and so on. Organising numeric or alphanumeric codes in this way facilitates retrieval of data at different levels of generalisation. In the example given, it would be possible to retrieve all roads from a database by searching for data with classification codes beginning with the digits 51.

Layers and coverages

The common requirement to access data on the basis of one or more classes has resulted in several GIS employing organisational schemes in which all data of a particular level of classification, such as roads, rivers or vegetation types, are grouped into so-called *layers* or *coverages* (Figure 2.9). The concept of layers is to be found in both vector and raster technology GIS. Typically the layers can be combined with each other in various ways to create new layers that are a function of the individual ones. However, it may be that the layer approach does not rest easily with an entirely object-based modelling approach, since the definition of new individual objects in terms of existing objects can become quite a complicated procedure.

A characteristic of each layer within a layer-based GIS is that all locations with each layer may be said to belong to a single areal region or cell, whether it be a polygon bounded by lines in a vector system, or a rectangular grid cell in a raster system. The idea of directly neighbouring, but non-overlapping, regions is described as *planar enforcement*. It may be contrasted with the possibility of independently digitised areal regions that could turn out to be overlapping in space. The latter type of data are referred to as *spaghetti data*. It should be remarked however that though areal regions cannot overlap in a planar enforced layer, it is possible for each region to have multiple attributes corresponding to multiple perspectives on the meaning of that region.

Vector data models

Vector data models treat phenomena as sets of primitive and composite spatial entities. In 2D models the primitive entities are points, lines and areas, while in three dimensions surfaces and volumes are also employed (Figure 2.10). Which individual primitives are used depends upon the scale of observation or level of generalisation. In a small-scale representation, phenomena such as towns may be represented by individual points, while roads and rivers are represented by lines. As the scale of representation increases, so the real-world areal extent of phenomena must be taken into account. At a medium scale a town might be represented by an areal primitive recording the boundary of the town. At larger scales the town would be presented as a complex set of spatial primitives representing the boundaries of the buildings, the roads, the pavements, the parks and the other physical and administrative phenomena of which it is composed.

The expression *vector model* arises from the fact that the primitive spatial entities are themselves usually defined in terms of coordinates. The location of a point is described by a single set of coordinates in two or three dimensions. A line is usually defined by an ordered sequence of two or more sets of point coordinates. The path of a line between the specified coordinates is then implied as a linear function or a higher order mathematical function, which itself may be evaluated at a set of intermediate points. An area is typically defined by a boundary consisting of one or more lines that form a closed loop. If the area has holes in it then more than one such loop may be used to describe it.

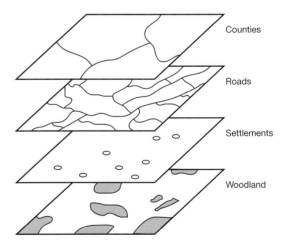

Figure 2.9 Some GIS organise spatial data into layers. Typically layers represent information belonging to particular classes, and can be combined to create new layers containing selected information specific to a particular query on the GIS.

There are requirements for a variety of 3D models, depending upon the type of application. Terrain modelling applications require either simple, single-valued surfaces (i.e. one value for each point in 2D space), which may represent only the ground elevation, or they may be combined with topographic features 'draped' on the terrain surface. In landscape architecture it is necessary to combine terrain surfaces with 3D representations of the features, such as buildings and vegetation, located on them. For cartographic purposes, the traditional method of representing terrain surfaces is by means of contours, but as indicated earlier in this chapter, contours are not a convenient representation for purposes of analysis. If surfaces are originally sampled as contours (perhaps having been digitised from a map), they will usually be transformed to one or other of the most common GIS-based terrain representations, which are the regular grid and the triangulated irregular network (TIN).

The regular grid, or matrix, of point values may be derived directly from an original regular sampling scheme or, as is often the case, by interpolating from an irregularly distributed set of values, which might include digitised contours and spot heights. TINs are characterised by the fact that they retain original irregularly sampled data values. A triangulation procedure is then used to construct a surface of triangular facets which connect these original points.

The facets of a TIN are by default regarded as planar, but curved surface functions may also be used to interpolate between the vertices.

If a TIN is used to represent a single-valued surface (whether of terrain data or otherwise), then in combination with an interpolating function, it provides a digital representation of a 2D field. Similarly, if a grid of point samples is accompanied by a function to interpolate between the points, then it also implements a field model. The representation and manipulation of single-valued surfaces is an extensive subject in its own right, being relevant to terrain modelling specifically and to any other data that are regarded conceptually as a field. It is addressed in more detail in Chapter 12.

If volumetric objects are to be stored in vector-based GIS, they are usually defined by one or more surfaces that bound them. Surfaces can be defined by polygonal facets bounded by 3D lines. The set of lines and constituent points, or vertices, defining such a surface is referred to as a polygonal mesh or polygon network. The facets of the mesh may be regarded implicitly as planar or they may be curved. In both cases a mathematical function must then be used to infer the location of the surface between the discrete coordinates. If a smooth surface is required, mathematical surface functions may be fitted through the vertices of the polygon mesh. The computer graphics display of such a surface could then be achieved by

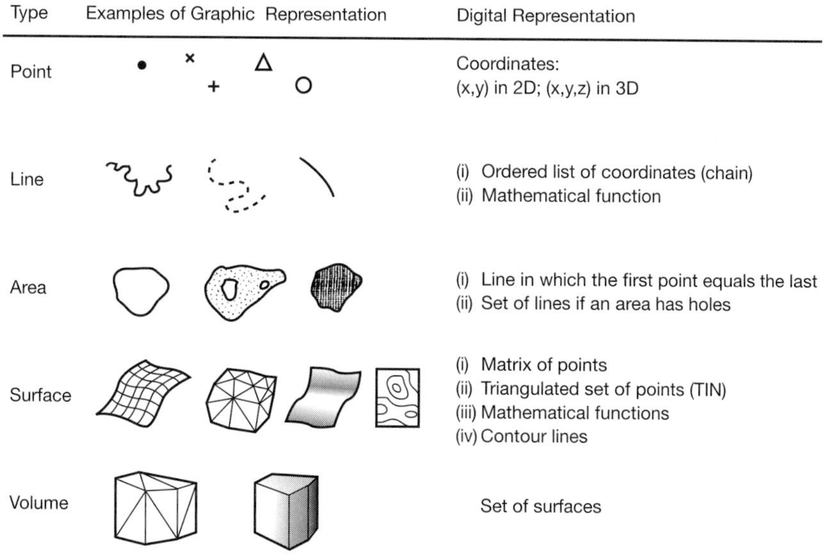

Type	Examples of Graphic Representation	Digital Representation
Point	• × △ + ○	Coordinates: (x,y) in 2D; (x,y,z) in 3D
Line		(i) Ordered list of coordinates (chain) (ii) Mathematical function
Area		(i) Line in which the first point equals the last (ii) Set of lines if an area has holes
Surface		(i) Matrix of points (ii) Triangulated set of points (TIN) (iii) Mathematical functions (iv) Contour lines
Volume		Set of surfaces

Figure 2.10 Geometric primitives can be categorised as point, line, area, surface and volume.

subdividing the mathematical surface with very small planar facets. Examples of mathematical surface functions are B-splines and non-uniform rational B-splines or NURBS (Foley *et al.*, 1990). These types of function provide considerable control over the relationship between the known control points and the fitted surface, including the degree of the surface and how tightly it fits the control points.

Topology

An important aspect of vector-based models is that they enable individual components to be isolated for the purpose of carrying out measurements of, for example, area and length, and for determining the spatial relationships between the components. Spatial relationships of connectivity and adjacency are, as we noted earlier, examples of topological relationships and a GIS spatial model in which these relationships are explicitly recorded is described as topologically structured. In a fully topologically structured data set, wherever lines or areas cross each other, nodes will be created at the intersections and new areal subdivisions defined. In two dimensions, this may be regarded as part of the process of planar enforcement referred to previously.

Vector-based spatial data models that are topologically structured are often described in a terminology of topological objects or primitives, which are classified in terms of topological dimensions, where an n-cell topological object is of n-dimensionality. Typically topological primitives in three dimensions (Figure 2.11) consist of polyhedra (3-cells) bounded by faces (also referred to as areas), and polygons (2-cells) built up of one or more arcs (1-cell), also called edges and links, joined together at nodes (0-cells). Two-dimensional topological objects consist only of polygons (faces or areas), arcs and nodes. For areal information, adjacency between the areas can be recorded in terms of feature codes associated with the left and right sides of arcs. The expressions 'left' and 'right' are given meaning in this context by specifying the direction of the arc in terms of a 'from node' and a 'to node'.

A hypothetical map that includes polygons with holes is illustrated in Figure 2.12. Note that all arcs have their direction specified. The composition of each polygon can be defined by listing the component arcs, including a negative sign where necessary to ensure consistency in arc direction. In order to distinguish between external and internal boundaries, a

convention of clockwise for the former and anti-clockwise for the latter (or vice versa), can be adopted. For the purpose of network analysis, each node may be associated with a list of the arcs which it bounds. The list of arcs connected to each node will generally be in a predetermined order, namely clockwise or anticlockwise.

Topological structure is important in keeping track of the components of complex objects and in determining the spatial relationships of connectivity and adjacency between recorded phenomena. Thus if two lines cross each other they will share a common node. If two areas are adjacent to each other, such as two neighbouring counties, they will share a common boundary arc. If the boundary of a county coincides with the path of a river they might also share the same arc. The inclusion of one area in another, such as a specific type of forest within a county, will result in their sharing common polygons.

The presence of these various spatial relationships can be determined by relatively simple comparisons of the identifiers of their topological components, rather than requiring possibly computationally demanding geometric calculations based on coordinates. It may also be noted that because shared spatial objects are only stored once, though perhaps referenced many times, storage space is saved by avoiding duplication of the same geometric data. This in turn assists in the maintenance of the integrity of the database by avoiding the possibility of two different versions of the same geometric components.

The relatively high degree of topological structuring described with reference to Figure 2.12 is not to be found in all vector-based GIS. Thus there is often a greater emphasis on multiple, topologically structured and feature-specific layers, in which only objects of particular classes are recorded. If this is the case, then arcs might be duplicated between layers if they represent the boundary or the path of more than one class of phenomenon. As indicated earlier in the context of feature identifiers, lines representing the same location in space might be present in multiple versions, possibly having been digitised more than once for the different purposes. When analyses that combine attributes from different layers are required, it would then become necessary to overlay the individual layers and find intersections between the arcs by geometric processing. The result of overlaying all layers in a layer-based GIS would be to produce a fully topologically structured representation, comparable to that described in Figure 2.12.

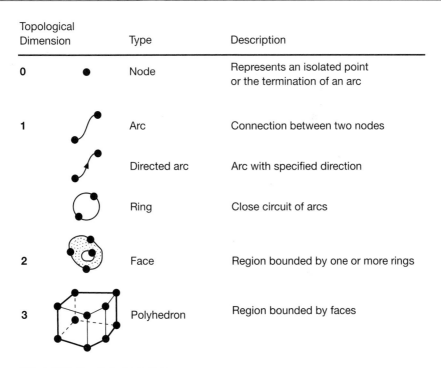

Topological Dimension		Type	Description
0	●	Node	Represents an isolated point or the termination of an arc
1		Arc	Connection between two nodes
		Directed arc	Arc with specified direction
		Ring	Close circuit of arcs
2		Face	Region bounded by one or more rings
3		Polyhedron	Region bounded by faces

Figure 2.11 Topological primitives.

Raster data models

Raster-based spatial models regard space as a tessellation of cells, each of which is associated with a record of the classification or identity of the phenomenon that occupies it (Figure 2.13). The term 'pixel', meaning picture element, is commonly used to refer to a cell, and indeed both it and the term 'raster' derive from the context of image processing in which individual images may be created by a raster scanning process of the sort used to create video images in television cameras. Raster data in GIS are frequently derived from scanning devices used in remote sensing from aircraft and satellites and from those used in digitising documents.

Information systems that use the raster model usually apply the layered approach described in a previous section. In each layer the raster cells record the presence of phenomena of particular classes. The value of the individual cells indicates the categories of phenomena within the given class. In a soils layer, for example, the category might be the dominant type of soil within the cell.

Since the cells are of a fixed size and location, rasters tend to represent natural and human-made phenomena in a blocky fashion. Thus boundaries between categories are forced to follow grid-cell boundaries, as each cell of a raster layer is usually allocated to a single category. The extent to which this may misrepresent the distribution of a phenomenon depends upon the size of the cells relative to the feature of interest. If the cells are sufficiently small relative to the feature, the raster may be a particularly effective means of representing the often somewhat random distribution of the boundaries of natural phenomena that may tend to merge gradually into each other, rather than being neatly delineated.

If each cell is confined to a single classification, the raster model may still fail to represent adequately the transitional nature of change of some natural phenomena. Unless sampling is reduced to a microscopic level, many classes of data, such as soils, sediments and vegetation, are in fact mixtures of categories. Such fuzzy characteristics can be represented more effectively in a raster by means of *mixed pixels*, in which the component categories are represented by measured or expected percentages of the total composition of the cell. It should emphasised though that, in general, the cells of a raster are allocated only a single value.

For the purposes of GIS data processing, an important property of the raster model is that because the cell locations in each raster layer are predetermined, it makes it very easy to carry out overlay operations to compare attributes recorded in different layers. Because the cell locations are predetermined and are usually the same for different layers in a particular

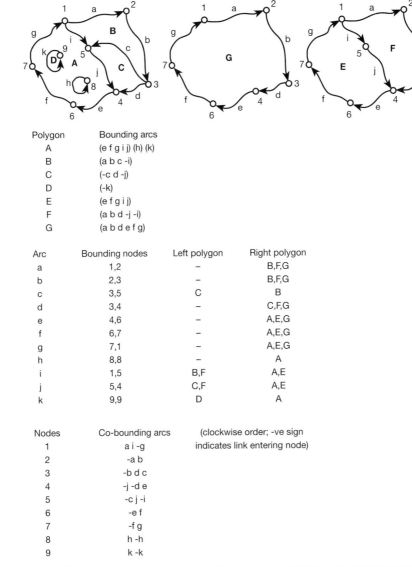

Polygon	Bounding arcs
A	(e f g i j) (h) (k)
B	(a b c -i)
C	(-c d -j)
D	(-k)
E	(e f g i j)
F	(a b d -j -i)
G	(a b d e f g)

Arc	Bounding nodes	Left polygon	Right polygon
a	1,2	–	B,F,G
b	2,3	–	B,F,G
c	3,5	C	B
d	3,4	–	C,F,G
e	4,6	–	A,E,G
f	6,7	–	A,E,G
g	7,1	–	A,E,G
h	8,8	–	A
i	1,5	B,F	A,E
j	5,4	C,F	A,E
k	9,9	D	A

Nodes	Co-bounding arcs	(clockwise order; -ve sign
1	a i -g	indicates link entering node)
2	-a b	
3	-b d c	
4	-j -d e	
5	-c j -i	
6	-e f	
7	-f g	
8	h -h	
9	k -k	

Figure 2.12 A topologically encoded map. Polygons (also called faces in topological terminology) describe areas bounded by lists of arcs (sometimes called links). Each directed arc is defined by its bounding (terminal) nodes and by the polygons to the left and to the right. Notice that in the map shown there is a hierarchical structure and hence some arcs are the boundaries of more than one polygon. Typically GIS only store one left and one right polygon identifier. Each node may also be associated with the identifiers of each arc that is bounded by the node.

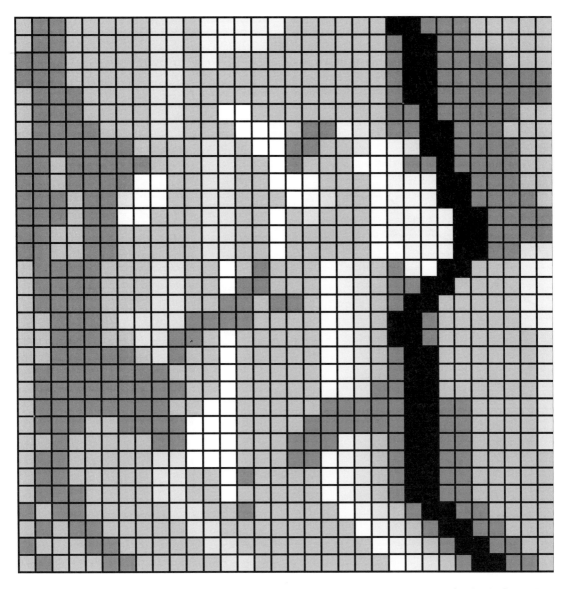

Figure 2.13 The raster model represents the two-dimensional location of phenomena as a matrix of grid cells. Each cell stores a data item defining the identity, the class or the value of the represented phenomenon.

application, each attribute can be combined logically or arithmetically with attributes in corresponding cells of the other layers to create a new attribute value for the resulting overlay. This differs from layer-based vector models in which the areal units of the layers are independent of each other, such that direct comparisons between layers cannot be made without further processing to identify the nature of the overlap (as indicated at the end of the previous section).

Voxels

The grid-cell based representation of raster GIS can be extended to three dimensions to create a voxel model in which cells consist of rectangular, typically cubic, volume elements (Figure 2.14). Some types of geoscientific data, such as that concerning subsurface geological bodies, are not always easily modelled by boundary representations, since the data values may relate to an attribute that may vary from place to place without clearly understood boundaries (e.g. ore grade or porosity). A more appropriate model for this type of data is that of the voxel model.

Voxel models have been widely used in medical imaging, where they have been derived from computer-aided tomography (CAT or CT) and nuclear magnetic resonance (NMR) scanners. They are good for representing gradual and very localised variations and are suitable for creating cross-sectional views of such variations.

Multimedia models

The increasing storage capacity of modern computers has led to a greater diversity in the types of data that are stored. A notable trend in recent years is towards *multimedia* representation of image, audio and video media items in addition to the more traditional textual, numerical and graphics-based representations. This can be very useful in allowing the user to see a

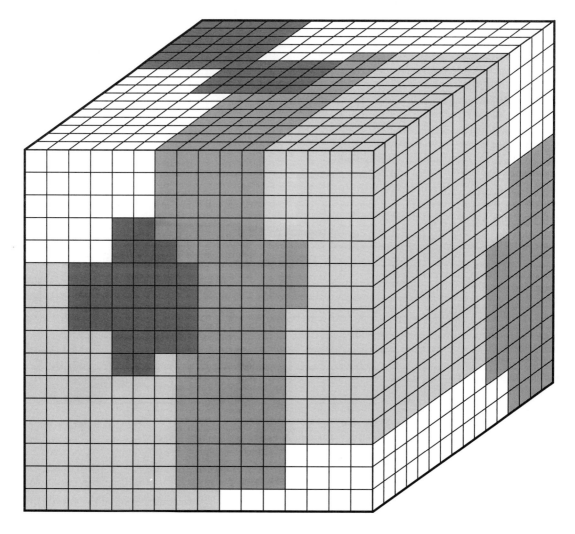

Figure 2.14 A voxel model consists of a three-dimensional matrix of cells in which each cell stores a data item defining the identity, the class or the value of the represented phenomenon.

photograph or a video (e.g. of a house or a landscape) that corresponds with other stored data.

One characteristic of digital media objects is that they are usually somewhat less structured than other data, their interpretation being to a large extent dependent upon the human senses of sound and vision. Clearly, visual sense is required to appreciate the content of a graphical display of a map or a graph, but usually in these latter cases the components of the graphic are individually identified within the computer and to varying degrees structured, as described in the previous sections. A scanned photographic image of a street scene or a video sequence of a forest fire may have a great deal of useful information in it, but it is often the case that the computer 'knows' relatively little about it, other than perhaps a few key words relating to the subject matter.

The benefits of multimedia databases have resulted in research into methods of linking together related types of media objects through semantic relationships that depend upon knowledge of the information content. Interactivity with images can be enhanced by designating features in the image as buttons or hot spots that the user can point to in order to obtain more information, which might be provided by text, sound or structured graphics, or by displaying another image or video sequence.

Summary

In this chapter we reviewed the main types of information that are represented in GIS and we have introduced the primary spatial data models that are used for representing the spatial components of that information. The main information types have been categorised at the highest level as spatial and non-spatial, the latter consisting of semantic and statistical aspects, and we have noted that temporal data provide an essential additional dimension. Semantic information gives meaning to spatial data and is concerned particularly with identification, with names and unique identifiers, and with classification of what are usually hierarchically structured, often overlapping concepts. Associated with the miriad classes of data are statistics which describe the nominal, ordinal and numerical values of particular attributes. Spatial information can be categorised into that concerned with locations, shape descriptors and patterns, relationships (topological, proximal and directional) and correlations and interactions between phenomena. Locations can be described either by name, or quantitatively in terms of geometric primitives, such as points, lines and areas, referenced to a coordinate system.

Three conceptual models of spatial information relevant to GIS have been identified. These are the object-based, network and field models. Object-based models emphasise discrete objects which may be described in detail usually in terms of their boundaries and in terms of other objects from which they may be composed or to which they are related. Network models represent interactions between specific objects, such as flows of water or of traffic. The field model represents data that are regarded as continuously variable either in two or three spatial dimensions.

GIS packages are commonly classified in terms of vector and raster models, where vector models adopt a largely object-based view. Raster GIS represent space as a regular tessellation of grid cells, which may be regarded as sampling either a field or an object-based model. Vector GIS, in following the object-based model, facilitate recording the structure of discrete phenomena, the spatial relationships between them and the performance of spatial analyses that depend upon an explicit record of the structure and connectivity of objects. They often include a capability for implementing network models. The raster data model is advantageous for representing the spatial distribution of phenomena, particularly natural features, that may be imprecisely defined. Raster models are appropriate for handling data acquired by remote-sensing techniques involving scanners that create an array of pixels that must subsequently be classified. Multimedia models have also been introduced in this chapter as a hybrid representation that enables the maintenance of image, sound and video media.

Further reading

Many of the representations of geographic information used in GIS may be regarded as based on a variety of spatial structures and processes documented in textbooks such as Haggett *et al.* (1977), which arose from the quantitative revolution in geography. The use of topological data structuring in GIS may be traced back to the work of Corbett (1979) at the US Bureau of Census. Early efforts to define spatial models in the specific context of GIS are found in Peuquet (1984, 1988). Discussions of the object-based and field views are to be found in, for example, Goodchild (1992) and Couclelis (1992). For a review of some of the various concepts of geographic space, including a discussion of the treatment of space in GIS, see Nunes (1991). Several conference proceedings from the early 1990s provide rich sources of articles discussing concepts and representations of geographic space. In particular, see Mark and Frank (1991), Frank *et al.* (1992), Frank and Campari (1993) and Frank and Kuhn (1995).

CHAPTER 3

GIS functionality: an overview

Introduction

In the previous two chapters we have reviewed the main applications of GIS and have summarised the characteristics of geographically referenced information. In this chapter we complete our introduction to GIS by providing an overview of the capabilities that can be expected of commercial GIS packages. Most commercial packages provide a range of functionality, which we categorise here into the five areas of (1) data acquisition; (2) preliminary data processing; (3) data storage and retrieval; (4) spatial search and analysis; and (5) graphical display and interaction. The available GIS packages differ considerably in their strengths and weaknesses in these areas and they also differ in the techniques that they use for implementing the functions.

Figure 3.1 illustrates the relationship between these broad categories of function and the different representations of information upon which they operate. Data acquisition functions derive data from observations of real-world phenomena and from existing documents and maps, some of which could already be in digital form. The resulting data are described here as raw data, implying that they may not be directly usable for purposes of query and analysis. Preliminary data-processing functions transform the raw data to structured data which can be operated upon directly by functions for spatial search and analysis. The results of particular queries or analyses are referred to here as interpretations, being specific combinations, subsets or transformations of the structured data. Facilities must be provided to store data at all levels, from raw data to interpretations, and hence functions for storage and retrieval operate upon all of these representations. Similarly the functions for display and graphical interaction must be available at all stages.

The remaining sections of the chapter are organised according to the fivefold subdivision of functionality indicated above. The longest section in this chapter is that on spatial search and analysis. It is here that we review the fairly wide range of tools that enable users of GIS to retrieve and analyse their data in a manner that is distinctly characteristic of GIS technology. This overview should help to set the scene of much of the rest of the book, in which the chapters focus on concepts, techniques and methodologies that underly the various functions performed by GIS.

Data acquisition

Data acquisition refers to the process of obtaining data in a form that can be input to a GIS. At its simplest level of implementation it could consist of the facility to interpret the format of some digital data sets that were supplied from an external source. The effectiveness of this type of facility depends upon it being able to recognise a variety of standard or *de facto* standard formats for data exchange, such as DLG, DXF and NTF, and the 'export formats' of widely used GIS packages such as ARC/INFO and IDRISI. GIS may also be able to import images, such as scanned photographs, which may be stored in formats such as TIFF and GIF. There are now

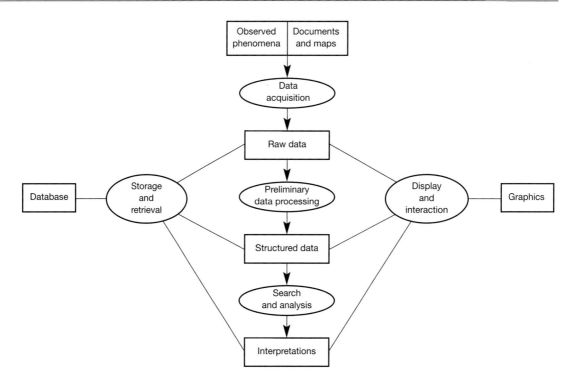

Figure 3.1 Overview of the five categories of function to be found in GIS (text in ovals) and the different representations (in rectangular boxes) upon which they operate.

many sources of spatial data and some organisations may be able to meet their requirements fully by obtaining such data. Frequently, however, organisations have to accumulate their own digital data sets relating to subject matter specific to their business or research activities.

The survey techniques which may be employed for *primary data acquisition* range between ground-based studies of topography, geology and vegetation, airborne surveys using satellites and aeroplanes, and socio-economic studies that could involve interviews and transcription of written documents (Figure 3.2). Ground-based studies often result initially in hand-drawn maps and written reports which must subsequently be digitised. It is common practice though, particularly in engineering surveys, to use hand-held computers, or data loggers, for recording information directly in digital form. The use of remotely sensed data from satellite-based instruments and from aerial photography has assumed

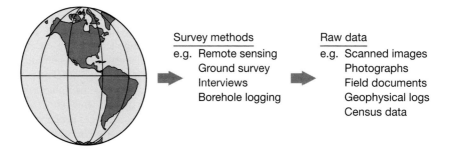

Figure 3.2 Primary data acquisition generates raw data from observation and survey.

major importance in many studies of natural resources and in topographic mapping. Much of the satellite imagery used in GIS has a ground resolution of between 10 and 80 m, which makes it very useful for studies of environmental variables such as vegetation and hydrology, but its use for general topographic mapping is limited to smaller scales such as 1:50 000 or less. Photogrammetric interpretation of aerial photographs, on the other hand, can produce highly accurate data which, can be used for creating and updating large-scale maps, such as at 1:500 or 1:1000 scales.

A major source of spatial information is published and unpublished maps in analogue form. To make use of this information requires processes of *secondary data acquisition* (Figure 3.3). This may involve manual digitising, in which an operator traces the shape of features on the map and separately enters attribute data describing the features. More automated approaches involve the use of raster scanning digitisers which produce a complete digital image of the map. The output from raster scanners is a 2D array of pixel values which for most practical purposes must be vectorised to generate an explicit digital representation of the point-referenced, linear, areal and text features on the map. This process of vectorising may sometimes be performed interactively with a screen display of the scanned map. Such on-screen digitising may be assisted by pattern-recognition software that can follow lines automatically, converting them to vectors in the process. Some systems also provide automatic text-recognition functions.

Digitising procedures must include *data coding*, *verification* and *error correction* if the resulting data are to be used reliably in a GIS. Data coding is the process of attaching descriptive feature attributes to all geometric objects, consisting of points, lines and areas. This is done typically at the same time as digitising the geometry, by typing at a keyboard or selecting from a menu.

Verification entails comparing plots of the digitised data with the source document. Care must be taken to ensure that all required features have been digitised and that they have been done so to acceptable levels of accuracy. An important part of verification and error correction for geometric data is ensuring that lines which are supposed to join up actually do so precisely. The precise fitting together of connected components of a map is a necessary precursor to topological data structuring, introduced in Chapter 2.

In general, data acquisition is a critical stage in many geographical information tasks. This is partly because it may be the most time-consuming and expensive phase of an operation. Equally importantly, it is at this stage that many aspects of the quality of data are determined, in terms of accuracy and in particular the association of information which records the source of data and the error inherent in it (see Chapter 7).

Preliminary data processing

Major aspects of preliminary data processing include creating topologically structured data and, in the case of remotely sensed data, classifying the features in scanned images in terms of phenomena of interest. As we saw in Chapter 2, there are several clearly different conceptual models of spatial information, namely object-based, network and field-based. Analyses based on these different views require the data to be represented and organised appropriately. It is important, therefore, to provide facilities to enable the GIS user to change the structure and sample schemes of data in order to adapt them to different requirements. This requires functions not just to create topologically structured vector data and raster data models in the first place, but to change between representations, to

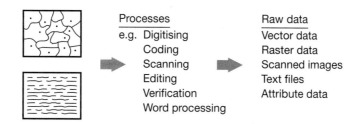

<table>
<tr><td></td><td>Processes</td><td>Raw data</td></tr>
<tr><td></td><td>e.g. Digitising</td><td>Vector data</td></tr>
<tr><td></td><td>Coding</td><td>Raster data</td></tr>
<tr><td></td><td>Scanning</td><td>Scanned images</td></tr>
<tr><td></td><td>Editing</td><td>Text files</td></tr>
<tr><td></td><td>Verification</td><td>Attribute data</td></tr>
<tr><td></td><td>Word processing</td><td></td></tr>
</table>

Figure 3.3 Secondary data acquisition involves deriving data by digitising existing maps and documents.

modify classification and sampling schemes, to simplify or generalise data, and to transform between different coordinate systems and map projections (Figure 3.4). Such operations may be regarded as preceding operations for spatial analysis. The stage at which the preliminary data processing tasks take place may vary somewhat according to the purpose of a GIS. Thus in a well-defined, single application context, all such manipulation may take place in conjunction with the data acquisition stage, prior to permanent storage of data in a chosen format. In other circumstances the acquisition of data from a range of sources, intended for a variety of applications, may require much more flexibility to modify data on an *ad hoc* basis.

Certain types of analysis tool are heavily dependent upon the use of a raster data model (Tomlin, 1990). The required conversion to a raster from a vector data model, in order to use such a methodology, is relatively straightforward. It can be done with rasterisation algorithms, which were developed originally for use in computer graphics systems and are indeed essential for most computer graphics display techniques used in computer cartography (Figure 3.4c). The converse transformation, from a raster to a vector model, termed vectorisation, was referred to earlier in the context of automated scanning and is not so straightforward as rasterisation (Figure 3.4d).

We may note the fact that rasterisation is a process of subdivision of a given, well-defined linear or areal object into cells, effectively by chopping up simultaneously along horizontal and vertical slices. Vectorisation, on the other hand, requires reassembly of the pieces, i.e. the pixels, often without prior knowledge of what piece belongs to what object. If the raster model was originally derived from a well-structured vector model, it may be fitted back together fairly reliably. If the raster data were not well structured, perhaps due to their having come from a scanned photograph or satellite image, it is a complex task of pattern recognition. Typically the process involves many ambiguous situations which may require human intervention to resolve. Chapter 8 describes the techniques involved in transforming between raster and vector representations, while Chapter 6 includes a summary of the image processing procedures required to process satellite imagery.

Although the transformation of discrete objects from vector format to the rectangular cells of a raster model is a well-defined procedure, transformations of arbitrarily located point-referenced data to a regular grid is a less precise procedure (Figure 3.4e).

Re-sampling of originally irregularly distributed points may be done by a variety of interpolation techniques which attempt to model the variation of the statistic at unsampled locations as a function of known sample points in the vicinity. An alternative to creating a regular grid is, as mentioned in the last chapter, to create a triangulated irregular network or TIN (Figure 3.4f). Interpolation techniques fall into the broad field of surface modelling, which is an important aspect of spatial analysis. Although interpolation to a regular grid may be an objective in its own right in some circumstances, it may also be treated as an intermediate representation from which others, such as area-class maps and contour maps, may be derived. The principles of interpolation from irregular point samples to a regular grid, interpolation between areal units, and the construction of TINs are described in Chapter 12.

When comparing data from a variety of sources, a common problem is the use of two or more classification or coding schemes referring to the same phenomenon. This might be due to international differences or to changes in time as schemes are improved. Provided the different schemes refer to the same phenomenon, re-coding from one to the other may be a fairly mechanical process. Where one is more detailed than another, or recognises different aspects of the phenomenon under consideration, it may be necessary to introduce approximations or simplifications to transform to a common scheme.

A relatively simple type of re-coding is that of reclassification whereby several classes are combined to form a less-detailed, or generalised, class (Figure 3.4g). Thus on a geological or soil boundary map, certain internal subdivisions could be removed to produce larger polygons identified by a higher-level classification. This process, when applied to polygons, is also known as line dropping and map dissolve.

A major problem of data integration, particularly in an international context, is that map data may be recorded on a wide variety of coordinate systems based on various map projections. Data from such disparate sources cannot be integrated on a single map without transforming to a common coordinate system. The first stage of coordinate transformation is that from the original recording or digitising system to the real-world units of a map projection. This may be regarded as a necessary part of data acquisition and, in the case of digitising existing maps, it depends for its success upon the quality of the data available to describe the map projection. When the details of the original map projection are

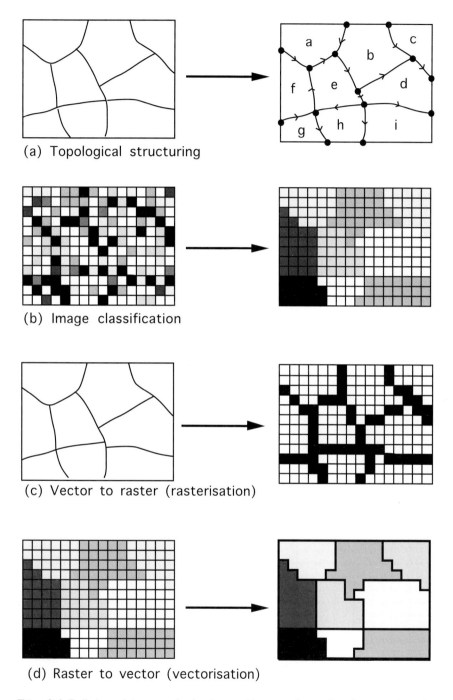

(a) Topological structuring

(b) Image classification

(c) Vector to raster (rasterisation)

(d) Raster to vector (vectorisation)

Figure 3.4 Preliminary data processing involves a wide range of operations for structuring, classifying and transforming representations of data to make them suitable for subsequent query and analysis.

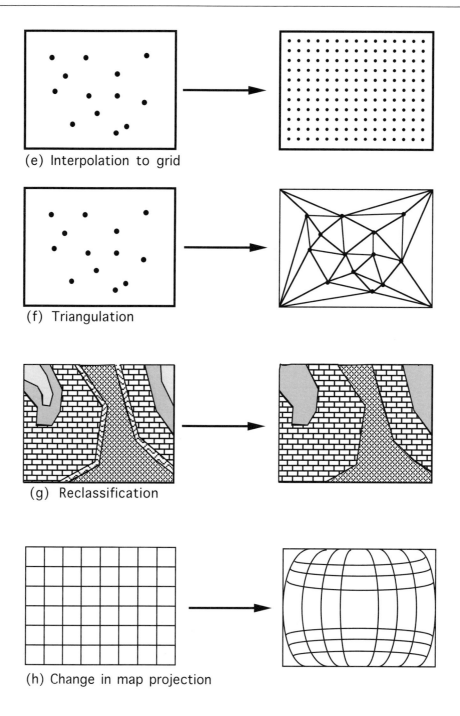

(e) Interpolation to grid

(f) Triangulation

(g) Reclassification

(h) Change in map projection

Figure 3.4 (cont.)

known, the transformation to a corresponding map grid system and between map grid systems can be performed, albeit often involving quite complex mathematics. Problems arise, however, when not all such details are known, in which case they may need to be guessed, or, when this is not feasible, 'rubber sheet' transformations (see Chapter 4) may be used to force (or 'morph') one undefined coordinate system to fit into another.

Data storage and retrieval

The function of data storage concerns the creation of a spatial database. In practice the contents of this database may consist of a combination of vector and (or) raster spatial data, and of attribute data which identify spatially referenced phenomena, attach meaning, via classification codes, and record textual and numerical statistics. The attribute data are usually stored in tables (files) which, in the case of object-based GIS, include a unique identifier for the corresponding spatial object, accompanied by the various attribute data items. The unique spatial object identifier serves as a link between the attribute data and the corresponding spatial data. Sometimes the data items in the attribute tables include spatial data values such as area and length that may have been derived from the geometric data representations (Figure 3.5). For raster data, attribute files typically include data relating to classes of phenomena, such as soils or vegetation, rather than discrete objects. The choice of whether the spatial data are organised according to a vector or a raster data model is often made at the data acquisition stage, since each model corresponds to a different approach to information sampling and description. However, as previously indicated, many GIS databases enable the maintenance of both spatial models. When building a spatial database it may be necessary to link together tables of data relating to the same phenomena. This is described as a join operation, and is described in Chapter 10.

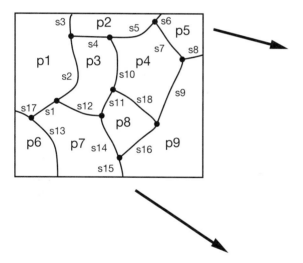

Non-spatial data

Polygon id	Parcel no.	Owner	Date
p1	856	F. Hollins	1976
p2	455	D. Newson	1942
p3	152	L.U. Doyne	1899
p4	357	F. Futter	1902
p5	358	F. Futter	1903
p6	480	R.F. Mutter	1930
p7	390	O.R. Doyne	1900
p8	840	P. Gween	1970
p9	362	F. Futter	1902

Geometric (spatial) data

Polygon id	Line segment ids
p3	s2, s12, s11, s10, s4
p4	s7, s5, s10, s18, s9
p8	s11, s14, s16, s18
. . .	. . .

Line seg id	Line segment coordinates
s1	(x1,x1) (x2, y2) (x3,y3). . .
s2	(x7,y7) (x8,y8). . .
s3	(x10,y10) (x11,y11) . . .
. . .	. . .

Figure 3.5 Tables of attribute data can be linked to geometric data by storing in the table the unique identifiers of the corresponding geometric objects (typically points, lines or areas).

In the terminology of database management systems, raster and vector data models may be seen as examples of *conceptual models,* which describe real-world application-related concepts that we wish to represent in the database. Conceptual models may be described at different levels of abstraction. Vector and raster data models are, as indicated in Chapter 2, at a somewhat lower level of abstraction, being nearer to computer data representations, than the object-based, network and field-based models. The term *logical data model* is used to refer to the way in which the database management system organises the conceptual models in terms of database-specific concepts such as files, records and indexes. The primary examples of logical data models are *relational, network, hierarchical* and *object-oriented.* We have already indicated that, at least until recently, conventional database technology was not well adapted to geographical data handling. Several widely used GIS, such as Arc/Info, are based on a combination of the relational model, for handling non-geometric attributes, and special-purpose, non-relational schemes for storing and manipulating the spatial data. Some GIS exploit facilities of relational database storage schemes to handle both the spatial data and the attribute data.

Much interest has been shown recently in the object-oriented data model as providing the potential for a single database foundation that can be adapted to the special requirements of GIS. The reason for this is that the database development environment combines both data storage and special-purpose procedures which operate on the data. This results in considerable flexibility, enabling the maintenance of a variety of different types of spatial data models and media, each of which may require different low-level processing functions to maintain them. Object-oriented techniques are thus well suited to the implementation of *multimedia databases* for GIS that store not just spatial models and associated non-spatial attributes, but also sound, video and static images.

Another reason for the interest in object-oriented databases for GIS is that information can be represented in a manner which reflects, to some extent at least, the real-world modelling concepts of the object-based view of spatial information. There is a contrast here between object-oriented databases and the relational database management approach, which enforces a manner of data storage, consisting of fixed tables of rows and columns, and associated operators. However, geographic data are often regarded as con-sisting of complex objects (such as a topologically structured network, or an entire image) that need to be treated as a whole, rather than disaggregated into their lowest-level components which often cannot be usefully manipulated by the standard operators provided by relational databases, intended originally for commercial and business applications. Object-oriented database methods, on the other hand may be regarded as reducing the impedance mismatch between the conceptual model and the lower-level database model (Worboys, 1994).

Retrieval facilities in a GIS database do, however, need to include the types of facility available in standard relational databases. This includes selecting stored data on the basis of exact matching of specified attributes such as names and classes of phenomena (e.g. towns, administrative areas, types of roads and pipeline, etc.). It also includes being able to construct queries that find information in which the data values are equal to or lie within specific numerical ranges (e.g. population exceeding 50 000, or inhabitants over the age of 60). The distinguishing characteristic of queries to a GIS database, however, is the need to be able to specify data in terms of metric location (based on coordinates) and spatial relationships.

Because of the need to retrieve on the basis of location, which is defined in at least two dimensions, spatial databases usually incorporate specialised spatial indexing methods. A typical spatial query is to find stored objects that lie within or that cross a rectangular spatial window, i.e. a rectangular region of space. To help meet such requirements, the spatial index partitions the contents of the database, either on the basis of regular cells, or irregularly in a manner determined by the location of objects or the bounding rectangles of areal and linear objects. Spatial indexing techniques are discussed in more detail in Chapter 11. Although retrieval on the basis of location and spatial relationships may be regarded as basic to data access in a GIS database, the distinction between retrieval and some types of analysis of spatial data can become somewhat blurred. We will therefore include spatial database queries in the context of spatial search and analysis in the next section.

Spatial search and analysis

The acquisition and storage of data in a GIS is normally motivated by a desire to use the data to solve a problem or make decisions relating to a particular application. Many GIS installations include or are associated with data modelling and statistical functions which may be quite specific to their application or organisation. However, there is a common body of GIS functionality which many more specialised functions depend upon and which in a number of cases may be adequate to meet all the needs of an application. We now summarise these various types of function.

Containment search within a spatial region

A relatively straightforward spatial analysis is to find those features, or parts of features, that lie within a given region of space. At its most trivial, the region may simply be a rectangular window defined by four coordinates, or two opposite corners (Figure 3.6). As has just been indicated, this may be regarded as a basic spatial database query rather than an analysis. What raises it above the level of a conventional database query at least is that it requires the capacity to clip spatial objects consisting of linear and polygon components at the boundary of the spatial window, a task that cannot be performed using a relational database query language such as SQL, without additional specialised processing.

Other spatial containment searches may specify the region in terms of an existing areal object. A typical example would be an administrative area such as a county, or a planning regulation zone such as a conservation area (Figure 3.7).

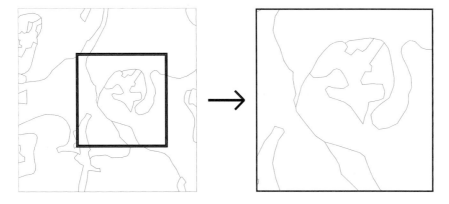

Figure 3.6 Spatial containment search with a rectangular window.

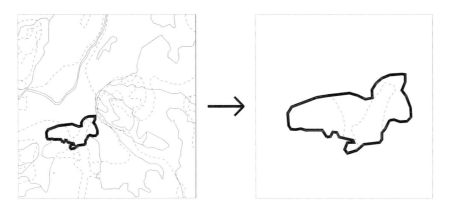

Figure 3.7 Spatial containment search based on an existing object.

Proximal search

We can identify several types of proximal search. The first may be regarded as an extension of the spatial containment search, in that a region of interest is defined in terms of its proximity to an existing phenomenon. Such searches are distance-constrained in that they define a zone that extends some specified distance from the specified object. In terms of GIS functionality, the zone is described as a *buffer* and a distinction may be made according to whether the buffer is constructed around a point-referenced, line-referenced or area-referenced object (Figure 3.8). In raster-based systems, the creation of a buffer may be referred to as a *spread* function.

A second type of proximity search is that which finds regions that are directly connected to a specified object. For example, it could be of interest to find all land parcels that were the immediate neighbours of a specified land parcel that was subject to industrial or building development. In a statistical analysis of the correlation of the incidence of unemployment or of measles between neighbouring areal sampling units (such as enumeration districts), a method is required to find the neighbours of any given unit. The topological data structures associated with vector GIS are specifically intended to facilitate answering queries concerned with determining such topological relations of connectivity.

A third type of proximal search occurs when it is necessary to find the regions of space that are nearest to each of a set of irregularly distributed sample locations, which are typically point-referenced. Such searches underly the creation of *Thiessen polygons,* which define the zones of space around each point that are closer to that point than to any other point. Thiessen polygon zonation schemes, also called *Voronoi diagrams*, have been used to create soil maps from isolated soil samples, on the assumption that nothing is known about the unsampled space between the sample points, and thus for adjacent samples with different soil types, the boundary line is chosen arbitrarily as half-way between them. In Figure 3.9 a Voronoi diagram has been constructed from the point locations of a set of retail outlets, indicating the areas for which each particular shop is the nearest.

A related type of proximal search is that of nearest-neighbour analysis, where the intention is to find, for each of a set of locations, those nearby (but separate) locations that are nearest to it. A common reason for wishing to do this is to interpolate

between data samples. For point locations, the Voronoi diagram indirectly provides a solution to the nearest-neighbour problem in that each edge of a Voronoi polygonal region marks the boundary between a pair of neighbouring points. When each pair of neighbours is connected, the resulting graph is a Delaunay triangulation, which is described as the dual of a Voronoi diagram, meaning that each is derivable from the other.

Phenomenon-based search and overlay processing

Phenomenon-based search techniques may be categorised according to whether the search is for a single type of phenomenon, irrespective of other phenomena, or a search for regions that are defined by combinations of phenomena.

The former case is very much the simpler, since unless any other spatial constraints are specified, the search consists of retrieving spatial objects only on the basis of specified attributes. This might just be the category, as for example in searching for all regions for which the land-use class is 'urban' (Figure 3.10a). Such a search might be further refined by specifying that only urban regions that exceeded a certain area were to be retrieved (Figure 3.10b). Another example would be to find all census enumeration districts in which the unemployment levels exceeded some particular level. In practice, searches of this sort can often be expected to be constrained by location, as described above. Thus the search may be confined to a particular rectangular spatial window, or an administrative area, or it may be subject to a proximal constraint, such as, in the case of the urban areas search, being within 500 m of a particular road.

The facility to search for locations that are characterised by a combination of phenomena is often regarded as fundamental to GIS functionality. One of the early motivating factors in developing GIS technology was the need to perform *suitability analyses*, in which the objective is to find sites or regions that satisfy particular criteria of land usage and terrain characteristics. This process had been performed in the past by overlaying thematically specific maps representing individual factors of interest, such as vegetation type, soil type, slope, elevation and land subject to particular planning regulations. The manual technique of overlay analysis to determine suitable sites for particular purposes was described

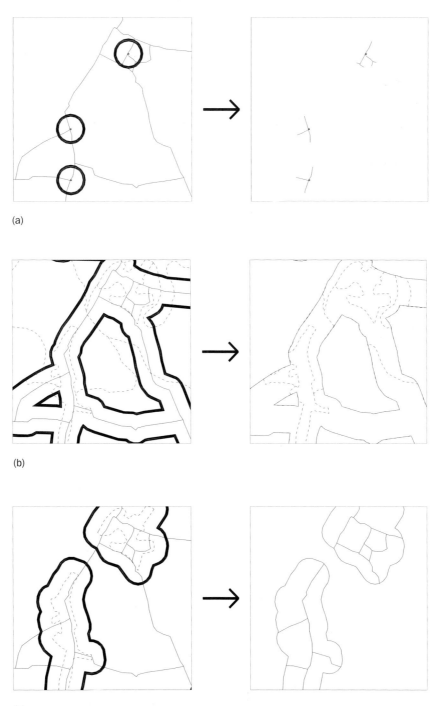

(a)

(b)

(c)

Figure 3.8 Distance-constrained proximal search. Buffer zones of specified radius, equal to the distance constraint, may be constructed around (**a**) point-referenced, (**b**) line-referenced and (**c**) area-referenced features.

and illustrated by McHarg (1969) in his book *Design with Nature*. The automated version of this technique is now widely available in GIS packages, whether based on raster or vector data models. The fact that it is generally referred to as overlay analysis accords well with those systems (the majority at present) that adopt a layer-based approach to data storage. Even when layers are not emphasised in a GIS, as in object-oriented systems, the principle of identifying regions on the basis of combinations of classes of phenomena and their attributes is still fundamental.

The primary types of operator used in overlay analysis are Boolean logical operators and set operators (Figure 3.11); and relational and arithmetic algebraic operators (Table 3.1) applied to the attributes within the layers. Boolean logical operators, consisting primarily of AND, OR, NOT and exclusive OR (XOR), produce results of either true or false. For example, if one layer defines regions where it is either true or false that the shallow geology consists of alluvium (which might be subject to mineral extraction), and another layer has true and false values for the presence of agricultural land of grade 3 (one of the medium or lower quality grades), then the

two layers could be combined using the AND operator to find regions in which alluvium and grade 3 agricultural land coincided (Figure 3.12), and hence could be considered for possible mineral exploitation. Alternatively, use of the OR operator would return all regions that were alluvium and the regions that were grade 3, whether or not they coincided. The NOT operator can be used to reverse (or complement) the meaning of false and true. Thus NOT applied to the alluvium layer would identify all regions that had a geology other than alluvium. The XOR operator would identify as true those regions that were either alluvium or grade 3 land, but not those regions where they coincided.

Note in Figure 3.11 that, when applied to spatial regions that are regarded as either present or absent (or true or false), the set operators of intersection and union are equivalent to the AND and OR operators. The difference set operator provides further functionality that is equivalent to subtracting one region, or set of regions, from another.

Boolean operators can be combined with algebraic (arithmetic and relational) operators to construct more logical expressions to specify the required

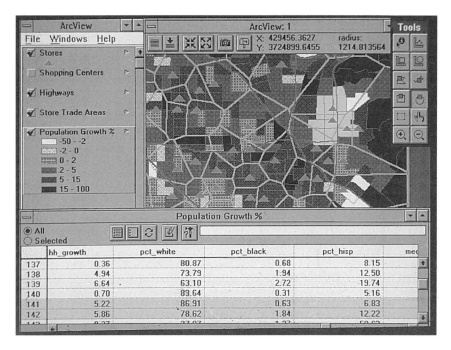

Figure 3.9 The Voronoi diagram defines polygonal regions of space that are nearer to each point location than to any other of the point locations. Courtesy ESRI.

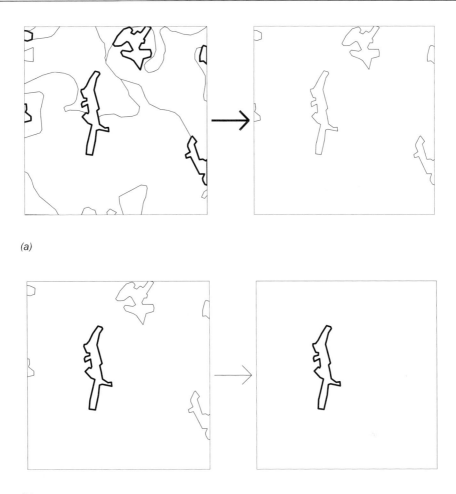

(a)

(b)

Figure 3.10 Phenomenon-based search, using a single theme or class, selects spatial objects that (**a**) belong to particular categories or (**b**) meet specified constraints on their spatial and non-spatial attributes, such as exceeding a given area.

result. In doing so, logical expressions can be applied to layers that include many attributes with various data types, rather than simply true or false, in order to find combinations of attributes that meet some search criteria. Referring to the previous example, we could therefore assume that we had one layer containing cells or regions categorised according to their geology, and another layer containing land categorised according to agricultural land use, of which grade 3 was just one of several categories. We might also wish to specify that only regions that exceeded a particular area were of interest. Alluvium and grade 3 land areas greater than 5000 m^2 might therefore be found using a

logical expression, containing the Boolean operator AND and the relational algebraic operators greater than (>) and equals (=) in the following form:

$$\textit{Select} \text{ geology} = \text{Alluvium } \textit{and}$$
$$\text{landclass} = \text{Grade 3}$$
$$\textit{where} \text{ area} > 5000$$

A notable characteristic of this expression is that it refers to what in a layer-based GIS system would be the contents of two separate layers. In a raster-based GIS it might well be possible to issue a command or commands equivalent to this, and the result would be a third layer which the user would need to specify. In

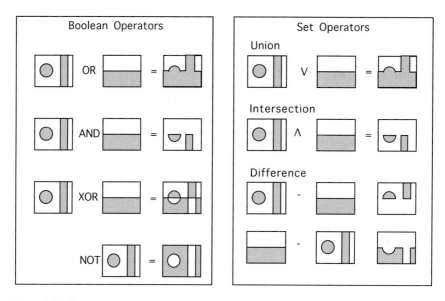

Figure 3.11 Boolean and set operators can be used to combine spatial regions in overlay analysis. The regions may be regarded here as true (shaded) or false (clear) in the sense of representing space that contains or does not contain some specified phenomena.

a vector-based GIS, however, before being able to issue the command, the two layers would usually have to be combined to create a third layer that included attributes from both layers. This might require issuing an overlay command which would combine the polygons in the respective layers to produce a new layer that contained all possible areal subdivisions resulting from superimposing one set of polygons on the other. Figure 3.11 illustrates differ-

Table 3.1 Relational and arithmetic operators may be used to construct logical expressions which specify the phenomena of interest in a spatial query.

Relational operators	
$a > b$	Greater than
$a < b$	Less than
$a = b$	Equals
$a \geq b$	Greater than or equal to
$a \leq b$	Less than or equal to
$a \neq b$	Not equal

Arithmetic operators	
$a + b$	Addition
$a - b$	Subtraction
$a \times b$	Multiplication
$a \div b$	Division
a^b	Exponentiation

ent ways in which two layers could be combined using the set operators of union, intersection and difference. If a variety of logical expressions are to be applied to the data from a pair of layers, then the union operator may be the most appropriate, since it retains all regions from both layers.

Overlay techniques are often used in combination with spatial constraints of containment or of proximity. Continuing with the above theme, an example that uses proximity constraints (see Figure 3.13) would be:

Find regions of alluvium that coincide with grade 3 agricultural land, that are more than 500 m from urban boundaries and more than 100 m from main roads.

Ideally, one would hope to be able to issue a GIS command similar in content to this. In practice, in layer-based GIS, it is generally necessary to carry out several subtasks in order to achieve the required result. Often there are multiple routes to the same result. One approach to performing the above query, in a vector-based GIS, would be to select grade 3 land regions and alluvium regions from each of their respective geology and land-use layers, then to intersect the layers to find all regions in which both types of region coincided. Buffer zones could then be constructed from separate layers containing the urban

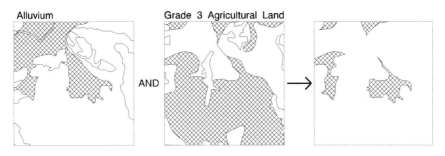

Figure 3.12 Example of a phenomenon-based search combining themes: find regions of alluvium geology that coincide with agricultural land of grade 3.

boundaries and the roads respectively, and each of these buffer zones could be subtracted from, or 'cut out' of the combined alluvium/grade 3 layer, using the equivalent of the *difference* set operation.

Performance of overlay analyses is computationally more efficient when the data are stored in raster layers, since the process can be reduced to logical comparisons between equivalent cell locations. The result will of course be subject to the limitations of the raster model. When the data are defined by polygons, or other vector objects, the computation is considerably more complex since, as indicated above,

before the logical comparisons can be performed, all data must be reduced to a single set of polygonal cells resulting from the intersection of the individual layers. This requires a combination of geometric and topological processing. First, the intersections of lines belonging to all overlaid polygons are found, and the relevant lines are segmented at the intersections. Secondly, the newly segmented lines are assembled to define the boundaries of all polygonal regions resulting from the overlay. These polygons must then be classified with respect to the combination of properties of the original polygons.

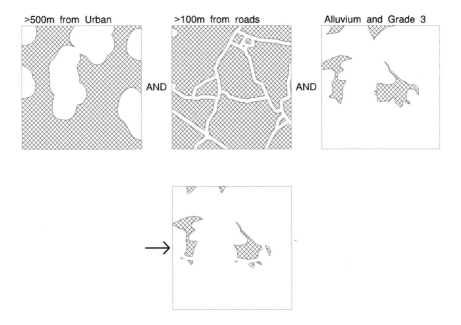

Figure 3.13 Example of a phenomenon-based search that combines multiple themes with a proximal spatial constraint: find regions of alluvium geology that coincide with agricultural land of grade 3, that are more than 500 m from urban boundaries and more than 100 m from main roads.

The techniques of overlay analysis summarised above can be regarded as quite crude for many analytical purposes, in that they do not take account of the relative importance of the different phenomena involved in the overlay. This issue can be addressed in overlay processing by attaching weights to the different layers. As the layers are combined, so the weights can be accumulated. The use of weighted combinations of overlays is discussed in Chapter 13.

Interpolation and surface modelling

The need for functions to perform interpolation arises when survey data are regarded as sampling a field; in other words, a variable that has a value at all points in space. When data are interpolated from the sample points to a regular grid of points, a frequently used technique is to estimate each grid point as a weighted sum of the sample points in the vicinity of the grid point. *Geostatistical techniques* adopt a relatively sophisticated approach in which the weights are calculated in an optimal manner based on a prior analysis of the correlation between sample values as a function of distance. An alternative technique is to triangulate the sample points and to estimate unknown points either through linear interpolation within each triangle of the triangulation, or by fitting a continuous mathematical surface through the triangulation. These mathematical surface functions, called splines, can also be fitted through a regular grid of points.

Surface modelling techniques applied to representation of physical surfaces such as terrain, and geological strata, cannot always be based upon an assumption of relatively smooth and continuous variation. In the case of terrain modelling, the surfaces should be able to incorporate ridges and valleys, while for geological modelling it is necessary to be able to incorporate discontinuities representing faults.

Best path analysis and routeing

Path analyses address problems of finding the shortest or, in some other sense, *least-cost route* between two locations (Figure 3.14). Solutions to problems of this sort can be found using network data models and using grid-cell-based raster data models. Network data models are used to represent relatively precisely defined sets of paths such as roads and rivers. Raster data models are used when the problem

Figure 3.14 Example of a shortest path through a network.

is to find a path across terrain that may not have any predefined paths. In both cases movement, either from node to node of a network, or from cell to neighbouring cell of a raster, is subject to impedances which are measures of resistance. In a road network this could relate to speed limits and to quality of the road surface, while in a terrain model the impedances might relate to the type of land cover and the presence of obstacles such as swampy ground, or somebody's backyard. Techniques for network analysis are described in Chapter 13.

Spatial interaction modelling

GIS are widely used now for purposes of identifying optimal locations of facilities such as supermarkets and fire stations. The definition of optimality varies according to the application, but in general it is possible to cast the problem as one of analysing flows, or interactions, of people, vehicles or money between service centres, such as a shop, that provide a facility, and demand centres, such as people's houses, where the people requiring the service are, at least temporarily, located. In locating a fire station, the technique of *allocation* can be used to determine all locations on a network of roads, that are within a given travel time. This is another example of the application of network models. Location of fire stations and schools may be regarded, fairly simplistically, as one of *districting*, in which the problem is to to define a set of districts in which everybody is guaranteed to be served by the relevant facility within a given cost of travel.

The situation becomes much more complicated when searching for the optimal location of a facility, if account is taken of the *attractiveness* of the facility gauged in application-specific terms that might, for example, take account of the spending power of the population at particular locations, the cost of travelling to the facility, and the attractiveness of competing facilities that may already be in use. Estimation of the relative attractiveness of a set of facilities relative to a set of demand centres is performed using the technique of *gravity modelling*, in which attraction is assumed to drop off as a function of distance or cost of travel. When several possible sites are to be considered with a view to finding optimal locations for what may be more than one facility, the problem is referred to as *location/allocation* analysis. It is potentially very demanding to take account of all possibilities and all factors that may affect the solution of such problems. It is commonly recognised that satisfactory solutions involving the use of computers are only likely to be obtained when there is considerable scope for interactivity to enable the application of expert, human judgement.

Correlations, associations, patterns and trends

There are some fields of study in which the power of a GIS to integrate information within a spatial context is potentially very useful in assisting in the search for causative links between events. A major area of application is in epidemiology, in which the objective is to find factors in the environment that are associated with the occurrence of poor health and particular diseases. Commercial GIS have frequently been criticised, however, for being poorly equipped to carry out such analyses in a reliable manner. Typically, where GIS packages have been used in detailed studies of correlations between environmental factors, they have been linked with specialised statistical packages. Thus some of the major commercial GIS packages do not include quite basic statistical tools such as that required to

perform a regression analysis to find the degree of correlation between two variables.

Examples of other analytical tools that would be useful in GIS but are not often to be found include cluster analysis, used to identify unusual groupings of phenomena such as a disease that might be correlated with a source of atmospheric pollution; and time series analysis, used to examine trends and periodicities in the nature of change of phenomena over time.

Map algebra with gridded data

A particularly flexible approach to manipulating layers in GIS is to be found in systems that provide implementations of map algebra, which was pioneered by Tomlin (1990). At present the term is generally applied to raster-based systems and is sometimes called *grid processing*. A grid (or raster) is regarded as consisting of a set of cells. The value of each cell may be linked to a value attribute table that associates attributes with each cell value. Groups of adjacent cells with the same values are described as *regions*, while all *regions* of the same value in a grid are described as a *zone*. A wide range of operators and functions may be provided, enabling many of the different types of analysis referred to above to be performed. Typically the result of applying an operator or function to a single grid or to a pair of grids is a new grid, or at least an update to an associated value attribute table.

Grid operators may fall into various classes, such as arithmetic (e.g. $+$, $-$, $\times$, $/$), relational (e.g. $>$, $<$, $=$), Boolean (i.e. and, or, not, exclusive or), and assignment (to update a cell value), as well as logical set operators and operators for manipulating binary numbers. Functions for grid-cell-based processing fall into the four categories of *local*, *focal*, *zonal* and *global*.

Local functions apply to individual grid-cell locations independently of other grid-cell locations. Thus, if operating on multiple grids, output at a grid-cell is a function of the values of the equivalent cell location in the input grid. Examples of types of local functions are trigonometric; exponential and logarithmic; reclassification, which may be may be via a table; selection according to a logical expression or geometric criterion; and determination of statistics for data from two or more grids (e.g. sum, maximum, minimum or mean). Using Boolean operators to combine two grids is equivalent to some overlay operations discussed earlier.

In focal functions the output at a grid cell is a function of the values of cells in an immediate neighbourhood. That neighbourhood is of regular form, e.g. a 3×3 set of cells or a circle centred on the cell. Functions include sum, weighted average, maximum, minimum, standard deviation and focal flow (which determines which cells of a 3×3 neighbourhood have higher values than the central cell).

When applied to a single input grid, zonal functions calculate some characteristics of the geometry of each of the zones on the grid, such as perimeter or area. With two input grids, one is a *zone grid*, which defines zones to be used in the function, and the other is a *value grid*, which contains data values that are to be examined on a within-zone basis. The output for each cell location is a function of the cell values that fall into the same zone as that cell location. Thus in the output grid, all cells from the value grid that lie in a particular zone will have the same output value. Functions that may be applied include mean, standard deviation, sum, range, minimum and maximum.

With global functions, output grid-cell values are potentially a function of all cell values in the input grid. A wide range of functions may be used. In a *Euclidean allocation* function, given a grid containing one or more source cells, all other cells are allocated to the nearest source cell. A *Euclidean distance* function, calculates the Euclidean distance from each cell to its nearest source cell. In a *cost allocation* function, each cell is allocated to the nearest source cell on the basis of weights associated with each cell that determine the 'cost' of passing through that cell. The path to the 'nearest' source cell must go via successive adjacent cells, moving vertically, horizontally or diagonally. The paths from each cell are calculated using a shortest-path procedure which moves out from the source cells selecting highest priority cells according to the nearness to their current respective source. Other functions may be used to record the cells that constitute the shortest, or least-cost, path from each cell to the nearest source cell, or to use costs associated with passing through each cell to determine a *corridor* (which could be several cells wide) from one source to another.

Graphics and interaction

A consequence of the essentially spatial nature of geographical information is that the specification of

queries and reporting of results can often be accomplished most effectively with the use of maps. Thus cartographic facilities of some sort can be expected within a GIS. These facilities should include options, often found in non-specifically cartographic drafting packages, for specifying the colour and style, or pattern, of point, line and area-referenced symbols (Figure 3.15). It should also be possible to annotate maps and their legends with text of different fonts with a choice of letter size and of orientation of the lettering. Many GIS packages also include facilities for changing between different map projections and for plotting the latitude/longitude graticule or a grid defining the map projection coordinate system. This may include the capacity to specify curves which the text should follow.

Recently there has been a renewed interest the use of cartograms to assist in data visualisation (Dorling, 1992). Here dimensions of the mapped features are modified in proportion to a mapped variable, such as population density or income. Other visualisation techniques can help with understanding the certainty of data, by changing colour density or tone and by means of fuzzy rather than precise delineation of boundaries. Representation of temporal change in a map can be a challenge. Sometimes this is done by marking the locations of phenomena at different times on the same map and connecting them by arrows. A more novel technique, which is not usually

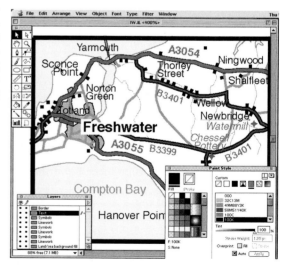

Figure 3.15 Example interface for interactive cartographic design using a desktop drafting package (Adobe Illustrator). Digital Map data Courtesy Automobile Association.

available with standard GIS facilities, is the generation of animated maps in which the changes are played out over a short period of time, possibly accompanied by associated graphics such as histograms or pie charts that could change in time with the map (Monmonier, 1992).

In a GIS with quite basic facilities, a map may take on a much more dynamic and flexible role than conventional paper maps. When enquiring about mapped data, the user can expect to point to features displayed on a computer screen or perhaps to draw a boundary on it to indicate the limits of a spatial search. The results of a query may then be displayed by adding features to, or modifying, a displayed map, or by generating an entirely new map. This could be printed on paper.

An important facility for some GIS applications is the capacity to view 3D scenes from different viewpoints in order to evaluate aspects of the landscape with regard to their appearance, and their visibility from different locations. The latter requirement results in the creation of viewshed maps (Figure 1.7 illustrates the results of a viewshed analysis). Three-dimensional visualisation is also required to assist in understanding the nature of phenomena in geoscientific studies of subsurface variation in geological properties such as porosity, and in oceanographic studies.

Computationally, 3D computer graphics is very much more demanding than working in two dimensions. However, the decreasing cost of powerful computer hardware, combined with increasing sophistication in techniques for visualising both synthetic and natural phenomena, have greatly raised the potential for using 3D graphics in geographical information systems. For example, technology that had previously been confined to expensive flight and ship simulators has now been transferred to moderately priced graphics workstations. This is particularly relevant in the context of applications concerned with assessing the impact on the landscape of new buildings or civil engineering works (see Plate 7). Looking further ahead, there is tremendous scope for programs which allow us to wander through present, historical and future worlds, using simulation based on varying degrees of factual, interpreted and imaginative representation. The term *virtual reality* refers to such interactive systems in which the user can experience stereoscopic vision, using goggles equipped with VDUs, as well as pointing to and manipulating objects within the virtual reality (see Chapter 14).

It is a frequent criticism of GIS that their automated map-making facilities barely deserve the description of cartography, because they are often so crude when compared with traditional cartographic products. The quality of automated maps is tending to improve however, with attention being paid to issues such as providing the user with guidance on the appropriate use of colour and map symbols. The automatic positioning of text on maps is an issue which has been the subject of a considerable amount of research, and it is now possible to generate maps automatically in which names are usually legible, do not overlap each other and are clearly associated with the feature they annotate. At the time of writing, however, it is still a rarity to find a GIS that incorporates such a facility.

A particularly challenging aspect of automated cartography is the process of generalisation, whereby information and its symbolisation are selected and modified in a manner that is appropriate to the scale. It involves processes such as the simplification, merging and removal of line and area objects and the displacement of map symbols to avoid their overlapping. Certain of these processes have been automated, but it is still not possible at present to take account automatically of all aspects of generalisation without human intervention, in order to produce a well-designed map. The result is that, although many maps are generated by computers, those that are produced entirely automatically are usually fairly simple in content, or else, if they have a very high data content, they are likely to suffer from obvious cartographic shortcomings of lack of clarity and cogency in the communication of particular messages.

Although it is easy to dwell on the limitations of computer cartography, it must be remarked that computing technology has rendered it relatively easy to produce maps which, because of the type of projection or the quantity of information, would have required enormous effort using manual methods. It is also possible to produce types of graphics, relating for example to the display of 3D landscape scenes, which were rarely attempted before the introduction of computers. The recent growth in multimedia and hypermedia technology is of particular relevance to GIS, since it provides the potential for combining computer graphics, photographs, sound and video to integrate a wide variety of information and to give the user very flexible means of interrogating it.

Summary

The purpose of this chapter has been to provide the reader with an overview of the facilities that are usually to be found in commercial GIS, along with an indication of types of functionality that, though not very common at the time of writing, may be expected to become more commonplace. We have made a broad division of GIS functions into the categories of data acquisition; preliminary data processing; data storage and retrieval; spatial search and analysis; and graphic display and interaction.

Data acquisition falls into two areas: primary data acquisition, whereby the information is collected directly via field observations; and secondary data acquisition, where the concern is to transform existing sources of information, typically found on maps and reports, to a digital form that can be manipulated by computing technology. Preliminary data processing covers a wide range of tasks addressed to making digital data usable for analytical and reporting purposes. These may include functions that apply changes of coordinate system, changes of classification and changes in data model.

Data storage entails organising data in a form that conforms with a particular logical database model. In GIS this often results in a hybrid approach involving relational database methods for the largely nonspatial attribute data, and specialised geometric data models for the spatial data. We saw that object-oriented and multimedia databases hold promise for handling geographically referenced information in a way that adapts to the complexity of some of the information modelling concepts and to the diversity of types of information that may be required.

The analytical facilities of GIS were presented here from the point of view of the techniques and methods, rather than the applications (which were reviewed in Chapter 1). There is no standard classification of GIS analytical functionality, but a distinction is made here between the following techniques: spatial containment search; proximal search; phenomenon-based search, including overlay analysis; path analysis; spatial interaction modelling; correlations, associations and trends; and map algebra which encompasses several of the other techniques.

Facilities for display and interaction normally include those for 2D viewing providing user control over symbology and map projections and those for 3D viewing that can enhance the sense of reality, as well as helping the user to understand and explore spatial relationships. Facilities for graphical user interaction constitute an essential aspect of GIS technology as they provide a means to control the spatial data handling and visualisation procedures by direct manipulation of graphic objects corresponding to the stored data.

Further reading

To gain a more detailed appreciation of the functionality of conventional GIS, it is necessary to study and work with actual systems. The user manuals of the ARC/INFO system give useful explanations of the wide range of functionality associated with this system, which is one of the longest-established commercial GIS packages. The IDRISI package is widely used in academic environments and the system is accompanied by several workbooks that give explanations of the functions provided by this largely raster-based GIS system. Most of the types of function referred to in this chapter, including those less well catered for in commercial GIS, are explored in more detail in later chapters of this book, where further references are provided.

PART 2 Acquisition and Preprocessing of Geo-referenced Data

CHAPTER 4

Coordinate systems, transformations and map projections

Introduction

The concept of position as defined by coordinate systems is essential both to the process of map-making and to the performance of spatial search and analysis of geographical information. To plot geographical features on a map, it is necessary to define the position of points on the features with respect to a common frame of reference or coordinate system. Having created such a frame of reference, it also provides a means of partitioning data for purposes of spatial indexing in a database. Thus the coordinate system can be used to guide a search through the database in order to determine which features occur in the vicinity of a point or a region expressed in terms of the coordinates. The coordinate systems that constitute the frames of reference necessary for mapping and searching geographical information allow us to specify position in terms of the distances or directions from fixed points, lines or surfaces (Figure 4.1). In *Cartesian coordinate systems,* positions are defined by their perpendicular distances from a set of fixed axes. The simplest and most familiar example is the case of two straight-line axes intersecting at right angles (Figure 4.1a). In *polar coordinate systems*, positions are defined by their distance from a point of origin and an angle, or angles, which give direction relative to an axis or a plane passing through the origin (Fig 4.1c,d).

Positions on the earth's surface are normally defined by a *geographical coordinate system* consisting of degrees of latitude and longitude. This is a form of spherical polar coordinate system in which two angles are measured with respect to planes passing through the centre of a sphere or approximate sphere (spheroid) representing the shape of the earth (Figure 4.1d). Distance is not specified in the coordinate system but it is implicit, being the radius of the earth at any given location on the surface. Because latitude and longitude refer to positions in 3D space, it is necessary for the purposes of cartography to transform them to a 2D planar coordinate system, or *map grid*. This type of transformation, which is called a *projection,* can be done in many different ways. The principal types of map projection transform from the earth's surface either directly to a plane, or to a cylindrical or a conical surface which, having been conceptually wrapped around the earth, can be unrolled to form a flat surface. When lines of latitude and longitude are plotted on the map they are referred to as a *graticule* (Figure 4.2).

All projections from geographical coordinates on the earth's surface to 2D map-grids involve some sort of distortion. The choice of projection is usually governed by a desire to minimise one or more of the distortions of either angles, linear dimensions or areas. For this reason it is important to appreciate the processes of map projection and the way in which they introduce internal changes in scale which give rise to these distortions.

A consequence of the variety of map projections is that there are numerous map-grid coordinate systems in use, some of which are unique to particular mapping organisations. This means that when compiling databases or maps with data from different sources, it will often be necessary to transform from one coordinate system to another in order to work within a single unifying framework. The use of computers has been most important in facilitating these often

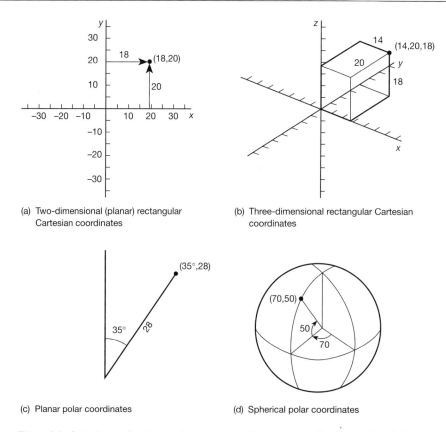

(a) Two-dimensional (planar) rectangular
   Cartesian coordinates

(b) Three-dimensional rectangular Cartesian
   coordinates

(c) Planar polar coordinates

(d) Spherical polar coordinates

Figure 4.1 Cartesian and polar coordinate systems. Cartesian coordinates consist of distances measured relative to fixed axes. Polar coordinates consist of a distance from a fixed origin and an angle or angles representing direction relative to a fixed axis or to a fixed place.

very complex geometric transformations which, when performed manually, may be extremely laborious.

We have already seen in previous chapters that the practice of computer cartography requires working with local coordinate systems which are specific to particular graphics display devices and to particular data acquisition systems. The use of technology for secondary data acquisition may require transformation from digitising table coordinates to geographical or map-grid coordinates, and again from the latter to plotting device coordinates. When, in the course of digitising, it is necessary to compensate for distortion in the source map, (e.g. due to paper stretching), then the transformations become more complicated. Problems also arise when the only coordinate system marked on the map is a non-rectangular graticule of latitude and longitude. In these cases it may be necessary to use interpolation procedures to deduce the map or geographical coordinates of points lying in-between known control points (such as at graticule intersections).

In the following sections of this chapter we start with a discussion of the way in which the shape of the earth is described, before examining the planar and spherical coordinate systems that are used as frames of reference. In the context of this introduction to coordinate systems we review methods for making measurements of length and area, before introducing the basics of simple geometric transformations of translations, scaling and rotation on the plane. The second main part of the chapter is concerned with map projections, including issues of the relationship between the sphere and the surface of projection, the concept of scale and way distortions are introduced in map projections. The following part of the chapter briefly reviews the problems of registering map data by means of rubber sheet transformations when not all projection parameters of existing maps are known. In the final section we introduce a relatively new type of coordinate system based on a tessellation of cells which approximate the shape of the globe.

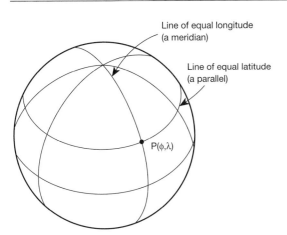

Line of equal longitude
(a meridian)

Line of equal latitude
(a parallel)

P(φ,λ)

Figure 4.2 Geographical coordinates are a special case of spherical coordinates in which a point P is defined by an angle of latitude which is measured relative to the plane of the equator (perpendicular to the axis of rotation) and an angle of longitude which is measured relative to a plane of a datum meridian passing through the axis of rotation. The network of lines of equal longitude (meridian) and of equal latitude (parallels) constitute the graticule.

The shape of the earth

Geographical maps are concerned with representing information that is spatially related to the surface of the earth. To represent the relationships accurately, maps should ideally be scaled down versions of the physical or cultural features of the earth's surface. Since the earth is three-dimensional, this implies creating 3D maps. Apart from the occasional use of globes and physical raised-relief maps, and some specialised 3D viewing systems, this is not generally practicable. It is necessary, therefore, to concern ourselves with the question of how to project the 3D world onto the 2D surfaces which characterise current graphics technology. The process of this projection depends very much upon the shape that we consider the earth to be.

A flat earth

In small regions, several kilometres in extent, the curvature of the earth's surface departs so little from a plane that it is possible to treat the earth as a flat surface on which local terrain can be measured as perpendicular variations in elevation. The projection of the lateral positions to a planar map is then a trivial matter, requiring only simple scaling of linear dimensions. For more extensive regions, the planar approximation becomes untenable for purposes of accurate locational mapping, since it is impossible to transfer measurements from what, in reality, is a curved surface to a flat one, without introducing distortions such as stretching and tearing.

Spheres and spheroids

For small-scale maps which cover large areas of the earth, it is appropriate to think of the earth as a sphere of constant radius. From a global point of view this is a very good approximation, as the radius actually only varies by about 10 km to either side of an average value of 6371 km (Maling, 1992). For larger-scale mapping of extensive regions, the spherical approximation is not adequate and account must be taken of the variation from a perfect sphere. Maling (1991) has pointed out that for many GIS applications, using data derived from secondary sources (i.e. digitised maps), the inherent inaccuracies of the data are such that the spherical earth assumption is often quite appropriate.

According to gravitational theory, if the earth were homogeneous in composition, its shape would be an ellipsoid of rotation, generated by rotating an ellipse about its shorter axis. Thus flattening occurs in a north–south direction along the axis of the earth's rotation. In vertical cross-section, the earth's shape would be an ellipse, with the major axis through the equator and the minor axis coincident with the rotational axis. Variations of relief due to mountains and oceans can be regarded as occurring above and below the surface of the ellipsoid, while the direction of gravity would always be normal to the ellipsoid surface, which could be defined as the horizontal at any given location. The shape of the earth represented by this gravitational equipotential surface is called the *geoid*.

The geological composition of the earth is such that there are both major and minor changes in rock density, which give rise to anomalies in the gravity field and hence in the form of the geoid. Satellite observations of the gravity field indicate that it can be represented by a surface that deviates somewhat from an ellipsoid in that it is dented at the south pole and squeezed in northern latitudes to produce something which, when greatly exaggerated, resembles a pear (Maling, 1992). These variations in the form of the geoid are, however, so minor that for the purposes of large-scale topographic mapping, it is sufficient to treat the earth as an ellipsoid. Because reference ellipsoids used to approximate the earth's shape are so similar to a sphere, they are often described as a *spheroid*, and in this context the terms ellipsoid and spheroid are frequently used interchangeably.

The reference ellipsoid is used to define the *geodetic datum* to which a geographical coordinate system may be linked. The dimensions of this ellipsoid can be defined in terms of the ellipse that generates it (Figure 4.3). The form of an ellipse is defined by the lengths of the semi-major axis (a) and the semi-minor axis (b). These values can be combined to define the degree of flattening, f, also called the ellipticity, oblateness or compression, where

$$f = (a - b)/a$$

The value of f is usually given as a fraction of the form $1/n$, where $n = a/(a - b)$. Several surveys have been made with the intention of estimating the values of a and b, and hence f, for the earth (see Snyder (1987) and Maling (1992) for lists of examples of officially accepted values). The results have varied somewhat as a function of which part of the world the survey data were obtained from, as is to be expected in view of the slight departures from an exact ellipsoid (as well as error in the survey methods). Several different values of f are in current usage, and most are in the range 1/297 to 1/300, though several larger values of the order of 1/294 are also in use. The measurements for the corresponding values of a and b only vary in general by less than a metre and are of the order of 6378 km and 6356 km respectively (Snyder, 1987).

Most of the ground-based measurements of the form of the earth have been based on the method of astro-geodetic arc measurement. The method involves determining various radii of curvature of the earth from which the form of the overall ellipsoid may be deduced. The principle depends upon measuring the distance d between two distant points, along with their angular separation θ, the latter being determined with respect to astronomical bodies. Using the relationship between these values and radius R, given by

$$d = R\theta$$

the radius can be found ($R = d/\theta$).

Recent measurements of the shape of a reference ellipsoid have made use of data obtained from satellites. This is the case for reference ellipsoids such as the North American Datum 1983 (NAD 83) and the Geodetic Reference System (GRS) 80 and the World Geodetic System (WGS) 84 (Snyder, 1987).

Planar rectangular coordinates

When a planar rectangular (Cartesian) coordinate system is adopted for cartographic purposes, the x and y, or horizontal and vertical, axes are usually referred to as eastings and northings respectively. The coordinate system itself is called a grid, or map grid, since it is frequently represented by sets of intersecting horizontal and vertical lines, drawn at regular intervals. Examples of such map grids are provided by the Universal Transverse Mercator system and the US State Plane system (both consisting of a set of grids), and the British National Grid (a single grid), all of which calibrate the grids in metres.

Graphical display devices and digitisers employ rectangular coordinate systems, but there is considerable variation in the units of measurement and, to some extent, in the orientation of the axes. At the level of the hardware, graphics display screens consist of 2D matrices of pixels which are individually addressable by means of an integer coordinate system in which unity corresponds to the separation between two adjacent pixels. Hard copy plotters and digitising tables also have a finite resolution, but it is often possible to work directly with units of millimetres or centimetres, and fractions thereof, rather than just pixels.

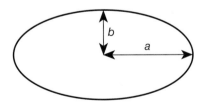

Figure 4.3 The shape of the earth can be approximated by an ellipsoid, obtained by rotating an ellipse about its minor axis (coinciding with the earth's axis of rotation). In the figure the difference between the semi-major axis *a* and the semi-minor axis *b* is greatly exaggerated compared with the difference between the values used for the earth.

Measurements with rectangular coordinate systems

There are obvious advantages with rectangular co-ordinates for mapping in that there is a direct relationship between coordinate units and distances on the ground. This statement is subject to the limitations introduced by projection distortions, but in the case of the types of map grid used by national mapping agencies, the distortions are sufficiently small to be ignored for many practical purposes. National mapping projections do usually involve some distortion in the representation of area but, taking the British National Grid as an example, the divergence from true areal scales may be no more than 0.1% (Maling, 1991). The width of the British National Grid is, however, only about 9° of longitude. The standard grids of the widely used Universal Transverse Mercator projection (to which the British system is closely related) are 6°. Distortion of area or of distance becomes more significant for more extensive regions.

Distance measurements

The shortest distance between pairs of points in a rectangular coordinate system is represented by straight lines which can be measured simply using Pythagorus's theorem. For example, given two points with easting and northing coordinates E1 = 302950, N1 = 2550802 and E2 = 315240, N2 = 2561844, the difference in eastings is E2 – E1 = 12290 and the difference in northings is N2 – N1 = 11042. Their distance apart D is therefore

$$D = (12290^2 + 11042^2)^{1/2} = 16521$$

When we wish to measure the length of a digitised line it is then only necessary to perform the same type of calculation for each successive pair of digitised points and to sum the results.

Area measurements with rectangular coordinates

Measurement of area on a rectangular grid is a relatively straightforward procedure. Assuming that the

region whose area is to be calculated is defined by a polygon consisting of digitised points, it is possible to express the area as the sum of the areas of a set of trapezia. A trapezium is a quadrilateral, *abcd* with two parallel sides. Referring to Figure 4.4, in which *ab* and *cd* are the parallel sides and *h* which is the distance between them, the area *A* is given by the formula

$$A = h/2 \, (ab + dc)$$

Considering the polygon in Figure 4.4, when vertical lines are drawn from each, clockwise ordered, vertex down to a horizontal line beneath the polygon, we can see that the area of the polygon may be expressed in terms of the differences in area between those trapezia in which the uppermost edge is directed rightwards and those in which the uppermost edge is directed leftwards. If we assume that the width of each such vertical trapezium (corresponding to *h* above) is given by the difference in *x* coordinates of successive vertices, we will find that some widths are positive and some negative, corresponding to the senses of right and left direction. This leads to the result that some areas calculated by formula *N* will be positive and some negative. Adding together all trapesium areas then gives the area of the polygon, since the positive areas correspond to the trapezia of the upper edges and the negative ones to those of the lower edges. The polygon area A_p is then given by the formula

$$A_p = \sum_{i=1}^{n-1} (X_{i+1} - X_i) \, ((Y_i - Y_b) + (Y_{i+1} - Y_b))/2$$

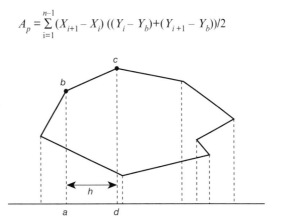

Figure 4.4 Measurement of the area of a polygon can be performed by summing the positive and negative values found for the areas of the individual trapezia (*abcd*) constructed relative to a horizontal line. The height of each trapezium is found by subtracting successive *x* coordinates on the boundary, represented by a list of clockwise, or anticlockwise, encoded vertices.

where X_i, Y_i are the coordinates of each vertex and Y_b is the y coordinate of a horizontal line. Since Y_b may take any value, including zero, the formula may be simplified to

$$A_p = \sum_{i=1}^{n-1} (X_{i+1} - X_i)(Y_i + Y_{i+1})/2$$

The computation required to find the area can be reduced if we express the formula as follows:

$$A_p = 1/2 \left[(X_n Y_1 - X_1 Y_n) + \sum_{i=1}^{n-1} (X_i Y_{i+1} - X_{i+1} Y_i) \right]$$

The centroid of an area

The centroid of an area is a representative point that is usually regarded as being a mean of all locations in the area. A simple approximation to this location could be found by taking the average of the co-ordinates of all points that defined the perimeter. However, if the density of points along the boundary was variable, this would result in skewing the location of the resulting centroid. A more accurate method is to triangulate the area and find the area-weighted mean of the centroids of the triangles. A triangle centroid is found at the intersection of the lines joining each vertex to the centre of the opposite side. For a concave area, the centroid, found by whichever method, may not be internal. If found to be external (using a point-in-polygon test, as described in Chapter 11), an internal centroid can be found by extending a line horizontally from the initial centroid and finding the centre of an adjacent pair of area boundary intersections.

Polar coordinates on the plane

An alternative to planar rectangular Cartesian coordinates is the polar coordinate system in which position is defined by the distance r from an origin and the angle O relative to an axis passing through the origin (Figure 4.1c). This type of coordinate system is of use in cartography when plotting certain types of projection in which position can conveniently be defined relative to a single, central point,

rather than a pair of axes. It is also relevant in general when it is appropriate to retain a sense of relative direction.

In mathematical usage, polar coordinate angles are conventionally measured anticlockwise from a horizontal axis (Figure 4.1c). In surveying and cartography, the angles are usually measured clockwise from a vertical axis, the angular units being either degrees (0–360) or grads (0–400), where 90° and 100 grads are equivalent to $\pi/2$ radians.

For a given origin, a cartographic polar coordinate of r,θ can be expressed in rectangular coordinates using the trigonometric formulae

$$x = r\sin\theta, \quad y = r\cos\theta$$

Conversely, the polar coordinates can be expressed in terms of the rectangular coordinates by finding θ from the relationship

$$\tan\theta = x/y$$

so that θ can be found using the appropriate inverse tan function on a computer. Knowing θ,

$$r = y/\cos\theta \quad \text{or} \quad r = x/\sin\theta$$

We may also note that

$$r^2 = x^2 + y^2$$

Spherical coordinates

We have seen that though planar coordinate systems are essential for constructing maps on flat surfaces, they cannot be used for representing extensive regions of the earth without introducing serious distortion in measurements such as distance and area. When high accuracy is not required these problems of distortion can be avoided by the use of a spherical coordinate system. This provides a single, consistent and relatively undistorted reference frame for recording positions and making measurements of the earth's surface. The coordinates can then be projected to a suitable planar coordinate system when a small-scale map of a particular region or aspect of the earth is required.

Any point on the surface of a sphere of given radius can be uniquely defined by the angles which the radius passing through the point makes with two reference planes passing through the centre (Figure 4.5). This is equivalent to a 3D polar coordinate system in

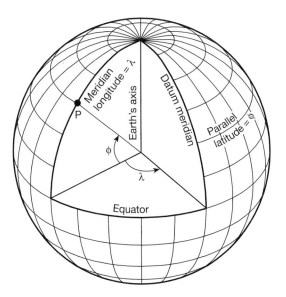

Figure 4.5 Cutaway view of the earth showing the angular relationship between a point *P* at latitude φ and longitude λ and the planes of the equator and the datum meridian. Note that φ is measured in the plane of the meridian which passes through *P*, and λ is measured in the plane of the equator.

which the distance from the point to the origin is fixed and hence the locus of all possible positions describes a sphere. On the earth, when it is treated as a sphere, the reference planes of the geographical coordinate system are the horizontal one perpendicular to the axis of rotation, which intersects the surface at the *equator*, and the vertical one which includes the rotation axis and intersects the surface on an arc called a *meridian*, which, for international purposes, passes through Greenwich, London. Angles measured in the vertical, meridianal planes relative to the equatorial plane constitute latitude, while those measured in the horizontal, equatorial plane relative to the plane of the Greenwich meridian constitute longitude. These two angles of latitude and longitude are also described as φ (phi) and λ (lambda) respectively. Although the meridian through Greenwich is the one most commonly adopted as 0° longitude, many national surveys measure longitude relative to a meridian which passes through their capital city.

The word meridian refers specifically to the semicircular arc formed by the intersection with the earth's surface of any plane which includes the axis of rotation. A single meridian is constituted by an arc

that extends from the north pole to the south pole where the word *pole* refers to the intersection of the rotation axis with the earth's surface. For any given meridian there exists an *antimeridian,* which is the arc in the same plane extending around the opposite side of the earth. Angles of longitude are conventionally negative when measured westwards of the zero meridian and positive when measured eastwards. Frequently the sign is omitted, direction being indicated by the specification of east or west.

Angles of latitude are defined either north or south of the equator, where northwards is conventionally treated as positive and south is negative. All points of a particular angle of latitude on the earth describe a circle, the plane of which is parallel to that of the equator. These circles, or lines of latitude, are referred to as *parallels*.

Great circles and small circles

In a spherical coordinate system, any line between two points must be curved, since it lies on a sphere. An important class of such lines is the circular arcs which result from the intersection of a plane with the sphere. Both meridians and parallels belong to this category. If the plane passes through the centre of the sphere, then the arc is of maximum radius, equal to the radius of the sphere, and it is termed a *great circle*. The shortest distance between any two points on a sphere is given by the length of the great circle arc which extends between them (Figure 4.6). Provided that the two points are not diametrically opposite, only one great circle will fit through them and the shortest distance will be that of the shorter of the two arcs which connect them.

It follows from the description of the meridian given above that all meridians are great circles. Thus all parallels are small circles, with the exception of the equator (which is a great circle). It should be noted that the shortest path between two points of the same latitude will not, therefore, be along the corresponding parallel, unless it is the equator.

Measurement along any great circle arc

In order to measure distance along any great circle arc, and hence be able to find the shortest distance between any two points on a sphere, it is necessary to make use of spherical trigonometry. Considering

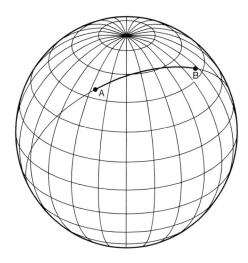

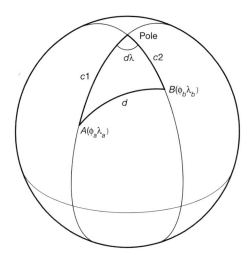

Figure 4.6 The shortest distance between two points A and B measured on the earth's surface, is given by the length of the great circle arc which extends between them. The great circle arc lies on a plane which passes through the centre of the earth. All meridians are great circles. The equator is the only parallel that is a great circle.

Figure 4.7 The shortest (great circle) distance between two points A and B can be found by constructing a spherical triangle between the two points and the pole. The angular distance can then be calculated, using the cosine formula, from the colatitudes c1 and c2 of the points, and the spherical angle $d\lambda$, which is the difference between the longitudes of the two points.

Figure 4.7, it is possible to regard the great circle distance between any two points $A(\phi_a,\lambda_a)$ and $B(\phi_b,\lambda_b)$ as the length of one side of a spherical triangle (i.e. a triangle on the surface of a sphere, as illustrated in Figure 4.8) in which the other two sides are the meridian arcs from the points to the nearest pole. If the latitudes of the two points are ϕ_a and ϕ_b, their angular distance to the pole will be $c1 = 90 - \phi_a$ and $c2 = 90 - \phi_b$, where $c1$ and $c2$ are called the colatitudes. The spherical angle opposite the unknown side is the difference $d\lambda$ in longitudes of the two points where $d\lambda = \lambda_a - \lambda_b$. It is a simple matter, therefore, to apply the cosine formula to derive the unknown angular distance d, where

$$\cos d = \cos c1.\cos c2 + \sin c1.\sin c2.\cos d\lambda$$

This formula can also be expressed in terms of the latitudes where

$$\cos d = \sin\phi_a.\text{son}\phi_b + \cos\phi_a.\cos\phi_b.\cos d\lambda$$

Having found the value of d, the angular distance in radians, the arc length s is then given by

$$s=R.d$$

where R is the radius of the earth.

Geometry of the spheroid

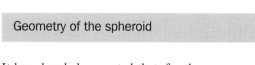

It has already been noted that, for the purposes of accurate surveying and mapping of the earth at large

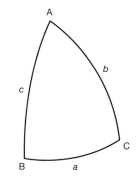

Figure 4.8 A spherical triangle. Each side is part of a great circle on a sphere. Each angle (A, B and C) is measured between the tangents of the two adjacent sides at the triangle corner. For a given corner, the tangents will lie in the plane which is a tangent to the sphere at that point.

or medium scales, the earth is treated as an ellipsoid of rotation. It is generated by specifying an ellipse that is rotated about its minor (shorter) axis. The minor axis is regarded as coincident with the earth's axis of rotation and thus planar cross-sections through the axis will always be elliptical in shape. Cross-sections perpendicular to the axis are always circular, so that, on the spheroid, the equator and all parallels are circular. Note that no other intersections of a plane with the spheroid will result in circles.

Just as on a sphere, measurements of distances between two points on the surface of a spheroid depend upon using the radius of curvature to calculate arc length. This introduces considerable computational problems for spheroidal calculations because the radius of curvature varies from one place to another on the surface as well as varying at each point, according to which direction it is measured in. Further explanation of the characteristics of the spheroid can be found in Maling (1992) and details of methods for measuring lengths of shortest distances between two points on the surface of a spheroid, i.e. the geodesic, are given in Maling (1989: 548–549).

Geometric transformations in rectangular coordinate systems

The basic geometric transformations of translation, scaling and rotation are essential requirements for computer visualisation and manipulation of map data. Combinations of these basic transformations are referred to as *affine transformations*. They are needed when changing between coordinate systems and when changing the location, orientation and size of displayed map symbols. The former requirement arises when data represented in 2D or 3D map-grid coordinates must be displayed on a computer device with its own coordinate system and, in the context of data acquisition, when data recorded initially in a local survey coordinate system must be transformed to a standard map grid. If a survey was recorded on a paper map, the latter transformation could include an intermediate representation in digitising table coordinates.

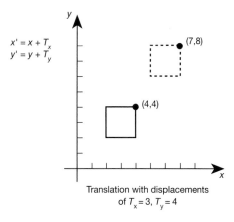

Figure 4.9 Translation transformation.

Translation

Translation in 2D (Figure 4.9) moves a point x,y to a new position x', y' by adding components T_x, T_y, i.e.

$$x' = x + T_x$$
$$y' = y + T_y$$

Scaling

To scale a point x,y to a new point x', y' we use the two scale factors S_x and S_y relating to scaling in each of the two (or three) dimensions (Figure 4.10). Thus for two dimensions

$$x' = x.S_x$$
$$y' = y.S_y$$

This scaling can be regarded as taking place relative to the origin of the coordinate system. When applied to an object defined by a number of points, the whole object is liable to be displaced, in the sense that the centre of the object will move by an amount determined by the scale factors. If it is required to keep one point of the object fixed, then that fixed point should be moved to the origin before applying the scaling, after which the 'fixed' point should be moved back to its original position. For a fixed point F_x, F_y the scaling therefore consists of the composite transformation of $(-F_x, -F_y)$ (S_x, S_y) (F_x, F_y). Hence to

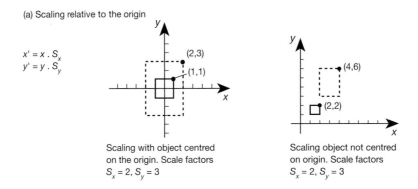

(a) Scaling relative to the origin

$x' = x \cdot S_x$
$y' = y \cdot S_y$

Scaling with object centred
on the origin. Scale factors
$S_x = 2, S_y = 3$

Scaling object not centred
on origin. Scale factors
$S_x = 2, S_y = 3$

(b) Scaling relative to a fixed point $F(2,2)$ using scale factors $S_x = 2, S_y = 3$, expressed
as a sequence of basic transformations

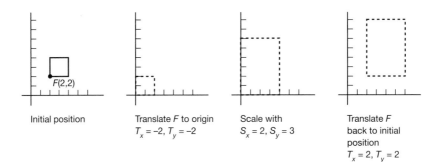

Initial position

Translate F to origin
$T_x = -2, T_y = -2$

Scale with
$S_x = 2, S_y = 3$

Translate F
back to initial
position
$T_x = 2, T_y = 2$

Figure 4.10 Scaling transformation.

find the transformed point x''', y''' from the initial point x,y, we can calculate as follows:

$$x' = x - F_x$$
$$y' = y - F_y$$

$$x'' = x' \cdot S_x = (x - F_x) S_x$$
$$y'' = y' \cdot S_y = (y - F_y) S_y$$

$$x''' = x'' + F_x = x.S_x + F_x(1 - S_x)$$
$$y''' = y'' + F_y = y.S_y + F_y(1 - S_y)$$

Rotation

To rotate a point x,y about the origin by an angle θ to a new position x', y' the formula is

$$x' = x \cos\theta - y \sin\theta$$
$$y' = x \sin\theta + y \cos\theta$$

The rotation, as defined above, is anticlockwise relative to the origin of the coordinate system (Figure 4.11). In order to rotate an object about an arbitrary

origin it is necessary to introduce translations relative to the given point before and after the rotation in a similar manner to that described above for scaling. The latter local rotations are relevant in computer cartography when manipulating individual map symbols such as arrows and items of text.

Changing between rectangular coordinate systems

To change from one coordinate system to another, one of the coordinate systems must be defined in terms of the other. Assuming, as we are here, that both coordinate systems are rectangular, we need to know a common point that can be defined in both coordinate systems, scale factors in each dimension (i.e. how many units of one coordinate system there are for each unit of the other), and the orientation of

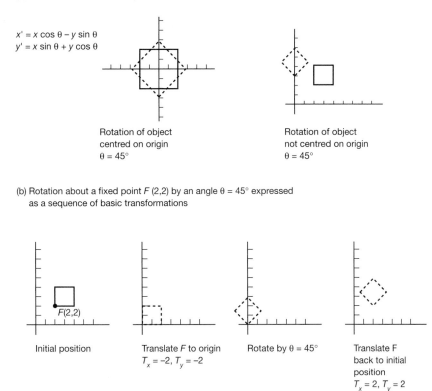

(a) Rotation relative to the origin (positive anticlockwise)

$x' = x \cos \theta - y \sin \theta$
$y' = x \sin \theta + y \cos \theta$

Rotation of object
centred on origin
$\theta = 45°$

Rotation of object
not centred on origin
$\theta = 45°$

(b) Rotation about a fixed point F (2,2) by an angle $\theta = 45°$ expressed
as a sequence of basic transformations

Initial position

$F(2,2)$

Translate F to origin
$T_x = -2, T_y = -2$

Rotate by $\theta = 45°$

Translate F
back to initial
position
$T_x = 2, T_y = 2$

Figure 4.11 Rotation transformation.

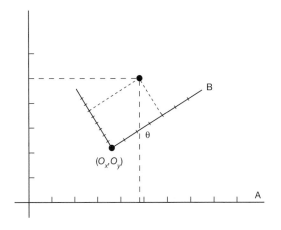

Figure 4.12 A change in coordinate systems from system
A to system B can be achieved by applying to the points in
coordinate system A those transformations required to align
the coordinate system of B with that of A.

(O_x, O_y)

B

A

θ

the axes of one system relative to the other. Let us
suppose, referring to Figure 4.12, that we wish to
transform points in coordinate system A to coordinate system B. The situation is particularly simple if
the common point is the origin of B. The transformation consists of applying to the points represented
initially in system A those transformations required
to align the coordinate system of B with that of A.
Referring to Figure 4.12, O_x, O_y is the origin of B
defined in coordinate system A, B_{sx}, B_{sy} are the numbers of units of B per unit of A in the x and y axes,
and θ is the angle between the x-axis of B and the
x-axis of A. The steps are then

1. Translate by $-O_x$, $-O_y$ moving the origin of B to
    that of A.
2. Rotate by $-\theta$, bringing the axes into alignment.
3. Scale by B_{sx}, B_{sy} to make the units of each dimen-
    sion equivalent.

If the common point used for translation in step 1 is
not the actual origin of coordinate system B, but is

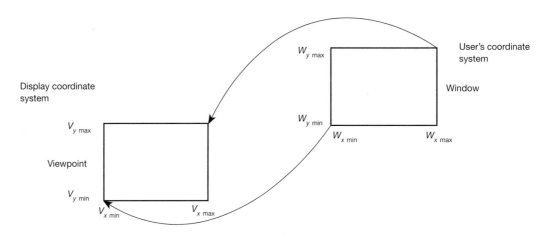

Figure 4.13 The window to viewport transformation in computer graphics. A change in coordinate systems is required to display a rectangular region of the user's (world) coordinate system, defined by a window, within a rectangular region of a display device, defined by a viewport.

some point b_x, b_y in the coordinate system of B, we need to introduce a fourth step to add on the coordinates of this point. The additional step is

4. Translate by b_x, b_y to add on the local origin.

A practical example of this change in coordinate systems would be where coordinate system B represents the grid system of a map being digitised on a digitising table represented by coordinate system A. If this was the case then the scale factors and angle might not be specified explicitly, but could be derived from three control points which the operator was required to specify. These points might be the origin of the map sheet and one point on each of the two adjacent corners of the map grid.

Another example of an application of the coordinate system transformation is that of plotting a digital map defined in its own coordinate system on a graphics display device. In computer graphics terminology, the transformation would be defined by a window on the user's coordinate system, i.e a rectangular area defining the region of the map to be displayed, and a viewport defining a corresponding rectangle on the display surface of the device in the graphics display coordinate system. Usually both rectangles are oriented parallel to each other, so that there is no need for the rotation step (Figure 4.13). The initial translation is given by the lower left corner of the viewport. The scalings are derived from the ratios of the sides of the window and viewport and the final translation is given by the coordinates of the lower left of the window.

Map projections

Map projections involve transforming a representation of the world in three dimensions to a representation in only two dimensions which can be plotted directly on a planar surface. There are many different ways in which the projection can be performed and, in order to understand the variety of map projections that result (Figure 4.14), it is useful to take account of several aspects of the problem. Of particular relevance are (a) the relationship of the planar surface to the global surface and (b) the nature of the distortion which the projection entails. The issue of distortion introduces in turn a requirement to consider variation in scale within a map and its specific association with the distortion of lengths, angles and areas.

Disposition of the plane of projection

The relationship between the map plane and the spherical or spheroidal surface gives rise to three distinct classes of projection: *azimuthal, cylindrical* and *conical* (Figure 4.15).

In the first case, of planar projection, the plane may be regarded as lying flat at a tangent to some point on the globe. This results in an azimuthal projection which takes its name from the particular case in which the plane is a tangent to the north or south

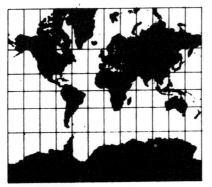

MERCATOR

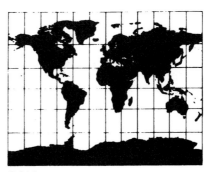

GALL

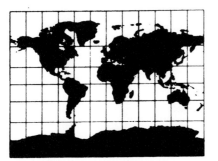

MILLER

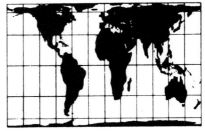

GALL-PETERS

MOLLWEIDE

ROBINSON

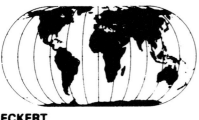

ECKERT

GOODE

Figure 4.14 Examples of world map projections. Reproduced by kind permission of the University of Wisconsin Cartography Lab.

pole from which meridians will radiate according to their respective azimuths. In the second, cylindrical, class of projection the plane is derived from a surface wrapped cylindrically around the globe, touching it along a single great circle. Note that unwrapping the cylinder into a genuine planar surface can be done

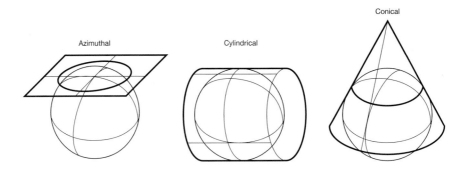

Figure 4.15 The three main classes of map projection.

without introducing further distortion. In the third class of projection, the conic, the map plane is derived from a surface that is wrapped around the earth in the shape of a cone touching the globe along a small circle or an ellipse. Projection surfaces such as the cone and cylinder, which can be unwrapped without distortion to form a plane, are called *developable surfaces*. It should be borne in mind when referring to planar, conical and cylindrical projections that ideas of the tangential plane, cylinder and cone are geometric concepts which are intended to help envisage the projection transformation. The projections are defined by mathematical functions which transform points on the globe onto the map surface along projection rays emanating in various ways from the globe to the map.

Each of the three forms of projection has modifications in which the projection surface intersects the earth rather than touching it tangentially at a point or along an arc (Figure 4.16). When the plane, cylinder or cone intersects the earth, it is described as *secant*. The secant plane intersects a spherical globe along a single circular arc. Both the secant cylinder and the secant cone intersect along two parallel

small circles. If the ellipsoidal shape of the earth is taken into account, the arcs of intersection will only be exact small circles if they lie in planes parallel with the Equator and hence perpendicular to the rotational axis.

Aspect

The appearance of a map projection, in terms of the form of the latitude–longitude graticules, varies according to the orientation, or aspect, of the plane, cylinder or cone of projection relative to the globe. The aspect of a projection is usually described as being either *normal*, *transverse* or *oblique* (Figure 4.17). The normal aspect of the azimuthal projection refers to the situation when the plane of projection is located at one or other pole, perpendicular to the earth's axis. It is characterised by radiating lines of longitude and concentric circles of latitude. The normal aspect of a conical projection has a similar pattern of latitude and longitude, and arises when the cone touches or intercepts the globe on a line or lines of latitude, corresponding to the

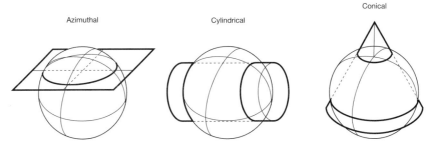

Figure 4.16 Secant projections are ones in which the plane, cylinder or cone of projection intersects the globe, as opposed to touching it tangentially.

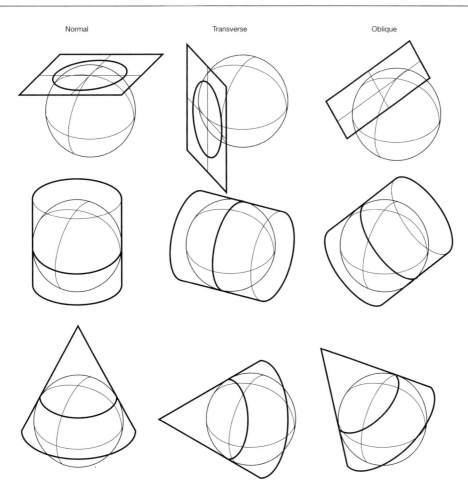

Figure 4.17 Aspects of azimuthal, cylindrical and conical projections. In the normal aspects of the azimuthal and conical projection, the plane and cone respectively may be positioned at the north or south poles. In the transverse aspects, the axes of the cone and cylinder, and the perpendicular to the plane may take on any horizontal (equatorial) orientation.

axis of the cone being parallel to the earth's axis. The transverse aspects of both azimuthal and conical projections occur when the orientation of the plane and cone is at 90° to that of the normal aspect. The appearance of the resulting maps are distinguished by the equator being horizontal while one, central meridian is vertical. Oblique aspects refer to all possible intermediate orientations of the plane and cone.

Considering the case of cylindrical projections, the normal aspect refers to the situation in which lines of latitude are horizontally orientated and lines of longitude are vertically orientated, though these lines may be curved, depending upon other properties of the projection. The normal aspect of cylindrical projec-

tion corresponds to the cylinder being orientated north–south parallel to the earth's axis. In the transverse aspect, the cylinder is orientated east–west and, when tangential, touches the globe along a great circle. The graticule includes a vertical central meridian and a horizontal equator. Oblique aspects of cylindrical projections refer to all other orientations of the cylinder relative to the globe.

Figure 4.18 illustrates the way in which the widely used Universal Transverse Mercator (UTM) projection system specifies 30 orientations for the horizontal cylinder, providing 60 UTM zones, each of 6° width (one on either side of the cylinder). The relationship between an individual zone and the corresponding grid system is illustrated in Figure 4.19.

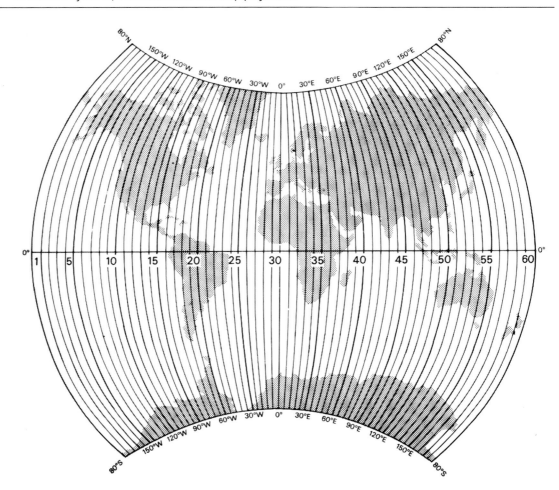

Figure 4.18 The UTM projection system employs 30 orientations of a cylinder, giving 60 zones each 6° wide. After Maling (1992).

Concepts of scale

Geographical maps can be thought of as scaled graphic representations of physical or abstract features of the earth. The map reader usually assumes that dimensions on the map can be related to their true dimension in terms of a scale value which may be expressed by the ratio between map dimension and actual dimension. When the scale value is given as a fraction in which the numerator is 1, it is called the *representative fraction*. If the representative fraction is relatively large, such as 1/1250 for a detailed urban survey, the map is referred to as *large scale*, whereas maps covering much larger areas in which the representative fraction is correspondingly smaller, such as 1/500 000, are referred to as *small scale*. In general, the meaning of large, small and medium scale will depend upon the conventions of particular applications.

The direct relationship between map dimension and actual dimension, implied by the representative fraction, is in practice somewhat misleading. This is because the distortion involved in transferring dimensions from the curved surface of the earth to a planar map prevent the possibility of a constant scale throughout the map.

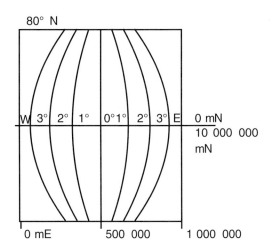

Figure 4.19 Each 6° wide UTM zone corresponds to a pair of map-grid systems for the north and south hemisphere respectively. Each grid extends from 0 to 10 000 000 in the north direction and 0 to 1 000 000 in the east direction.

Conceptually the production of a map requires that the earth be scaled down to a *generating globe* of a size that is compatible with the size of map plane onto which it is to be projected. The ratio between the radius of the generating globe and the radius of the real world equivalent is called the *principal scale*, and it is to this scale that the representative fraction refers. The representative fraction can only apply exactly to the scale at one or two specific points or lines on the map, which are positions of zero distortion. These locations are those of contact or intersection between the projection plane, cone or cylinder and the globe. Away from these positions of zero distortion, the scale varies in a manner that depends upon the type of projection. The variation of scale leads to the notion of particular scale, which refers to the scale in a specified direction at a specified point on the map.

Distortion

Understanding the way in which scale varies on map projections is equivalent to understanding the nature of the distortion that all map projections contain. One means of analysing distortion, devised by Tissot, is to examine, for corresponding points on the map and the globe, the way in which an infinitely small circle on the generating globe changes shape when plotted on the map. For positions on the map where there is zero distortion, the circle will remain a circle of the same radius. In all other cases, the circle becomes transformed in terms of either its size, its shape, or both. In the general case, the circle becomes an ellipse, the semi-major and semi-minor axes of which we may refer to as a and b respectively. The orientation of the two axes bears a special relationship to the globe, in that there is normally one pair of perpendicular directions at any given position on the globe which remain perpendicular after projection and assume the orientation of the axes of the ellipse. Tissot called the ellipse an *indicatrix* and its axes are called the *principal directions* at the given point. If the original circle is defined as being of unit radius, then the two semi axis dimensions, a and b, are equal to the maximum and minimum particular scales for that point on the map.

The relationships between the values a and b can be used to characterise the distortion properties of a projection. The product of a and b is related to the area of the ellipse and gives rise to a quantity known as the area scale s (also known as p) where

$$s = a.b$$

If the area of the ellipse remains the same as the original circle then $s = 1$. Maps in which this property holds are called *equal-area projections*, since they involve no distortion of area between the globe and the map. Maintenance of equal area is accompanied by considerable variation in the individual values and orientations of a and b.

The consequence of any departure from circularity of the original circle is the distortion of angles. Angular distortion can only be avoided by keeping $a = b$, i.e. making the ellipse circular. Maps in which $a = b$ at all points are known as *conformal projections*. It is important to realise that the two properties of equal area and conformality are always exclusive on a projection from a globe.

On maps which include angular deformation, it is of interest to be able to monitor the degree of deformation at different points on the map. For every point, there is a maximum value of angular deformation ω, given by the change in angle between two directions on the globe, each of which has undergone the maximum individual deflections for that point. The value of ω is given by the formula

$$\sin\frac{\omega}{2} = \frac{a - b}{a + b}$$

Choosing a map projection

When deciding on the most appropriate map projection for a particular purpose, one important choice is between equal area and equal angle (conformal) projections. When representing a very large area on a small-scale map, it will usually be the case that an equal area map is preferable, as it will result in a realistic representation of the relative size of different regions. Thus projections commonly used for world maps, such as the Mollweide and Goode projections, are equal area (Figure 4.14). Note that the Mercator projection, which is now rarely used for world mapping, results in gross distortions of the relative size of continents (Figure 4.14). For high- accuracy maps on which measurements of angle may be made, an equal angle projection is appropriate.

When choosing between planar, conic and cylindrical projections, and their respective aspect, the issue of distortion should be considered. It was pointed out earlier that the point or line of true scale on a map projection coincides with the point (or lines) of contact of the projection surface and the globe. Distortion increases with increasing distance from these locations. To reduce the maximum degree of distortion and to provide an even spread of distortion, the point or lines of true scale should be located as centrally as possible.

Thus for mapping a circular region, such as the whole of the Antarctic, a normal planar (aximuthal) projection is appropriate, since distortion will increase equally in all directions from the centre outwards. For an elongated region either a conic or a cylindrical projection may be appropriate, the objective being to locate the true scale line or lines axially relative to the elongation. For an east–west extended map, a normal conic projection is appropriate since distortion will be at a minimum along lines of latitude.

The use of a secant projection introduces two lines of true scale and hence reduces the distortion in a north–south direction for that projection. A north–south oriented region, such as Britiain, is appropriately mapped with a transverse cylindrical projection, since the true scale lines are oriented north–south. Transverse cylindrical projections such as the UTM and the British National Grid are secant and hence reduce the distortion across the map, when compared with a tangential projection.

Analytical transformations

Transformations from geographical coordinates to the grid system of a particular projection can be specified precisely using mathematical formulae. In a few cases the formulae are relatively simple, particularly when the earth is assumed spherical. Thus for the Mercator projection

$$x = R.\lambda$$

$$y = R \ln (\tan(\pi/4 + \phi/2))$$

where x and y are grid coordinates, λ and ϕ are longitude and latitude, R is the radius of the scaled sphere, and ln is the natural logarithm (i.e. to the base e). The equations are called the *forward solution*. The reverse transformation from grid coordinates to geographical coordinates is called the *inverse solution* and, in our simple example of the Mercator projection, is given by

$$\phi = \pi/2 - 2 \tan^{-1}(e^{-y/R})$$

$$\lambda = x/R + \lambda_o$$

where e is the base of natural logarithms and λ_o is the meridian passing through the origin of the geographical coordinate system.

For other map projections, forward solutions are often more complicated than the above example and become especially so when the transformation takes account of the non-spherical form of the earth. The inverse solutions often require an iterative form of solution. Details of the solutions for many important transformations are found in Snyder (1987).

Rubber sheet transformations

The use of the various well-defined formulae and procedures for converting back and forth between geographical and grid coordinates depends upon a clear definition of the projection in use. It is quite common, however, to encounter maps in which inadequate details of the projection are provided. If this is so then it may sometimes be necessary to guess appropriate projections and hope for the best. An alternative approach is to make use of so-called

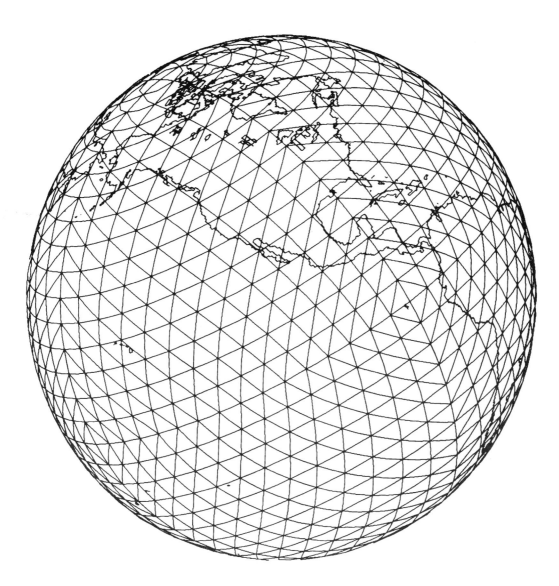

Figure 4.20 An example of a global tessellation, based on triangular faces. From Goodchild and Shiren (1992). Reproduced with permission from National Center for Geographic Information and Analysis.

rubber sheet transformations, which provide a means of converting a poorly defined coordinate system to a well-defined system, on the assumption that it is possible to identify some control points, the location of which is known in both coordinate systems. Such control points are likely to correspond to the location of features which can be clearly recognised in maps represented in both coordinate systems.

Rubber sheet transformations are also of use in situations other than when a conventional map projection is poorly defined. One important application is for registering satellite imagery with geographical or grid coordinate systems, the problem being that the geometry of the projection from the surface of the earth to the satellite's imaging system may not be precisely defined. Another application arises in the context of map digitising when adjacent or overlapping map sheets are found not to match up precisely, due to errors in digitising and due to distortion in the paper on which the map is drawn.

One technique for rubber sheet transformations makes use of polynomial equations to relate the coordinates in one map to those in the other. Rubber sheet transformations generally need to account for differences which are non-linear and thus require higher-order polynomials. Another approach, which is designed to ensure that the control points transform precisely from one coordinate system to the other, makes use of a triangulation scheme in which the control points are triangulated in the two maps and each equivalent triangle is transformed individually, with points internal to the triangles being interpolated in a linear manner (White and Griffin, 1985).

Global hierarchical tessellations

Interest in the development of global databases has led to efforts to design locational coding schemes (or geocodings) which allow spatial phenomena to be studied at different levels of detail in a consistent fashion across extensive regions of the earth. We have already seen that conventional projections can only retain simultaneously approximations to properties of equal area and a lack of shape distortion across

limited areas. Following Goodchild and Shiren (1992), we can identify four desirable properties of a global coding scheme for recording properties of the earth's surface in terms of finite spatial elements (or areal units):

1. The scheme should be hierarchical with elements at each level being subdivisions of elements at the next higher level.
2. Elements at any level of resolution should be approximately the same size, wherever they are located on the globe.
3. Elements at any level should all be approximately the same shape, wherever they are located.
4. The scheme should preserve topological relationships correctly, particularly adjacencies.

One of the simpler schemes which aims to meet these objectives is that described by Dutton (1989), which represents the globe in a hierarchical fashion by subdividing the triangular facets of an octagon, the six vertices of which are located on the earth's surface at the north and south poles and at 0°, 90°, –90° and 180° longitude. An octagon is one of the five regular polyhedra or Platonic solids, the vertices of which lie on a sphere. In Dutton's scheme, the triangular facets are divided recursively into four subtriangles. For a triangle with a horizontal base, the subtriangles are numbered 0 for the central one, 1 for the upper one, and 2 and 3 for the lower left and lower right one respectively. As triangles are subdivided, they are allocated numeric codes which Dutton calls QTM (quaternary triangular mesh) codes. Each time a new triangle is created its code consists of that of its parent with the addition, on the right, of 0, 1, 2 or 3, depending on its location within the parent. Clearly with each level of subdivision the triangles become smaller, and Dutton shows that at the 21st level of subdivision their size is approximately 1 m, going down to 17 cm at the 24th level (Figure 4.20 illustrates a level 4 subdivision). An important property of the QTM codes is the fact that the length of the code can be used to imply a particular level of locational accuracy. Thus individual points could be allocated the code of the smallest triangle they were known reliably to occupy. Similar objects of known areal extent could be allocated to the triangle that completely enclosed them.

Summary

Much of the data used in GIS are two dimensional, having been derived from maps which are based on projections from the 3D world to a planar map sheet. Locations on the earth's surface can be defined independently of a map by means of the geographical coordinates of latitude and longitude, which are a form of polar coordinates. When making measurements in terms of the geographical coordinates, the earth is assumed to be either a sphere or, when greater accuracy is required, an ellipsoid of rotation, or spheroid. These mathematical models of the earth's form are used when projecting to 2D maps. Projections are classed broadly as cylindrical, planar and conical, according to the type of surface onto which the projection is made, and as normal, transverse and oblique according to the aspect of the projection surface. All map projections introduce some form of distortion, which leads to a further categorisation into equal-area maps, which preserve measurement of area, and conformal maps which preserve angles in the projection. Computing technology has resulted in a recent interest in the possible use of global tessellations of the earth's surface, which provide a discrete, and in some cases variable scale, of cell-based locational referencing in three dimensions.

Further reading

A more detailed account of much of the material in this chapter concerning map coordinates, the form of the earth and map projections can be found in Maling (1989, 1992) and Snyder (1987), which have been referred to in course of the chapter. For a more extensive coverage of the transformations used in computer graphics, see Rogers and Adams (1990), in particular Chapters 2 and 3. Tobler has published several papers on the transformations used in cartograms, whereby the scale of a map is modified locally in proportion to a mapped variable such as population (see for example Tobler, 1979, 1986). Nyerges and Jankowski (1989) have developed an expert system for helping in selecting map projections.

CHAPTER 5

Digitising, editing and structuring map data

Introduction

Much of the computing technology for handling geographical data has been available, in varying levels of sophistication, since the mid 1970s, yet it is only since the late 1980s that it has come to be widely used. One of the reasons for this is the fact that the application of a GIS depends upon the existence of digital data, relating both specifically to individual applications and to topographic base maps to which they may be referenced. Although modern survey methods are able to produce their results directly in a digital format, the bulk of available topographic surveys were stored originally on conventional map documents, which must be converted to digital form to be of use for GIS.

The conversion of maps to create a digital database poses a major problem for national mapping agencies. In the USA there are about 54 000 of the basic scale map sheets, at a scale of 1:24 000, while the British Ordnance Survey, which was one of the first organisations to transfer to digital mapping, maintains a total of more than 220 000 maps at the basic scales of 1:1250, 1:2500 and 1:10 000. It was not until the early 1990s that sufficient of this basic scale mapping became available in digital form to satisfy the requirements of a significant number of map users, such as planners in local government and utility companies providing electricity, water, gas and telecommunications.

The process of digitising existing maps may be regarded as one of *secondary data acquisition*, in the sense that it is a transformation from one (analogue) form of information storage to another (digital). This may be distinguished from *primary data acquisition* whereby new data are obtained by surveying. Ground-based and remote-sensing surveying methods are now heavily dependent upon computing technology, with the consequence that the survey results are readily available in digital formats. Techniques of primary data acquisition are dealt with in Chapter 6.

The remainder of this chapter reviews the technology and procedures involved in digitising existing map documents. This will include descriptions of manual digitising, semi-automatic line-following digitisers, and scanning systems which require transformation from a complete raster image to a vector representation of the map features. The issues of data verification and correction, and of data structuring to produce topologically structured databases will also be considered.

Manual digitising

Manually operated digitisers probably provide the most widely used means of converting pre-existing maps into digital form. The digitising systems usually include facilities for recording both spatial and non-spatial information. Spatial data are recorded in the form of single coordinate pairs representing points, and series of coordinates representing lines and area boundaries. Non-spatial data, which describe the spatial data, consist of feature attribute codes, defining classes of feature, text used for naming individual

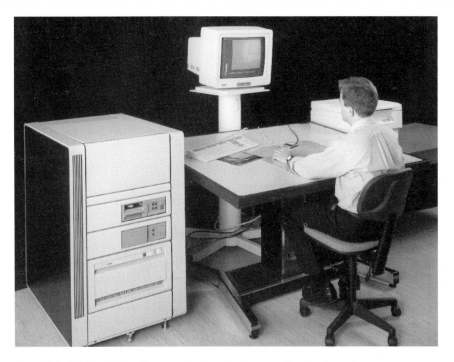

Figure 5.1 Digitising table with cursor, keyboard and screen. Courtesy Laser-Scan.

features, and numbers that quantify the entity represented (e.g. contour elevations, populations, geochemical measurements).

The main components of a *manual digitiser* are a flat surface, ranging in size between small tablets about 30×30 cm to large tables 120×80 cm or more in dimension; a hand-held puck or cursor, used by the operator to indicate positions to be recorded; and a keyboard for entering alphanumeric data and possibly commands (Figure 5.1). It is the larger devices that are generally of most use in cartography. Exact positioning of the puck is made possible by a cross hair mounted within a flat glass panel, which may sometimes include a magnifying lens. Also mounted on the puck are buttons that may be used for controlling data entry.

The most commonly used technology for the digitising tables is electromagnetic, in which a table inlaid with a fine grid of wires is associated with the puck which contains a metal coil. The grid of wires in the table and the coil in the puck act either as transmitter and receiver, or receiver and transmitter, respectively. If the puck is the transmitter, the position of the cross hair is found by scanning the x- and y-coordinate grid wires to identify those nearest the puck. The exact position is then found by interpolat-

ing between the adjacent wires on the basis of the nature of the signals received. Small-format, lower-resolution digitising tablets may sometimes use a stylus with a small coil in its tip, rather than a puck, as the locating device.

Operation of manual digitisers

Typically, the operating procedure for digitising a map starts by taping the source document firmly to the digitising surface. Grid or graticule reference points may then be digitised in order to register the map's coordinate system, before going on to digitise the map features themselves. An essential part of cartographic digitising is to ensure that the locational referencing information, represented by grids and geographic graticules on the map, is retained in the digital version of the map. Most medium- to large-scale maps (say larger than 1:50 000) include a rectangular grid which can be used as the basis of a linear transformation from digitiser coordinates to grid coordinates. If the map can be assumed to be undistorted, or accuracy is not a high priority, then this can be done by digitising at least three non-collinear (i.e. not on the same straight line) points on the map, the grid coordinates of which are known and may be typed in when digitising. Where higher standards of accuracy are required, and particularly when the source document is known to be distorted, a much higher number of coordinate control points should be digitised.

Sampling modes

Two distinct modes, point and stream, are commonly employed for digitising coordinates. Point mode entails explicitly entering single coordinates, with a button push on the puck. These may refer to point-referenced features such as trigonometric points, point elevations, or small-scale map town positions, or to linear or polygonal features such as rivers and county boundaries. In the latter case, the operator selects characteristic points along the path of the features. Stream mode is intended to speed up linear or polygonal feature digitising, by automatically recording points as the operator moves along the line, rather than obliging the operator to identify them individually (Figure 5.2). This may be done either on a time basis, in which case points are recorded at a specified time interval (e.g. every fifth of a second), or on a distance basis, whereby a point is recorded whenever the puck has moved a specified distance, in x or y or both, from the previous recorded point.

Time-stream digitising is characterised by the fact that fewer points will be recorded along relatively straight sections of line, which the operator may move quickly along, compared with complicated, wiggly sections which take longer to trace and which will therefore be recorded by more points, appropriate to their greater detail. Unless stream mode (time or distance) is set up to record very frequently, and hence often with great data redundancy, there is a danger that minor details such as sharp corners will not be accurately recorded. Because of this, some organisations work only in point mode and instruct the operators to ensure that all detail of interest is recorded accurately.

Chrisman (1987b) has pointed out that for inexperienced operators, distance sampling has the advantage of reducing the probability of undersampling, while taking away from the operator the burden of responsibility for selecting significant points. A side-effect of distance sampling is oversampling of straight sections. This can be compensated for by subsequent filtering of unwanted points. In the project described by Chrisman, soil maps (with wiggly boundaries) were distance sampled with a sampling rate of three times the line width. For land parcels which consisted typically of straight lines with distinct corners (i.e. sampling points) manual sampling was performed. He remarks that the parcel maps contained smooth arcs which should ideally be recorded with special-purpose digitising software tools.

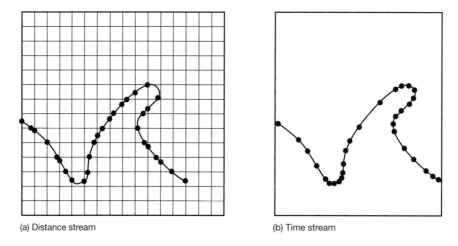

(a) Distance stream   (b) Time stream

Figure 5.2 Distance and time stream sampling modes. (**a**) In distance sampling a point is recorded whenever the cursor moves more than a specified distance in the x and y directions. (**b**) In time stream mode a point is recorded at regular time intervals. If the operator moves slowly, such as on intricate curves, points are recorded with higher density.

Feature coding for points, lines, polygons and text

For each point, line and polygonal feature that has its position digitised, additional data must be provided to indicate its meaning. This could consist of a classification, such as 'building' or 'primary road', or a specific identifier which could be supplemented by a specific name, giving for example a building's name or street number, or, for a road, the highway number or a street name. This information would normally be typed on a keyboard or selected from a menu, either immediately before or after digitising a location. However, it could also be done as a separate phase of encoding after precise locations have been entered. The latter approach may have the advantage of being a little less disruptive, obviating the need to keep alternating between the use of a line following digitising cursor or puck and the use of the keyboard. On the other hand, it may increase the risk of

error in associating the identifier with a specific graphical object. This could involve the user in 'clicking' on the object with a mouse or cursor before entering the identity. An alternative would be for the program to highlight graphic objects on the VDU screen, one by one, prompting the user to enter the appropriate data, in which case no physical pointing device would be employed.

Polygon encoding

Feature encoding of point and line objects is a relatively straightforward operation, but the process of encoding polygons is complicated by the fact that each line segment, or link, which bounds an individual polygon will normally belong to at least one other polygon. If each polygon were to be encoded as an entirely separate graphic unit, it would involve a great deal of duplicate digitising of the shared line

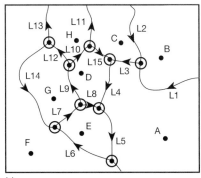

(a)

Seed point encoding
All links are digitised in one phase.
The coordinates of seed points
and the identifiers of their associated
polygons are recorded separately.
The seed points can be any point inside
their polygon.

Link	Left	Right
L1	A	B
L2	B	C
L3	A	C
L4	A	D
L5	A	E
L6	F	E
L7	G	E
L8	D	E
L9	G	D
L10	H	D
L11	H	C
L12	G	H
L13	F	H
L14	G	F
L15	C	D

(b)

Left/right attribute encoding
When each link is digitised, the identifiers
of the polygons on the left and right are
recorded, possibly in response to prompts
from the digitising program

Figure 5.3 Polygons can be digitised either using seed point encoding (spaghetti and meatballs) or left/right encoding. In (**a**) dots and capital letters indicate seed points, arrows indicate the directions of links (numbered L1 to L15), and circles indicate nodes. In (**b**) each link with its left and right polygon is listed. Completion of polygon encoding with both methods would result in identifying each link in terms of its terminating nodes.

segments. One method of avoiding duplicate line segment digitising is to associate with each line segment a pair of feature codes, describing the identities of the polygons on the 'left' and 'right' side of the line respectively. The terms left and right are given meaning by the direction in which the link is digitised (Figure 5.3). Having encoded the bounding segments in this way, the software can subsequently construct each polygon by searching for those links which reference it. The segments will be ordered, automatically, by matching equivalent start and end points. Note that while links belong to more than one polygon, it will not be necessary to store them twice. Provided each link is allocated a unique identifier, it is simply a matter of creating a list of link identifiers which will reference the relevant geometric data.

Another method of avoiding duplicate line digitising is referred to as the *spaghetti and meatballs approach* (Chrisman, 1987b), and is used in several commercial GIS, being somewhat simpler than explicit recording of left and right polygons. It involves an initial phase of digitising all the line segments of a polygon map in any order. This is the spaghetti. The operator then digitises at least one point inside each polygon and associates with it (e.g. by typing at the keyboard) the identifier of the polygon. These points are the so-called meatballs, also referred to as *seed points* (Figure 5.3). It is then left to the software to organise the spaghetti into discrete links which are used to construct polygons (as lists of relevant links). The software uses the seed points to deduce the identity of each polygon by performing point-in-polygon searches (see Chapter 11).

Text encoding

When digitising existing cartographic products, the operator can usually determine the identity of map objects either from their shape, their symbolisation (e.g. colour or line style), or from textual annotation on the map. In some organisations, it has been common practice when digitising maps to record text as independent graphic objects. This may be acceptable if the sole purpose of digitising is to enable a precise computer graphic reproduction of the source document, but if it is with a view to performing any analysis of the geographical information, or to creating maps of different scales from the original, items of text floating unattached in the original map space will be of limited value.

If text is present, it is generally desirable to attempt to associate it directly with a spatial object, so that when a database is constructed it will serve as an attribute of that object. It will then be possible to use the name as a means of referencing the associated geometry and, conversely, to retrieve the name when the geometry is accessed, in order to annotate a map.

Error detection and correction

All digital map data can be assumed to include errors of some sort. They arise from a combination of inaccuracy in the source data and from limitations of the digitiser operator and the computer system in use. Certain types of error, such as that due to inaccuracy of the original surveyed data, are implicit in the map source and cannot be rectified without obtaining another source of data. Other errors arise due to the operator failing to position the cursor accurately over the graphic object to be digitised or, more blatantly, missing objects from the map or wrongly entering the identity of an object. As a general rule, if an error can be detected at the time of data acquisition it will be easier to correct it then than if it is detected later, when someone is attempting to use the data. It is also the case that errors will usually be much easier to detect at the time of acquisition. Undoubtedly many digital data sets contain errors that may never be detected. In particular, locational errors often do not become apparent until similar data from different sources are combined.

Visual feedback and check plots

Modern digitising workstations commonly provide a visual display of the features that have been digitised, on a screen adjacent to the digitising table. Examination of this screen display may reveal relatively major errors such as operator-introduced knots or spikes in a linear feature, or the failure to complete a linear or polygonal object. Errors may then be corrected by interaction with the screen display and, if necessary, re-digitising or further digitising of the source document. A more thorough method of detecting positional errors and missing objects is to plot the newly acquired data at the source scale on a transparent film which is overlaid on the original document. Where discrepancies appear, the transparency, known as a check plot, can be marked to indicate where existing digitising should be deleted, repeated or inserted.

Errors in linear and polygonal feature digitising

Chrisman (1987b) has discussed typical errors which arise in digitising data that are to be structured into polygons. Specific errors include the following:

1. dangling chain (undershoot and overshoot),
2. unlabelled polygon,
3. conflicting labels for a polygon,
4. chain (line segment) with same left and right labels.

In a network of line segments, or chains, constituting a polygon map, it is essential that the chains terminate precisely at node points that are shared with connected chains. When digitising lines that meet at junctions, it is often found that the end of one line terminates either short of (*undershoot*) or beyond (*overshoot*) a line or lines with which it should be connected (Figure 5.4). These so-called *dangling chain* errors may be detected visually on a screen or check plot and may be corrected interactively by moving the end points of lines, often in combination with a *snap* function in the software which can unite two or more points (or a point and a line) that are separated by less than a specified tolerance distance. Alternatively,

having identified an undershoot, for example, an *extend* function may be used to automatically extrapolate the line onward to a neigbouring line.

It is also possible for these corrections to be carried out automatically. Thus a minimum length of dangling chain may be specified to cure overshoots, and a general tolerance may be specified which has the effect of merging any pair of points or a point and a line. Care should be taken in selecting an appropriate tolerance and in verifying the result of using it, since there may be situations where linear features are genuinely dangling, in the sense that they are unconnected at one or both ends. Examples of the latter would be roads that are culs-de-sac (no through way), and fences and walls that do not close off an area. On a hydrographic map, many stream sections would be unconnected at one end (their source).

Another problem which can arise when using tolerances to snap points together is that of *creep*. Here a point which has been snapped onto another point within the tolerance distance may then be snapped onto another which is within the tolerance distance of the second point but beyond that of the first. Thus the first point may effectively be displaced by a distance greater than the specified tolerance (Figure 5.5). This can happen within a cluster of points, causing significant 'illegal' displacements.

Errors of unlabelled polygons could be due to a failure to digitise a polygon seed point (a meatball), but they could also arise as a consequence of duplicate digitising of a single chain. Small discrepancies between the two versions of the chain would generate sliver-shaped polygons for which no seed point iden-

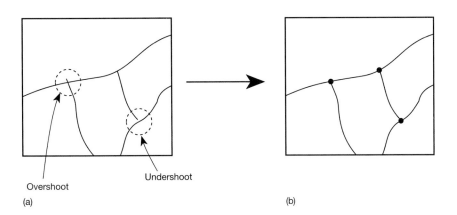

Overshoot

Undershoot

(a)

(b)

Figure 5.4 (**a**) Undershoot and overshoot digitising errors. Overshoot errors require the removal of a dangling chain and undershoot errors require extending the dangling chain to produce the 'clean' data in (**b**). The assumption is that the digitiser operator has not terminated lines exactly at intended junctions.

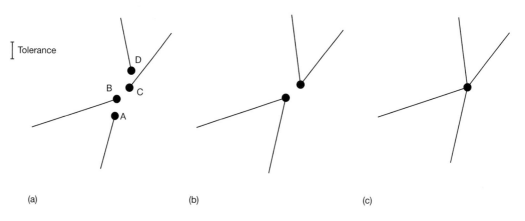

(a)   (b)   (c)

Figure 5.5 Creep can result in points being joined together that are further apart than a specified tolerance. (**a**) Each pair of points (AB), (BC) and (CD) is within tolerance, but a sequence of snap operations in which (**b**) is an intermediate state, result in the situation in (**c**) in which A has become connected to D, despite the fact that they were not originally within tolerance of each other.

tifier would have been digitised (Figure 5.6).

The occurrence of conflicting labels for a polygon might be due to operator error in assigning labels to seed points in situations where multiple seed points were digitised (intentionally or otherwise) for a given polygon. The use of multiple seed points may be regarded as advantageous in helping to reduce coding errors, since it effectively requires the operator to repeat an individual coding operation. Discrepancies would thus highlight errors which might not otherwise be detected if only a single point is used.

Identical left and right chain labels can occur if two adjacent polygons have been mistakenly coded with the same identifier. In some circumstances it may not in fact be an error if a chain is genuinely unattached, as mentioned above.

Human factors in positional error

In a discussion of the effects of what he terms human frailties on map digitising, Jenks (1981) has distinguished between psychological, physiological and logical errors. He illustrates the results of an experiment in digitising which suggest the possibility of very significant positional errors arising from working in stream digitising mode. He identifies latitudinal errors (lateral deviations from the true path of the source line) as due to pyscho-motor error in which the operator fails to locate the cursor accurately over the centre of the line to be digitised. Longitudinal errors, approximately parallel to the path of the line, are attributed to 'involuntary muscular spasms' resulting in spikes and polygonal knots in the digitised line.

In the context of point-mode digitising, Jenks introduces the notion of logical error, reflecting poor judgement on the part of the operator in selecting critical points which best represent the character of the line. He also reports the work of C. T. Traylor which demonstrated that human digitising error in general can be reduced by about 50% by training. It was found in particular that the direction of motion of the digitising cursor could have a major effect on the incidence of error.

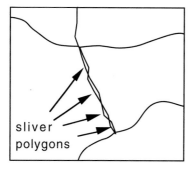

Figure 5.6 Sliver polygons can arise as a consequence of duplicate digitising of the same feature, in which the digitised lines differ slightly.

Hardware factors in positional error

All coordinates returned by a digitising table are limited in accuracy by the resolution of the table, and by the physical accuracy of the device in consistently returning the coordinate closest in position to the cross hair of the cursor. Both of the values are typically about one-tenth of a millimetre, which is within the width of the lines on most maps. However, because of the prevalence of the use of distance stream digitising mode, the opportunity for error is considerably greater than this figure might suggest. Thus the sampling interval is usually set to be much larger than the resolution of the device, in order to avoid generating large quantities of redundant data. The consequence is that any features occurring between the sampling interval will be lost. A typical effect of this would be that the tip of a spikey feature would be reduced slightly in the digitised representation.

Integration of digital map data from multiple maps and sources

Edge matching

Systematic digitising of a large geographic area will often require the use of several original map sheets that meet along their boundaries. When digitising sets of map sheets, problems often occur due to the fact that features in adjacent map sheets may not match exactly along their corresponding boundaries. Thus when the digitised data in the region of the boundary are plotted, certain features that cross the boundary may have small discontinuities, reflecting either inaccuracy in the original documents or digitising error. Such discontinuities must be removed if a seamless spatial database is to be created, i.e. one that is not constrained in any way by the boundaries of the source map data. To overcome the problem the digital map data must be edited in the boundary zones to ensure that there are no breaks in features that should be continuous (Figure 5.7). This process is referred to as 'edge matching' or 'zipping' and it can be automated to some extent using rubber sheet transformations that pull corresponding points together, while ensuring that other map features in the vicinity are modified slightly in location to retain the original relationships (Beard and Chrisman, 1988).

Conflation

Merging of data becomes more complicated when the source map sheets are overlapping to some extent. This may arise when wishing to integrate a set of surveys of the same area, perhaps carried out at different times or by different organisations. When such data sets are combined, or *conflated*, certain features may

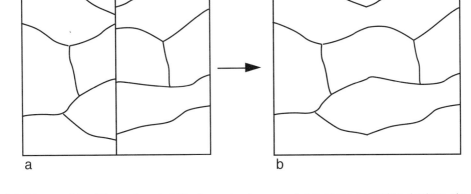

Figure 5.7 Edge matching. When adjacent digitised map sheets are merged to create a seamless database, the matching line segments may need to be shifted slightly to ensure continuity. The differences between matching lines may be due to inconsistency in the original surveys or to digitising error.

be duplicated between the maps, in which case one version of the feature will usually be retained, while the other or others are discarded. In doing this, care must be taken to ensure that associated map features, previously connected, or adjacent, to the discarded feature, become correctly merged with the retained feature, such as to ensure correct topological structure. The process of *conflation* is a difficult task to automate, since it involves matching equivalent features, or parts of features, which due to inherent survey and digitising errors may be in slightly different positions (Saalfield, 1988). The merging process may involve the use of rubber sheet transformations. If the merged data sets are at different levels of generalisation, requiring retention of the more detailed version, then the matching becomes particularly demanding due to the possibility of significant differences between generalised representations of the same real-world feature (Jones, 1996).

Semi-automatic line-following digitisers

Because manual digitising is such a time-consuming procedure, considerable effort has gone into attempts to develop automatic digitisers. This has resulted in the development of automated line-following devices, which in practice are only semi-automatic in that they must be positioned manually at the beginning of each linear feature to be digitised. They may sometimes also need to be guided manually when they encounter junctions between two or more linear features. More fully automatic devices which scan entire documents in a single pass are described in the next section.

The earliest line-following digitisers were based on mechanical designs which typically involved a flat surface on which the document lay and over which moved a sensing device mounted on a cross slide or gantry (Petrie, 1990a). These were superceded by laser-based technology, exemplified by the Laser-Scan company's Fastrak and Lasertrak systems. Here the source map was represented on a transparent sheet. A laser beam, deflected by mirrors, executed a local raster scanning pattern over a portion of the line to be digitised and recorded the image on film negative. This was automatically analysed to determine the path of the centre of the line. The resulting coordinates then served the additional purpose of

helping to determine where the next local scan should be centred in order to follow the line.

The Laser-Scan company improved considerably on the efficiency of the laser-based line-following digitiser with the introduction of the VTRAK system which works entirely in raster mode, starting with a fully scanned image of the map to be digitised (see next section). The operator is able to monitor the progress of the digitiser on a conventional screen display and for this reason the process is sometimes referred to as *head-up digitising*. Lines that have been digitised are marked automatically by a change in colour or symbolisation. Tasks to be performed by the operator include selecting features to be vectorised, and describing them by means of codes and attributes. Guidance of the process may sometimes be needed where the automatic procedure is unable to follow the line successfully. The latter situation could arise due to poor definition on the source document or, more specifically, where lines are irregularly discontinuous due to annotation or to the presence of map symbols which interrupt the line. Contours, for example, are often broken due to text giving height values, and where many contours merge together due to high topographic gradients. In the case of lines which are regularly discontinuous, due to a dashed symbolisation, it is possible to introduce sufficient 'intelligence' in the controlling software to cross the gaps automatically.

The effectiveness of line-following digitisers is a function of the frequency with which the operator must intervene to guide it. Thus they are more inefficient on detailed urban maps, which have many line junctions at small intervals on the map. Conversely they have proved most useful for digitising contour maps with long, relatively discontinuous and generally smooth lines. Ideally, for contour maps, the document to be digitised should be a separate sheet, with no other map data, apart from the contour values.

Full-document scanners

Scanners that scan entire documents are designed to create a digital representation of the source in the form of a 2D array of pixel values. For the majority of cartographic and GIS applications, this array or raster must then be analysed to derive a vector

Figure 5.8 Document scanner. Calcomp scanner. Courtesy Laser-Scan.

representation of the geographical features and of the annotation as indicated above for the **VTRAK** system. In respect of their primary function, commercial scanners have usually been very effective in generating high-resolution rasters of either binary, or grey scale or colour values. It is only recently, however, that raster to vector conversion software has become sufficiently sophisticated to be able to identify and correctly digitise (in vector format) a significant proportion of the graphical and textual information found on topographic maps. Even so, considerable effort must often be expended in validation, feature coding and interactive graphical editing of the vectorised data.

Early application of scanner systems was confined to digitising simple high-quality line work, such as the colour separates of contours of published maps. Current systems use pattern recognition techniques to distinguish between, and hence provide feature identification codes for, a variety of point, line and area symbols. They can also interpret, to a high degree of

success, both printed and hand-written text. A discussion of the details of raster to vector conversion will be deferred until Chapter 8. It may be appreciated that the latest scanning systems have enormous potential for dealing with the backlog of traditional 'analogue' maps which must be digitised by many of the organisations wishing to take advantage of GIS technology. With the recent improvements in processing of scanned data it can be expected that these systems will continue to become more widely used.

Raster scanners are usually based on either drum or flat-bed design (Petrie, 1990a). Drum scanners, such as those manufactured by Optronics, Scitex and Tektronix, wrap the document on a drum that is rotated adjacent to a photo detector head which moves incrementally along the length of the drum. The documents may be monochromatic requiring a single detector, or full colour, in which case several photodetectors may be used simultaneously, each with their own colour filter (Figure 5.8). As an alternative to the relatively large and expensive scanners, there are available small-format devices which use movable linear or area arrays of charged couple devices (ccd). The ccds can be combined with a lens to form a camera which, in the case of the linear arrays, moves the length of the document in a single scan.

Summary

Data acquisition through digitising of existing map documents has been, and in some organisations remains, a major stumbling block to the successful introduction of GIS technology. At present the most common means of digitising is still that of manually operated digitising tables and tablets, in which the operator traces out and identifies the point, line and areal features of the map. The process is essentially a tedious one and brings with it the responsibility for ensuring that adequate standards of accuracy are maintained. This occurs in the feature-tracing process in which the operator must take care to position the digitiser cursor accurately and to ensure that all features are recorded. It also occurs in the associated procedures of feature attributing which is required in order to link the geometric data to other application-specific data and to create topologically structured data. Issues of errors and data quality are pursued in more detail in Chapter 7. Semi-automatic procedures for digitising are becoming more widespread with the introduction of so-called head-up digitising whereby the operator controls line-following and feature-recognition software that operates on a scanned (raster) representation of the map. There have been several attempts over the history of digital mapping to introduce fully-automatic digitising, again based on scanned maps, but software for line following and particularly feature and character recognition is still not sufficiently intelligent to obviate the need for considerable interactive post-processing by an operator.

Further reading

Dangermond (1990) has provided a short review of the various data types and some of the practical issues involved in secondary data acquisition. For a review of digitising technology, see Jackson and Woodsford (1991). For a practical introduction to the wide range of facilities provided for digitising maps in the ARC/INFO system, see ESRI (1995). The papers by Chrisman (1987) and Jenks (1981), cited in the text, are recommended for a discussion of several practical issues in map digitising. As indicated in this chapter, fully automatic digitising is still a very challenging task. Some of the technicalities of automatic digitising are dealt with further in Chapter 8. Quality of digitised data is a major issue, which has been touched upon in this chapter, but is pursued in more detail in Chapter 7.

CHAPTER 6

Primary data acquisition from ground and remote surveys

Introduction

The digitising techniques described in Chapter 5 depend upon the presence of maps which record the results of earlier surveys. When no suitable maps exist, or when those which do exist need to be updated, it is necessary to resort to primary sources of information. These take the form of ground-based surveys and remotely sensed surveys and may initially be in either analogue or digital form, depending on the survey method. Ground surveys involve logging of field observations directly in digital form or in a manner that must subsequently be digitised. This may be appropriate for localised surveys and, in particular, ones requiring a resolution or a measure of ground truth which cannot be obtained in any other way. In many situations, however, the information required can be obtained more economically from remotely sensed surveys in which imagery, based on visible or non-visible electromagnetic radiation, must be interpreted by eye, or automatically by means of a variety of image-processing techniques.

Ground survey procedures often involve the use of electronic instrumentation which records measurements on digital media. Ultimately a ground-based survey provides the potential for obtaining the highest level of detail for topographic mapping. For some purposes, such as obtaining socio-economic data, verifying particular species of vegetation or studying very detailed geological structures, it may be the only currently practicable approach. In the context of topographic mapping, however, aerial photographic surveys often provide a more cost-effective means of collecting data. This is especially so in more inaccessible regions where ground-based surveying may be difficult and hence very time-consuming and costly. Digital data acquisition from photographs involves photogrammetric instruments which are used for interpretation of the photographs and recording of the features directly in digital format (older instruments output plots which can subsequently be digitised).

In the case of non-photographic remote-sensing surveys, employing electro-optical and microwave sensors, digital data are now accumulating in great quantities, raising major problems in storage and interpretation. These types of non-photographic sensors are effective in distinguishing between phenomena of the earth's surface and atmosphere, due to their differing physical and chemical properties, and have consequently been very useful in monitoring environmental resources such as crops, soils and geology, and in weather forecasting. Satellite surveys of natural resources started in 1972 with the American Earth Resources Technology Satellite (ERTS-1), which was the first of the Landsat series of satellites. As the spatial resolution of the data has improved, notably with the introduction of the SPOT satellites, it has now become feasible to attempt more detailed topographic surveying from the satellite data. In less-developed countries which have not previously been surveyed in detail, these data assume particular importance in their potential for providing basic (though relatively small-scale) topographic mapping. In regions where cloud cover is normal for much of the year, microwave sensors are becoming of special value, due to their relative insensitivity to the atmosphere. Interpretation of the satellite images is still a challenging task however, and the development of techniques for automatic recognition of ground features from the imagery is an active research field.

Ground-based surveys

Ground-based surveys are usually either concerned with recording general topographic information, such as the location of roads, rivers, buildings and the elevation of the ground, or with more thematically specific information, such as geology, soils, water quality and vegetation. The latter, more specialised surveys are based on topographic maps if they exist at an appropriate scale. If they do not, then it may be necessary to carry out some local topographic surveying to provide the context for other survey information.

Traditionally, topographic surveys have been conducted with a combination of theodolites, for

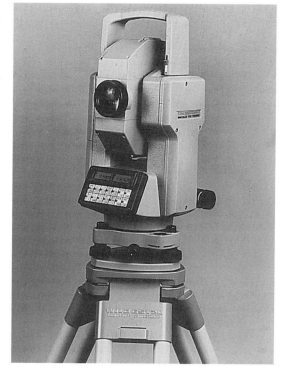

Figure 6.1a An electronic tacheometer used for measuring distances and angles (Zeiss).

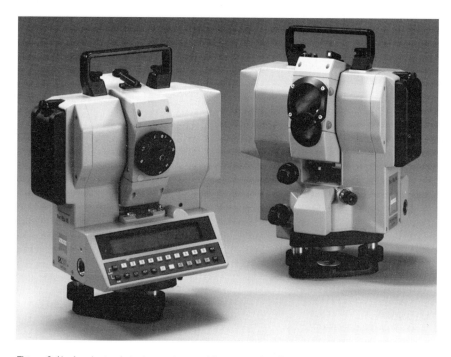

Figure 6.1b An electronic tacheometer used for measuring distances and angles (Zeiss).

measuring angles, and tapes and chains for measuring distances. The results of these surveys would normally be plotted in the field on maps which could subsequently be digitised on return to an office with computing facilities. The entirely manually operated tools are being replaced with electronic devices for measuring distances and angles, referred to as *electronic distance measurement* (EDM) and *electronic theodolites* respectively. The two can be combined within an *electronic tacheometer* (Figure 6.1) which, in addition to measuring angles and distances, can include a capacity for computation, based on the measurements, and for data storage. These devices are also called *total stations* and may consist of a physically integrated device or of a modular construction of flexibly linked components (Kennie, 1990).

When angle and distance measuring instruments are linked with a data-recording device, called a *data logger* or *data collector*, the measurements can be recorded digitally without the need to key in manually, thereby reducing the risk of human error due to misreading or mistyping.

The global positioning system (GPS)

A major development in ground-based surveys has been the introduction of the global positioning system (GPS). This enables positioning accuracy varying from about 100 m, down to a few millimetres, depending upon the way in which it is used (Colwell, 1991). The system is dependent upon a set of satellites, called NAVSTAR, which are intended to number at least 24 under a full implementation. This then provides instantaneous positioning for any location on the earth. The satellites each carry highly accurate clocks which are used in the transmission of signals picked up by GPS receivers on the ground (or on aircraft). A single hand-held receiver (Figure 6.2) can provide instantaneous accuracy well within 100 m, while differential GPS receivers provide relative positional accuracy of 3–5 m, with respect to a control point, the location of which must have been determined independently.

Higher accuracy, at the millimetre level, can be achieved, at greater cost, using two or more receivers simultaneously, one of which is positioned over one of a set of known locations, referred to as fiducial stations, while the other is placed at the unknown locations (Ashkenazi and Ffoulkes-Jones, 1990). The high accuracy is the result of complex computations which take account of the characteristics of the

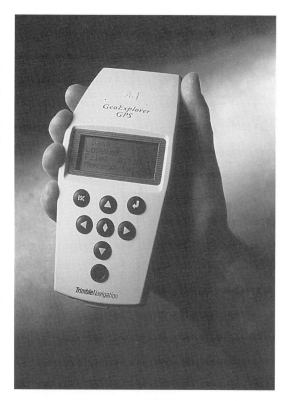

Figure 6.2 Hand-held GPS receiver (Trimble).

orbits of the satellites involved. The fiducial GPS measurements are sufficiently accurate to be used in monitoring movements of the earth's crust (Dong and Bock, 1989), and can be used as a cost-effective alternative to the highly accurate satellite laser ranging (SLR) and very long base interferometry (VLBI) techniques.

GPS coordinates can be calculated in three dimensions, thus allowing them to be used for elevation measurements. However, the accuracy in the vertical direction is poorer than in the horizontal.

Although GPS has an obvious value in surveying and is becoming an alternative to EDM and electronic theodolites, it has particular value in navigation. It was originally designed for navigation in a military context, but is becoming used increasingly in commercial and private vehicle navigation systems and in aircraft navigation. At the time of writing, a shortcoming of GPS is that the signals are subject to selective availability (SA) whereby they are degraded intentionally by the military, significantly reducing the positional accuracy obtainable from the (lower cost) navigational GPS systems to about 100 m, when 16 m would otherwise have been obtainable (Anon., 1990).

As with electronic tacheometers, GPS receivers can include storage facilities which enable recording of data for subsequent transfer to a GIS package.

Use of general-purpose hand-held computers

Survey data may be recorded in the field using purpose-built hand-held computers such as the Husky Hunter, or pen computers (Figure 6.3) in which data may be entered with a pointer device. Easily portable versions of PC-compatible and Macintosh computers are now widely available. These devices are very versatile in that the formats for data entry can be modified and special-purpose programs can be incorporated for local processing of the data, including digital map data. They may also be used for basic computing tasks such as word processing. Data entered on these instruments may be associated with field sketches illustrating point observations and strings of coordinates representing traverses or profiles, or in the case of topographic surveys, breaks of slope, ditches and road edges. The sketches would normally include codes or identifiers which link the observations to the stored data (Kennie, 1990).

Sampling patterns

For topographic surveys, sampling is usually on the basis of critical points relating to, say, the corners of buildings or to changes of slope and ridge lines. This type of sampling is appropriate where the features to be surveyed have an obvious structure that can be characterised by the critical points and lines. Where the distribution is less obviously characterised, sampling may be done on a regular-grid basis. Difficulty of access may dictate more irregular sampling pat-

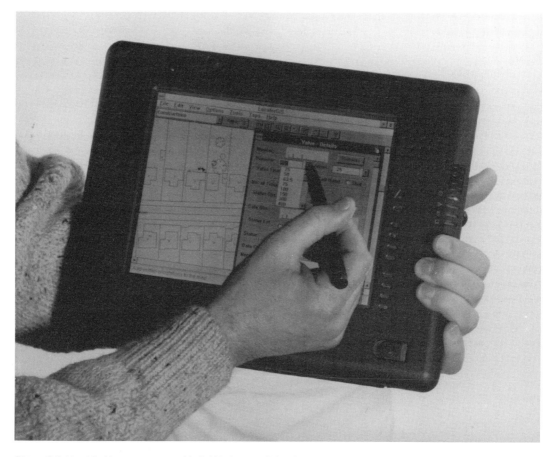

Figure 6.3 Hand-held pen computer with Sokkia Locator GIS software.

terns. Geological surveys, for example, are greatly affected by the presence of rock outcrops that are not covered by soil and vegetation.

Socio-economic data are frequently referenced to administrative areas such as census enumeration districts, electoral wards, local government districts and counties. The aggregation approach to data presentation can be a major source of uncertainty in the application of the information. This is because it may hide more detailed and often important spatial variation within the areas. Problems with uncertainty arise particularly when data from different-shaped sampling areas are combined, since in the overlapped areas estimates of statistics relating to the subareas may depend upon unreasonable assumptions about the uniformity of distribution within their parent sampling areas (see Chapter 12, page 211).

In general, it is preferable to obtain data initially referenced to the smallest available sampling areas. In this way data can then be used, according to the application, for local estimates of statistics and for aggregate or generalised estimates. In the latter case, the users of the data will have control over the way in which the aggregation is performed. For census data, localisation of data is facilitated by the use of postcodes which can be referenced to representative geographical coordinates within the postal zone. In some situations, data may be referred to individual houses, but for many purposes this is likely to create problems of lack of confidentiality.

Remote sensing

Remotely sensed data can be classified at one level by the wavelength of the radiation that is detected (Figure 6.4a). The ranges of electromagnetic wavelengths that

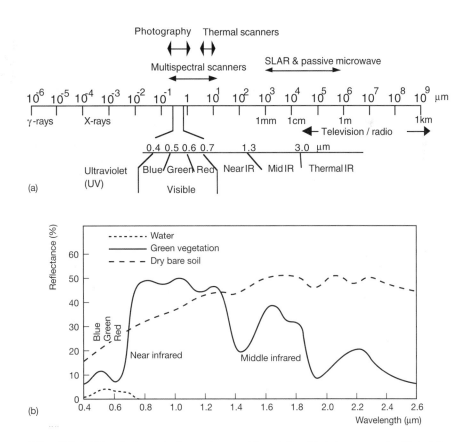

Figure 6.4 (**a**) The electromagnetic spectrum and the relationship between wavelength and the applicability of remote-sensing systems. (**b**) Typical reflectance spectra for water, green vegetation and dry soil.

are commonly used for remote-sensing surveys are 0.4–0.7 micrometres (μm), which is the visible range; 0.7–3 μm, called the reflected or near- and mid-infrared; 3–14 μm, the thermal infrared; and 5–500 mm, referred to as microwave radiation (Curran, 1985). Figure 6.4(b) illustrates typical reflectance spectra for water, green vegetation and dry soil. It is the characteristic differences between such spectra that enable phenomena on the earth's surface to be distinguished in remotely sensed images. Note, for example, the local peak in the green vegetation spectrum in the green wavelengths, and the much higher reflectance in the infrared wavelengths.

Figure 6.5 illustrates a number of images of the same location recorded in different wavebands. In this figure the changing spectral response of ground-surface phenomena between different wavelengths can

be clearly distinguished. Note, for example, the marked increase in value (lightness) of certain pixels in the near-infrared images. This is indicative of green vegetation, corresponding with the characteristic high near-infrared reflectance illustrated in Figure 6.4(b). Note also the high values of reflectance associated with buildings in the thermal infrared waveband. The capacity to distinguish between ground-surface phenomena is also well illustrated in Plate 9 in which several wavebands are combined to create a false colour image of southern California, which shows mountain ranges, with red indicating vegetation, and the city of Los Angeles, with blue tones corresponding to buildings. Plate 10 illustrates an image of the sea (off the east coast of North America) in which several narrowly defined visible wavebands have been combined to create a false colour image that distinguishes

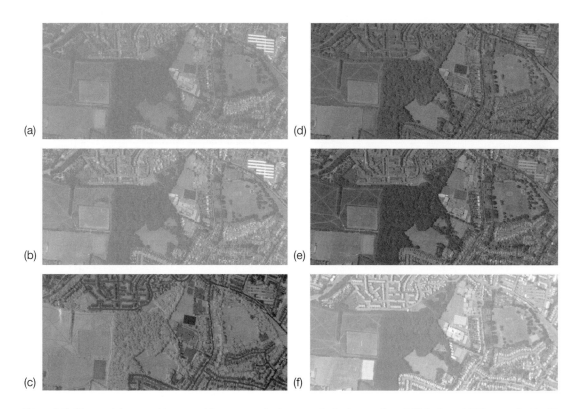

Figure 6.5 Six remotely-sensed images of the same suburban area in Orpington, Kent (UK) in June 1992, recorded in different wavebands of (**a**) green (0.52–0.59 mm); (**b**) red (0.6–0.7 mm); (**c**) near-infrared 0.76–0.9 mm); (**d**) middle infrared (1.55–1.75 mm); (**e**) middle infrared (2.08–2.35 mm); (**f**) thermal infrared (8–13 mm). Note the higher (lighter) pixel values of large areas in the centre of the near-infrared waveband image, corresponding to green vegetation. The distinct stripes in the upper right, apparent in the green and red wavebands, correspond to bodies of water. The very dark rectangle to the upper right of the centre of the near-infrared image represents an area of tarmac (road surface). Note similarly low reflectances for the roads in the scene. Note also the high reflectances of buildings and roads in the thermal infrared image. The images were recorded with the Daedalus AADS 1286 multispectral scanner with an original spatial resolution of 2 m. Data copyright NERC (1992).

between levels of phytoplankton, as well as showing sea currents and the land–sea boundary.

A second means of classifying remotely sensed data is according to the physical form in which it is acquired, which relates closely to the survey technology employed. Photographic surveys, from aeroplanes and satellites, record on film which must subsequently be interpreted, with varying degrees of automation, by means of photogrammetric instrumentation. Depending upon the type of film and on associated filters, photography can record in the ultraviolet (0.3–0.4 µm), visible (0.4–0.7 µm) and part of the near-infrared (0.7–0.9 µm) wavebands.

Multispectral scanners can detect several, and sometimes hundreds of, different wavebands simultaneously on an appropriate set of detectors. Multispectral images are usually recorded entirely in digital format, which may be stored on board the aircraft or satellite, or transmitted directly to ground stations. Video cameras detect energy in the visible and infrared wavebands and create a raster image that can be recorded on analogue video tape or be converted to digital form. They may also be viewed on a television screen at the time of recording or soon afterwards. Thermal infrared scanners record in a waveband beyond the capabilities of photographic film. Signals from the scanner can be recorded directly and may also be used to drive a cathode ray tube, the raster image from which may be recorded photographically.

Microwave sensors detect by means of antennae. They may be classified as *passive*, which record emitted radiation and those which are *active* in the sense that they send out microwave energy pulses, the reflections of which are subsequently recorded. The active method is that of *radar* (derived from Radio Detection And Ranging).

Having distinguished broadly between the classes of remote-sensing data, we will now examine in more detail the characteristics of photographic surveys and their associated photogrammetric technology, before going on to describe the non-photographic remote-sensing survey methods.

Aerial photography and photogrammetry

Aerial photography is an essential source of information in many disciplines. Notable fields include those of topographic surveying (for general-purpose base maps and for civil engineering operations), geology, hydrology, agriculture (soil and crop surveys), and urban and regional planning.

The detail that can be obtained from these photographs depends on several factors which include the flying height of the aircraft on which the camera is mounted, the focal length of the camera lens, the properties of the film and, for ground-elevation measurements, the degree of overlap between successive photographs. High-quality mapping cameras use lenses which minimise distortion and hence reduce the amount of geometric correction required to obtain measurements from the photographs. Focal length, i.e. the distance from the lens to the film, is typically 150 mm. The scale of the resulting photograph is given by the ratio of the focal length (f) to the height of the camera above the ground (h). Thus

$$S = f/h$$

For the simple example of a height above the ground of 7500 m and $f = 150$ mm $= 0.15$ m,

$$S = 0.15/7500 = 1/20\,000 = 1{:}20\,000$$

The relationship between focal length, the ground area photographed, and flying height is illustrated in Figure 6.6. Derivation of data from the photographs requires the ability to recognise the feature depicted, which can then be recorded either as point observations or, more generally, by tracing the path of linear features and the boundaries of areal features. When high accuracy is not required, this could be done directly on individual photographs using a digitising tablet of the sort described in Chapter 5.

For most practical purposes, however, it is necessary to take account of the changing elevation of the land surface, since the position of any point in an aerial photograph is a function of the ground elevation, as well as the horizontal ground position. As is illustrated in Figure 6.7, the image is a perspective projection of the land surface due to a cone of rays which converge at the camera lens. For a given horizontal ground location, the corresponding projection point in the image varies such that increasing ground elevation results in increasing displacement from the centre of the image.

Although there are techniques for measuring height and compensating for its effect in a single photograph, accurate photogrammetry requires working with overlapping pairs of aerial photographs for which stereo images can be obtained (Figure 6.8). Three-dimensional coordinates of a point on the ground can be calculated from a measurement, along the line of flight, of the change in photographic position of the point (relative to a fixed reference

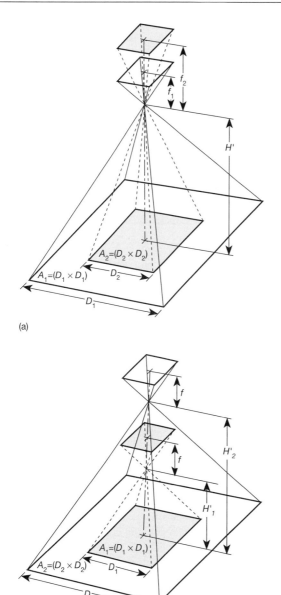

(a)

(b)

Figure 6.6 The relationship between areal coverage of a photograph, camera focal length and the aeroplane flying height. Modified from Lillesand and Kiefer (1994). (**a**) The larger the focal length, the smaller the areal coverage. (**b**) The higher the aircraft is, the larger the areal sample.

position) between successive overlapping photographs. This measurement, called *parallax*, is used, in combination with the known original separation of the photographs and the height of the aircraft, to

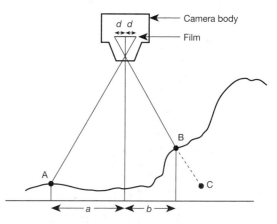

Figure 6.7 Effect of relief on the lateral position in a photograph. The point at raised elevation B projects to a position on the film which gives a false impression (C) of its horizontal ground location relative to point A because the two different ground dimensions, *a* and *b*, project to the same distances from the centre of the film.

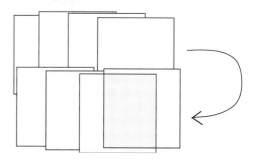

Figure 6.8 Aerial photographs are overlapped to generate stereo images. The shaded area shows a region of endlap (typically about 60% along the direction of flight) between successive images in a strip. The sidelap between images in adjacent strips is typically about 30% of the width of images.

determine the ground coordinates. Complications in the calculations arise from non-vertical photography and from changes in the height of the aircraft between succession photographs.

Stereoscopes and their associated instrumentation provide a low-cost means of creating a 3D view of the ground surface and of using the effects of parallax to record, by means of parallax bars, the 3D coordinates of points on that surface (Figure 6.9). The manually recorded values for individual points can be used to calculate heights and distances between points and hence the correct 3D coordinates (see Curran (1985) and Lillesand and Kiefer (1994) for details).

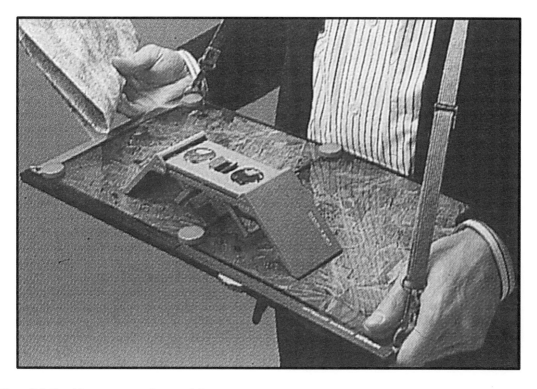

Figure 6.9 Portable stereoscope. Courtesy Leica.

Stereoplotters

Stereoscopes and parallax bars are, in practice, very laborious means of obtaining digital data. A more mechanised approach is provided by the use of stereoplotters equipped with instrumentation capable of digitally recording specific features and contours observed by the operator (Plate 8). A stereoplotter effectively reconstructs a scaled view of the original terrain by reversing the original projection on a pair of photographs. This requires reconstructing the original geometry of the projections, which must take account of the camera location, height and orientation for each photograph.

The main types of stereoplotter are optical, mechanical and now primarily analytical (Figure 6.10). Optical devices project the two photographs onto a table which the operator views stereoscopically. The features are then recorded in three dimensions using a tracing device. In the mechanical (or optical/mechanical) instruments, binoculars allow the operator to view the photographs directly, and the features are traced using mechanical rods. Digitisa-

tion can be achieved in both cases by linking the tracing mechanism to shaft encoders which record the motion of the x and y cross slides and of a vertical (z-axis) column. Digital data may be recorded on disk or tape for subsequent processing or may be processed on-line by computer, in which case there could be interactive facilities that allow error correction in the course of digitising, rather than in a later validation and editing session. Analytical stereoplotters perform the transformation from photographic coordinates to 3D ground coordinates by means of a digital computer. This is in contrast to the optical and mechanical stereoplotters which are essentially analogue devices, any digital output being due to sensing the mechanical motion. The analytical devices obtain digital results simply by storing the results of their calculations. The increasing use of computer techniques has culminated in *soft photogrammetry* in which all processing is performed digitally, starting with a pair of digitally scanned images. A digital photogrammetric workstation is illustrated in Figure 6.10(b).

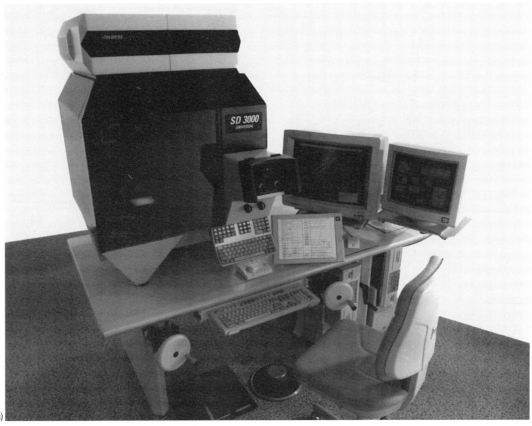

(a)

(b)

Figure 6.10 Analytical stereoplotter workstations. (**a**) Leica SD 3000 Stereo Analytical Plotter; (**b**) Leica Helava Digital Photogrammetric Workstation. Courtesy Leica.

Figure 6.11 Orthophotomaps superimpose cartographic data on photogrammetrically processed scanned photographic images in which the original photographic distortions have been removed to produce a planimetrically corrrect image. From Dahlberg (1993).

Orthophotography

Orthophotographs are geometrically corrected photo-graphs derived from stereo pairs of photographs by means of stereoplotters. The operator systematically scans the viewed stereo model while the machine (which may be entirely digital) simultaneously creates the orthophotograph, or ortho-image, in which positional displacements, due to ground relief and the effect of camera tilt, are removed. The primary purpose of the process is to generate composites of orthophotographs, called orthomosaics, which may then be enhanced with some graphic symbolisation to produce *orthophotomaps* (Figure 6.11). These are used in some countries as the basis of national map series (Petrie, 1990b). The significance of orthophotography from the point of view of digital mapping is that the process of scanning the stereo model necessarily generates height profiles which may be recorded digitally to form a terrain elevation model in the form of a sequence of parallel raster scans.

Correlators

The process of scanning for the purposes of ortho-photography, as described above, has been automated by the use of a correlator, or image-matching device. Here corresponding sections of the pair of scanned photographs, or originally digital images, are systematically compared in software to produce measurements of parallax which can then be used for height calculation. This approach results in an automatically generated digital terrain model.

Multispectral scanners

Remote-sensing surveys from satellites are typified by the use of multispectral scanners which record several wavebands of the electromagnetic spectrum. Interpretation of land cover often involves creating images which combine data from the different wavebands. In the early Landsat satellites the bands corresponded to the green (0.5–0.6 μm), red (0.6–0.7 μm), near-infrared

Table 6.1 Summary of remote-sensing satellites and sensors (Delbaere and Gulinck, 1994).

Sensor	Platform	Spectral range	No. of bands[a]	Resolution (m)	Image frame (km)	Period or launch	Target	Origin
HRV-XS	SPOT 1,2,3	Visible, near infrared	3	20	60 km	1986–	Land	CNES
HRV-P		Panchro./stereo	1	10				
Thematic mapper	Landsat 4, 5	Visible, near, middle infrared	6	30	180	1982–	Land	NASA
		Thermal infrared	1	120				
MSS	Landsat 1, 2, 3, 4, 5	Visible, near infrared	4	80	180	1972–	Land	NASA
AVHRR	NOAA (various missions)	Visible, near and middle infrared, thermal infrared	5	1100	2400	1982–	Atmosphere, oceans, land	ESA
SAR	ERS-1 ERS-2	Microwave	C (1)	30	100	1991–1995	Ocean, land	ESA
SAR	JERS-1	Microwave	L (1)	18	75	1992	Ocean, land	Japan
OPS		Visible to middle infrared	7					
MESSR	MOS-1	Visible to middle infrared	4	50	185		Atmosphere, ocean, land	Japan
ASAR	ENVISAT-1	Microwave	C (2)	13.5 to 30	55 to 400	1998	Coast zones, oceans, land	ESA
VEGETATION	SPOT-4 SPOT-5	Visible to middle infrared	5	1100	2200	1996 1999	Continental biosphere	CNES
ETM	Landsat-7	Visible, middle and near infrared	6	30	180	1998	Land	NASA
		Thermal infrared	1	120				
		Panchromatic	1	15				
SAR	RADARSAT	Microwave	C,X (5)	6 to 20	18 to 63	1995	Snow, ice land, oceans	Canada
LISS	IRS-1c	Visible to near infrared	4	72	148×174	1994	Land	India
MOMS-2	Space shuttle	Visible/stereo	3	4.5 / 13.5	78	1993	Land in intertropics	Germany
		Visible to NIR	4	13.5				
SIR-A		Microwave	L (1)	40	50	1981	Land, sea	NASA
SIR-B				30	20 to 50	1984		
SIR-C			X,C,L (3)	30	15 to 65	1992 1993 1995		
KEA-1000	KOSMOS	Photographs	5		80	1986	Land	Russia
MK-4		(visible and	8			1987		
KWR-1000		near infrared)	2		40	1993		
TK-350				10	180×270			

[a] For microwave specification with conventional code, within the number of polarisations in parentheses.

(0.7–0.8 µm and 0.8–1.0 µm) parts of the spectrum and produced images with a ground resolution of about 80 m (Plate 9). These Landsat images are usually referred to as MSS (multi-spectral scanner) images. The later thematic mapper (TM) scanner images (Plate 11) from Landsat provided a spatial resolution of 30 m, while the SPOT satellites carry high-resolution visible (HRV) imaging devices which can generate 20 m resolution multispectral imagery, and 10 m resolution panchromatic (black and white) imagery (Plate 12). Higher-resolution images can be obtained from multispectral sensors on aircraft, such as those in Figure 6.5 which have a 2 m spatial resolution. A summary of satellites and their sensors is given in Table 6.1. We will now examine the multispectral scanning technology in more detail.

Whiskbroom scanners

The main functions of multispectral scanners are those of collecting, detecting and recording the radiation received from the ground. This received radiation is referred to as *radiance* and consists of a combination of reflected, transmitted and emitted radiation resulting from interaction between the Sun's energy and objects (vegetation, soil, rocks, etc.) on the earth's surface. Collection is performed by means of a telescope directed towards a mirror which is oriented at 45° to, and which rotates back and forth about, an axis parallel to the line of flight. This enables radiation to be collected from a scan line perpendicular to the flight path before being reflected into the optics, which include a prism, or diffraction grating, for dispersing the component wavelengths towards their respective detectors (Figure 6.12). The *instantaneous field of view* (IFOV) of the collector determines the area of the ground that can be sampled at any given instant. Referring to Figure 6.12, the IFOV is given by the angle β subtended by the cone of rays entering the telescope. The diameter D of the ground sample can then be calculated by the equation

$$D = H\beta$$

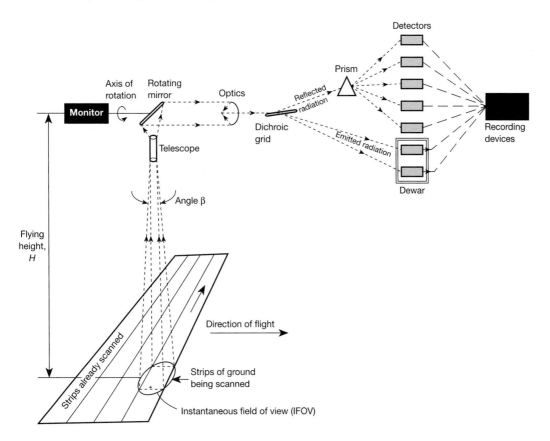

Figure 6.12 Diagrammatic representation of a multispectral scanner. After Curran (1985).

where H is the flying height of the aircraft (Lillesand and Kiefer, 1994). Taking an example of β 2 milliradians and a flying height H of 2000 m, D would equal 4 m. Typically, the mirror oscillates through an angle of approximately 90°, which is called the *field of view* (FOV). This in turn determines the width of the scan line on the ground, as a function of the flying height (see Figure 6.12).

The detectors are composed of various materials dependent upon the type of radiation to be detected. For example, thermal infrared may be detected by mercury-doped germanium (3–14 μm), indium antimonide (3–5 μm) or mercury cadmium telluride (8–14 μm). Thermal detectors must be cooled to very low temperatures of the order of –200 °C and below to avoid picking up radiation from the instrument itself.

The detected signals are amplified before being recorded, which, for digital purposes, requires an analogue-to-digital converter. The signals may also be used for analogue recording on cathode ray tubes (for temporary display) and on film. The digital values correspond to successive, adjacent and overlapping sample elements along the scan lines. The image is built up by recording successive scan lines along the line of flight, the width of this swath (i.e. the length of the scan lines) being determined by the FOV.

Geometric distortion of multispectral scanner images

Multispectral scanner (MSS) images suffer several distortions which must be corrected in order to be able to treat the digital image as a square matrix of pixels.

Lateral (tangential) distortion
As the scanner moves away from the vertical, the size of the ground sample element increases (Figure 6.13a). Thus, if the mirror rotates at constant speed and the detectors are sampled at equal time intervals, the resulting distance between samples will also increase in proportion to their distance from the centre of the scan line. This produces a decrease in the scale of the sample grid away from the centre. Further lateral distortions occur from an effect equivalent to the lateral displacement due to variable elevation in photographs. Radiation received from hills facing toward the scanner will map in the image to a horizontal location outward from the actual position (Figure 6.13b).

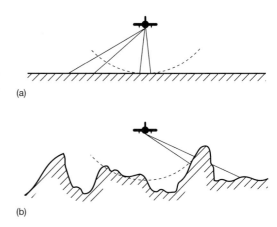

(a)

(b)

Figure 6.13 Lateral distortion in multispectral scanners. After Curran (1985).

Directional distortion
If the aircraft moves forward by a significant distance in the duration of a single scan line then the resulting line of pixels will not lie perpendicular to the line of flight. The problem can be obviated to a large extent with a combination of very rapid scan speeds and automatic instrument correction (Curran, 1985).

Aircraft stability
Any perturbations in the flight path of the aircraft will inevitably result in distortions in the scanning geometry. Specific causes of such distortion are changes in the disposition of the aircraft (pitch, roll and yaw), and changes in its direction, speed and elevation. Figure 6.14 shows the way in which uniform-sized ground cells may be distorted as a result of some of these factors.

Linear array (push broom) sensors

Many of the problems of geometric distortion of MSS images can be significantly reduced by the use of solid-state linear scanners. These devices, which are carried on SPOT satellites, use an array of CCD (charged couple device) sensors oriented perpendicular to the line of flight (Figure 6.15). The radiance from many hundreds of ground samples is detected simultaneously and hence there is no need for the oscillating mirrors of whisk broom scanners. As well as simplifying the MSS scanning mechanism, they also bring improvements in the signal-to-

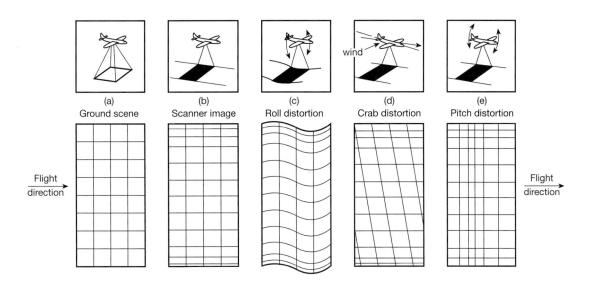

Figure 6.14 Scanner imagery distortions in size and shape of square areas on the ground resulting from (**a**) lateral distortion (see Figure 6.13) and (**c, d, e**) changes in aircraft orientation. Crab distortion (**d**) is equivalent to aircraft yaw. After Lillesand and Kiefer (1994).

noise ratio when compared with whiskbroom scanners, due to the greater dwell times possible. Because they record many samples simultaneously, the geometry resembles the perspective projection of photographic scanners and they are therefore well suited to creating stereo models. This aspect is exploited on the SPOT satellites by incorporating adjustment in the view direction of the array such that successive orbits can obtain stereo coverage of selected regions.

Imaging spectrometry

Improvements in multispectral scanning technology have led to the development of sensors that can record simultaneously in many very narrow wavebands, rather than a few relatively broad bands as, for example, in the Landsat and SPOT multispectral sensors. The Airborne Imaging Spectrometer (AIS) recorded 128 channels, each with a band width of about 9.3 nm. The Airborne Visible–Infrared Imaging Spectrometer (AVIRIS) records in 224 channels of about 9.6 nm width and with a spatial resolution of 20 m (Lillesand and Kiefer, 1994). The benefit of imaging spectrometers is that by producing a detailed spectrum for each pixel, they can discriminate quite effectively between types of minerals and species of vegetation. There has been particular interest in the monitoring of stress in vegetation due to pollution.

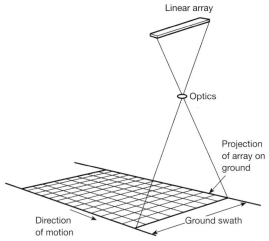

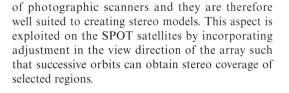

Figure 6.15 Linear array (push broom) sensor. After Lillesand and Kiefer (1994).

Figure 6.16 A metereological image, created using two wavebands from the multispectral sensor on the European Space Agency Meteosat satellite located 36 000 km above the earth's surface. Courtesy National Remote Sensing Centre.

This can result in a shift in the location of a characteristic steep slope in the reflectance spectrum called the 'red edge'. The red edge is located at a wavelength of about 0.7 µm (see Figure 6.4(b), which illustrates a typical reflectance spectrum for green vegetation). Since imaging spectrometers can be used to examine the detailed form of the reflectance spectra, the presence of such shifts can be detected.

Meteorological satellites

One of the main characteristics of metereological satellites is their very coarse spatial resolution. As illustrated in Figure 6.16, this can be exploited to create images covering very large areas of the earth. The Advanced Very High Resolution Radiometer (AVHRR) on board the NOAA (National Oceanic and Atmospheric Administration) series of satellites provides a ground resolution of 1.1 km, with data being recorded within a swath width of 2400 km on each orbit. The orbit, like Landsat and SPOT, is near-polar. The GOES series of meteorological satellites are geostationary, being positioned over the earth's equator. AVHRR images are used for a variety of applications in addition to meteorology, including vegetation mapping, water-surface temperature monitoring, geomorphology, soil-moisture

mapping and forest-fire detection. The main advantages offered by such sensors, when compared with earlier Landsat and SPOT systems, are greatly improved temporal resolution (several images may be produced in the course of a day) and the synoptic coverage provided by a large area.

Microwave imagers

Microwave imagers detect electromagnetic radiation in the relatively long wavelengths between about 0.1 and 100 cm (equivalent to 0.3–300 GHz in frequency). There are several commonly used spectral bands within this range. Microwave radiation is of special interest in remote sensing because of its capacity to penetrate cloud cover, which in many parts of the world can seriously hamper the use of visible and infrared wavelengths for the purposes of terrain surveys.

Naturally emitted energy in the microwave band is a function of temperature and of dielectric properties. Energy levels are, however, very low when compared with the visible and infrared, and devices for passive recording at these wavelengths are relatively expensive considering the limited advantages of the resulting surveys. Microwave surveys are most commonly performed by active imagers which record the backscattered radiation resulting from their own transmitted microwave energy. The returned signals are in this case dependent upon the roughness of the ground surface and its dielectric properties. As illustrated in Figure 6.17, this can result in images that give a clear impression of topographic and geological features.

Sideways-looking airborne radar (SLAR)

Active microwave imaging in remote sensing has developed from the use of radar, which is based on the principle of detecting the presence of static and moving objects by emitting pulses of radio energy from an antenna, and then recording the signals that are reflected from the objects back to the source (which may function both as transmitter and receiver). In sideways-looking airborne radar, the antenna is mounted on the base of the aircraft and sends out pulses of energy in a fan in the range direction. This is vertical and perpendicular to the flight path, which is termed the azimuth direction (Figure 6.18). The time taken for back-scattered energy to be returned to the receiver is a function of the distance it has travelled.

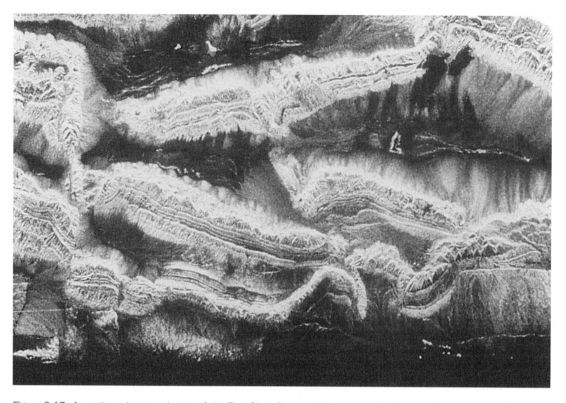

Figure 6.17 An active microwave image of the Tien Shan Ranges in China, acquired by the SIR-A imaging radar on the space shuttle Columbia in 1981. The radar beam was directed south, and hence north-facing slopes are brightly illuminated. Alluvial fans surrounding the hills are very prominent due to bright return signals from the gravels and boulders. Courtesy National Remote Sensing Centre.

Thus an image can be built up in the azimuth direction as a sequence of lines in the range direction, in which the amplitude of the returned signal is plotted

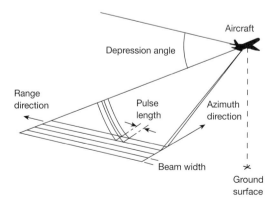

Figure 6.18 Sideways-looking airborne radar (SLAR). Modified from Curran (1985).

as a function of time and hence distance from the flight path (Figure 6.19).

For a horizontal ground surface, the time taken to receive signals from two points at a given distance apart increases with increasing distance from the aircraft. This produces a predictable outward reduction in scale, which can be compensated automatically when recording the signal. The recording method may use photographic film, analogue tape or digital media. The spatial resolution of SLAR images depends on the pulse length (i.e. the length of time for which a pulse is emitted), the depression angle and the beam width (Figure 6.18). Pulse length t affects resolution in the range direction. In order for two adjacent objects to be resolved, their returned pulses should not overlap. Thus the shorter the pulse, the higher the resolution. The corresponding distance on the ground is related to the depression angle γ, where decreasing γ results in decreasing ground distances. A consequence of this fact is that ground resolution in the range direction improves with

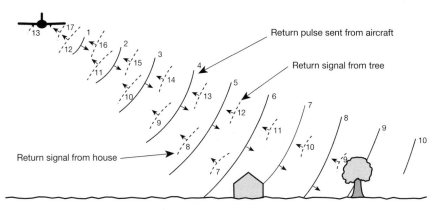

(a) Propagation of one radar pulse (indicating the wavefront location at time intervals 1–17)

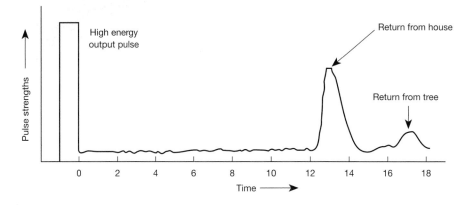

(b) Resulting antenna return

Figure 6.19 Radar pulses in sideways-looking airborne radar (SLAR). After Lillesand and Kiefer (1994).

increasing range. The range resolution Rr is given by the formula

$$Rr = ct/2\cos(\gamma)$$

where c is the speed of a microwave pulse (Curran, 1985). In contrast to this, resolution in the azimuth direction decreases with increasing range since the width of the beam on the ground increases with ground distance. For a beam width b and a ground range distance GR, the azimuth resolution Ra will be given by $Ra = bGR$.

If imagery is to be recorded in wide swaths, as is desirable for the purposes of reducing the number of flight paths across a given area, then the deterioration of azimuth resolution with increasing range becomes a significant problem. This is because for a given value of GR, the azimuth resolution can only be decreased by decreasing b, the angular beam width. The value for b is given by the formula

$$b = \lambda/A$$

where λ is the wavelength of the transmitted pulses and A is the length of the antenna. The wavelength cannot be significantly reduced without losing the advantage of microwave radiation for penetrating cloud cover, and the length of the antenna cannot

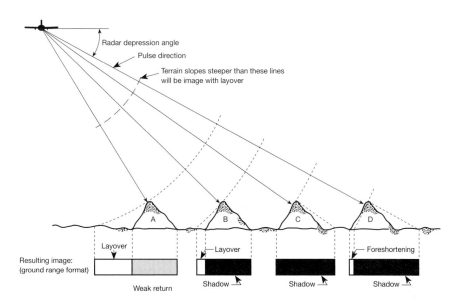

Figure 6.20 Effects of terrain relief on SLAR images. The term 'layover' refers to the arrival of reflections from elevated points before the arrival of lower elevation points which are physically nearer the aircraft. Shadows relate to parts of the terrain that are obscured to the pulses by raised relief. Foreshortening refers to the range-direction scale distortion resulting from raised relief, resulting in earlier arrival of pulses than their corresponding range points at lower elevation. After Lillesand and Kiefer (1994).

be increased beyond about 15 m without becoming physically unmanageable. These problems have been overcome by means of *synthetic aperture radar* (SAR) in which the effect of an ideally long antenna is synthesised by combining the signals from successive overlapping beams in the azimuth direction (Lillesand and Kiefer, 1994). The image in Figure 6.17 was obtained using SAR in the synthetic imaging radar (SIR-A) imaging radar on board the NASA Space Shuttle.

Effects of relief on SLAR imagery

The introduction of geometric errors in SLAR images due to increasing scale in the range direction has already been referred to. A more complex source of image distortion in the range direction is due to the fact that raised ground relief can result in the early arrival of returned signals relative to the corresponding ground range, illustrated in Figure 6.20. When the angle of the ground surface which faces the radar source is steeper than a line perpendicular to the direction of the radar pulse, then the reflection from the top of the feature arrives back before that

from the base. This displacement is called *layover*. If the angle of the facing slope is less than that of the perpendicular to the pulse direction, the reflection from the top will arrive after that from the base, but the difference may be quite small, resulting in a *foreshortening* of the true distance on the ground in the range direction. Note that these displacements are in the opposite direction to the relief displacement in aerial photography, in which the tops of mountains are displaced away from the direction of view, relative to the bases.

Another important effect of relief is the introduction of shadows behind raised features which block the path of the microwave pulse. These shadows can be of use to geomorphologists and geologists because they highlight the structure of the land surface.

The occurrence of approximately right-angled ground features can result in white spots in the image due to pulses being reflected directly back to the receiver following two successive 90° reflections. This effect, called *corner reflection*, may be particularly noticeable in urban scenes in which the angles are introduced between the ground and the walls of vertical buildings (Curran, 1985).

Image processing of remotely sensed data

The use of remotely sensed data for environmental studies depends upon the ability to refer data samples or pixels to known ground locations and to interpret the data values in terms of the phenomena on the ground. Determination of the relationship between original data samples and ground locations requires an understanding of the various data collection techniques and the geometric distortions which they introduce. Some of the factors that give rise to distortions have already been referred to. They include variations in height, orientation and velocity of the sensing device, the changes in the geometric relationship between the radiation direction and the earth's surface, atmospheric effects and changes in ground elevation.

Interpretation of pixel values requires *radiometric corrections* and *calibration* which compensate for non-uniformities and noise in the radiance signals, arising from the data collection process. This is followed by the application of various *image-enhancement techniques* which help in distinguishing the characteristics of the data between different locations. They involve transforming the data, sometimes by producing combinations of different images of the same area, in order to produce a clearer correlation between known environmental parameters and the recorded data. Enhancement techniques include *thresholding*, *contrast stretch* (or *contrast manipulation*) and *waveband combination* (or *ratioing*). Finally, *classification* processes perform the correlation between data values and the environmental phenomena. Techniques for classification may be categorised into those of spectral pattern recognition, which are based primarily on the radiance values, and spatial pattern recognition, which searches for geometric forms (shape, size, patterns) within the set of pixels as a whole (Lillesand and Kiefer, 1994). Some aspects of spatial pattern recognition are dealt with in more detail in Chapter 8, where we look at the identification of edges and linear features within a raster image.

Geometric restoration and correction

The coordinates of a matrix of pixel values produced by a remote-sensing device will not in general be aligned with any standard map coordinate system. Geometric restoration and correction are required to transform the sampled grid to a required coordinate system. Preliminary corrections may be necessary to compensate for changes in flying height of the aircraft or satellite and for characteristics of the sampling device which frequently result in overlaying of scan lines in the direction of flight. The latter would manifest itself as a repetition of the same ground locations in different scan lines of the image. Correction of such distortion is normally carried out before data are released for use (Curran, 1985).

Transformation of pixel coordinates to a standard coordinate system, such as the Universal Transverse Mercator, may involve a resampling procedure whereby locations in the image are correlated with known points on the ground, called *ground control points* (GCPs). The GCPs would normally correspond to features on the ground that could be recognised clearly in the image. Having recognised common control points on the ground and in the image, the pixel values can be used to estimate the values at regular intervals between the GCPs (Figure 6.21). This could be done, for example, by taking a weighted average of the nearest pixel values or by fitting a surface function to the local values and evaluating it at the required resampling positions (see Chapter 12, page 203).

Image enhancement

Contrast manipulation

The values for pixels in one waveband of a remote-sensing image typically range from 0 to 255, this being the range of possible values which can be represented by one byte of computer storage. The detectors on the sensing devices will be set such that the highest and lowest values for radiance or backscatter, for any part of the earth that the device will detect, will be recorded digitally within this range. It will often be the case that objects of interest lie within a limited range of these values. If the pixel values corresponding to these objects are all higher or lower than some value then *thresholding* can be used to produce a binary image, in which all values above the threshold are one and those below are zero. Such binary images, while useful in their own right in many contexts, can also be used as masks to segment other images on the basis of the binary classification, prior to subsequent processing. Figure 6.22(a) (from Lillesand and Kiefer, 1994) shows a histogram of a thematic mapper (TM4) image which displays a

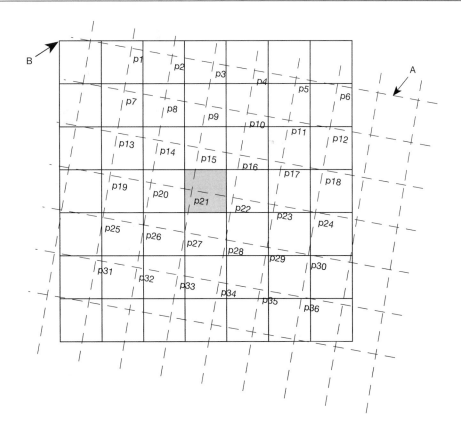

Figure 6.21 Resampling an image in a local coordinate system A to a standard geographical grid coordinate system B. The value of the shaded pixel in B is estimated as a function of the overlapped and adjacent pixels in A. If a distance weighting function is used, its value may depend not just on the overlapped pixels (p14, p15, p20 and p21) but also on nearby pixels such as p7, p8, p9, p10, p13, p16, p19, p22, p26, p27 and p28, depending upon the degree of smoothing or averaging to be employed. Using such methods, all pixels in the B grid could only be estimated to a uniform level of reliability if image A extended beyond the boundary of B.

clear bimodal distribution corresponding to water and land areas in the TM4 image (Figure 6.22b). Using a threshold value of digital number DN 40, a binary image (not illustrated) was produced, to distinguish only between land and water areas. This served as a mask to be superimposed on another thematic mapper image from a different (TM1) waveband (Figure 6.22c) in order to produce an image (Figure 6.22d), in which all land pixels have been set to a background value to simplify subsequent processing of the water areas.

For any one image, the majority of the actual data values recorded may be concentrated within only a part of the 0–255 range, depending on the characteristics of the objects in the image. This is the case with the image in Figure 6.22(d). A consequence of

this can be that small variations in pixel values, which may have scientific interest, may be difficult to detect visually because they correspond to only slight differences in grey scale or colour allocated to the pixel values. The visual contrast between different pixel values can be increased by re-allocating a range of pixel values of interest, say 80 to 150, to the entire range of possible pixel values (0–255). This process is called *contrast stretch*. Thus if 0–255 correspond in colour to shades of grey from black to white, the original grey value of 80 would become black, and 150 would become white, while values in-between would be interpolated linearly to appropriate grey shades. Figure 6.23 illustrates this process with part of the high-resolution imagery illustrated in Figure 6.5.

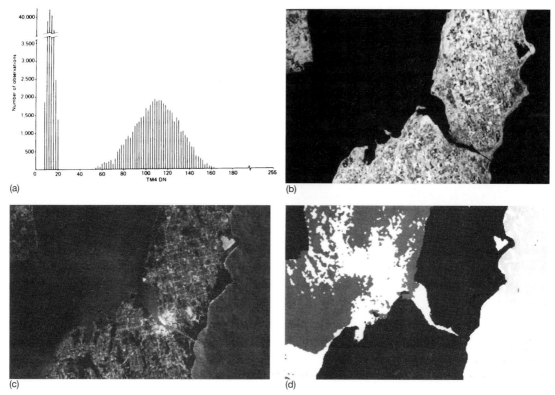

Figure 6.22 (**a**) Histogram corresponding to the thematic mapper image in (**b**) and showing a bimodal distribution. (**b**) TM4 image which was thresholded at DN level 40 to create a mask (not shown) distinguishing land from sea. (**c**) TM1 image with continuous distribution of grey tones (i.e. without the clear land/sea division). (**d**) Result of masking out the land areas on the TM1 image to enable the sea part of the image to be analysed. From Lillesand and Kiefer (1994).

There are several other ways of performing contrast stretch. Use can be made of different colours to distinguish different parts of a spectrum of values. If the user was interested in two parts of the spectrum, then two ranges of values could be contrast-stretched within two different colour ranges, e.g. red and green. Automatic procedures can be employed to identify minimum and maximum pixel values and to perform contrast stretch linearly between these values. Alternatively, using *histogram equalisation*, ranges of pixel values in the original image would be stretched over different ranges according to their frequency of occurrence. The effect is that a range of pixel values that were frequently occurring in the original image would be allocated to a wider range of pixel values in the output image, while adjacent ranges of infrequently occurring pixel values could be compressed to a single pixel value in the output.

Combining wavebands

Information about vegetation, soils and environmental change can sometimes be derived from remotely sensed data if images are constructed from more than one waveband. Thus, for example, red soils and iron ores have been detected from Landsat images by using the ratio of the green and red wavebands (Vincent, 1973). Curran (1985) points out that for detecting vegetation amount, Landsat MSS 5 (red) and MSS 7 (infrared) wavebands have been useful. Channels 1 (visible) and 2 (near-infrared) of AVHRR data have regularly been used to detect the presence of vegetation by calculating the normalised difference vegetation index (NDVI), where

$$NDVI = \frac{c2 - c1}{c2 + c1}$$

When vegetation is present the NDVI will produce a relatively high value because of the character-

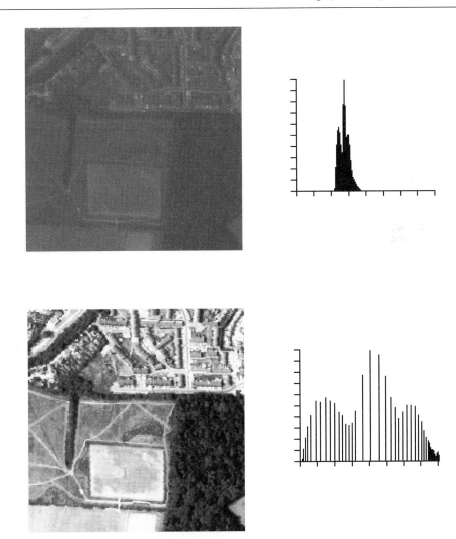

Figure 6.23 Contrast stretch. The upper image illustrates an image with low contrast due to the pixel digital numbers falling in a relatively narrow range, as indicated in the associated histogram. The lower image and its histogram show the effect of contrast stretch. Data copyright NERC (1992).

istically high reflectance in the near-infrared $c2$ wavelengths relative to that in the visible $c1$ wavelengths (see Figure 6.4b).

For the purpose of detecting change, such as flooding or forest-fire damage, an image can be constructed from the difference between two images of the same waveband recorded at different times. This process can require additional geometric correction to register the two images.

Ratioing and differencing of different wavebands are relatively simple methods of producing images

with specific information which may not be obtainable from the single waveband. A more complicated use of multiple wavebands, which can result in images that more clearly distinguish between a variety of environmental parameters, than could a single waveband, is that of *principal components analysis* and *canonical analysis*. When corresponding values for two or more wavebands are used to produce scatter plots, regions of which are then classified in terms of the objects of interest, it can often be seen that new axes can be fitted to the data such that when used

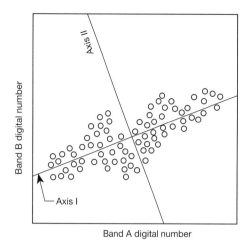

Figure 6.24 In principal components analysis, new axes are fitted to scatterplots such that new axes represent greater variance in the data than the original axes. After Lillesand and Kiefer (1994).

instead of the original values they differentiate more effectively between classes of object. The new axes are calculated on the basis of capturing the maximum variance in the data (Figure 6.24). Transformation of several wavebands to the primary axis obtained in this way can provide an effective way of producing an image which encapsulates the primary element of variation recorded by the different wavebands. Such methods may prove particularly useful when attempting to assimilate information from as many as seven simultaneously recorded wavebands, as in the case of Landsat thematic mapper (TM) data (Mather, 1987). Thus the main information content is represented by a smaller number of dimensions than those of the original data.

It was pointed out above that colour can be useful in helping to distinguish between different parts of a single waveband. Colour is also very useful for combining different wavebands, for the same image area, each of which could be allocated a range of red, green and blue colour intensities respectively. Each pixel value of the resulting composite image would have a red, green and blue component. If, as can often be expected, the original data wavebands did not represent the corresponding visual wavebands of red, green and blue, the resulting image would be described as a *false colour composite* (see Plate 9 for an example, in which vegetation appears as red).

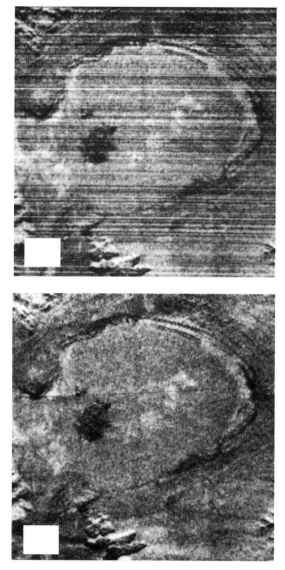

Figure 6.25 Noise elimination in the frequency domain. (**a**) Multispectral scanner image with noise, and (**b**) a reconstruction of the original image after filtering out the frequencies corresponding to the horizontal striping pattern. From Lillesand and Kiefer (1994).

Spatial filtering

Processes such as contrast stretch and ratioing of wavebands, discussed above, derive their value from emphasising or suppressing particular pixel values or ranges of values. When analysing images it can also be of interest to enhance or suppress structural features of the image represented by multiple pixel values, which together constitute visual patterns. Processes of spatial filtering are based on the assumption that an image can be regarded as being composed from a summation of sinusoidal functions, of different frequencies, representing change in image intensity (or value). The various spatial forms within the image are characterised by differing frequencies in different directions.

Using *Fourier analysis*, the frequency composition of an image can be examined and certain frequencies can be removed before reconstructing a filtered image. Removal of high frequencies, i.e. the application of a low-pass filter, could serve to reduce 'noise' consisting of speckle in radar images. Figure 6.25 illustrates the removal of noise which can be isolated in the frequency domain because it corresponds to high frequencies in a particular direction. Geological lineaments could be emphasised by filtering out frequencies corresponding to directions other than those of the lineaments. A high-pass filter (which suppresses low frequencies) could be used to enhance edges in an image, or to pick out small features which the remote-sensing detector had tended to suppress because they were small compared with the pixel sample size.

Transformation to a set of frequencies, some of which are removed before image reconstruction, is known as *frequency domain filtering*. Similar effects can also be obtained by applying a moving window operator, or filter, to the original image. The moving window operator, also known as a *kernel*, examines successive blocks of pixels in the image (say 3×3 or 15×15) which are used to re-estimate the pixel in the centre of the window as a function of all pixels overlapped by the window. The filter function could be weighted towards the centre of window, i.e. with a strong peak in the centre, in which case high frequencies would be emphasised (a high-pass filter), or it could be relatively flat in shape producing a high degree of averaging of pixel values and hence suppressing the high frequencies, i.e. a low pass filter.

Classification

The primary objective in processing remotely sensed data is to establish a relationship between components of the image and features on the ground. The processing techniques discussed so far have all been intended to facilitate the establishment of this relationship by transforming the original data to give it a simpler relationship with the real-world objects, and by enhancing or combining it to make aspects of the relationship clearer.

Procedures for classifying imagery operate on either a single waveband or on a combination of wavebands. In the former case the waveband is segmented into several ranges of pixel values, called *density slices*, each of which is classified in terms of the regions on the ground. The classification must be done with reference to a training area on which the land cover is known, so that relationships in this area can be used to infer the interpretation of images where the ground cover is not known. If there turns out to be a direct correlation between each class of ground object and a distinct range of values in a single waveband, then a simple density slicing procedure for correlating ranges of values with objects will be adequate for classification.

It is more commonly the case that no single waveband can distinguish all classes, but distinctions can be made when wavebands are combined to produce a scatter plot as referred to earlier. Once combined waveband values have been interpreted by reference to the training area, the scatter plot of interpreted values can be used to classify pixel values for unknown regions (Figure 6.26a). Three common methods of classifying, given the training data set, are as follows (Lillesand and Kiefer, 1994):

1. minimum distance to means;
2. parallelepiped;
3. maximum likelihood.

In the *minimum distance to means* classifier (Figure 6.26b), the means of all points for each class are calculated and an unknown point is then classified according to the mean value that it lies closest to. In the *parallelepiped* classifier, boxes are placed around each set of points of a given class, where the boundaries of the box correspond to the minimum and maximum waveband values of pixels falling into the given class (Figure 6.26c). Since boxes may be found to overlap, resulting in ambiguous classification, they may be modified in shape on the basis of a closer inspection of the training data. The third method, of *maximum likelihood classification*, is more sophisticated in that the scatter plot is used to construct probability contours from the probability density function of each object class. This will allow for the possibility of overlapping classification, but will be able to use the probability functions to estimate the most likely classification (Figure 6.26d).

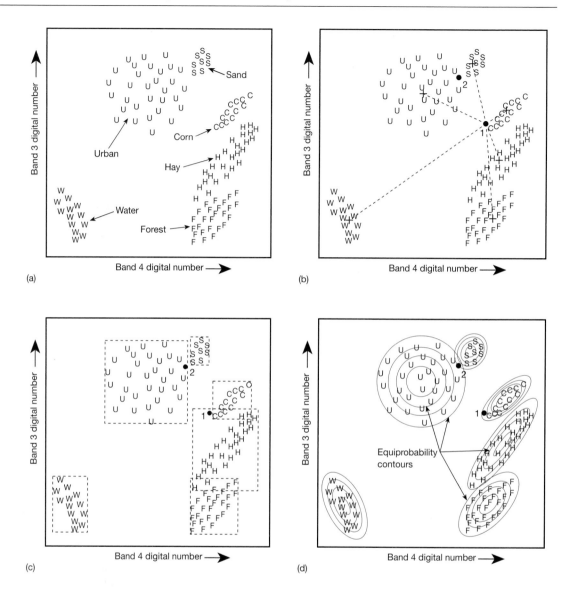

Figure 6.26 Classification methods. (**a**) A scatterplot of pixel values known from the training site to correspond to particular classes of phenomena can be used to characterise unknown pixel values. After Lillesand and Kiefer (1994). (**b**) Minimum distance to means classifier. The crosses represent the means of the various clusters of training site pixels. Point 1 is closest to the corn mean. The method is not always reliable. Note that point 2 is closest to the sand mean, though it lies within the urban cluster. After Lillesand and Kiefer (1994). (**c**) Parallelepiped classification strategy. A pixel is allocated to the box which encloses it. Ambiguity arises when a point lies inside more than one box. Also, if a cluster is diagonally elongated, inclusion in the box is not a reliable indicator of correct classification (see point 1). After Lillesand and Kiefer (1994). (**d**) Maximum likelihood classification uses the probability density function of each category to determine which category an unknown pixel most probably belongs to. Thus each category has its own function which is applied individually to the unknown pixel. After Lillesand and Kiefer (1994).

Having applied the results of a classifier, their verification will require ground sampling, following which it may be necessary to make adjustments to the classification scheme, in the light of unacceptable errors. In addition to recording the various types of ground cover, ground sampling could include on-the-spot measurement of radiometric properties of particular classes of object, which could then be correlated with the remotely sensed data.

Neural nets for image classification

Recently there has been considerable interest in the use of *artificial neural networks* or *neural nets* for the classification of remotely sensed images (Foody, 1995). Neural nets simulate mechanisms assumed to be used by the human brain to acquire and store knowledge. A neural net is composed of processing units referred to as *neurons*, or *simulated neurons*, which, by analogy with the neuron of an animal's brain, are connected to each other via *links*. Each neuron receives signals on input links (dendrites) which it reacts to by providing, on an output link (axon), a signal that is a function of the inputs. The outputs of neurons are connected (synapse) to the inputs of other neurons. Each of the input signals is assigned a numerical weight which is used by the neuron's *activation function* to generate an output signal. A typical function will generate an output value of either zero or one, depending on whether the weighted sum of the inputs (0s and 1s) exceeds a given threshold value. The neurons are arranged in layers consisting of an input layer, one or more internal hidden layers, and an output layer. The intention is that data enter at the input layers and are interpreted to provide a response at the output layer.

For a neural net to succeed in recognising or interpreting input data it must undergo a training process, the purpose of which is to allocate weights to the input links of neurons in all the layers, such that particular patterns of input result in predetermined outputs. The training data must consist of examples of input data and the required corresponding output data. For satellite imagery, this would be the image density values of however many channels were to be used, and the classified image. The power of neural nets lies in the fact that if training is performed adequately it will be possible to present examples of inputs that may differ slightly from any of the training examples, while still yielding an appropriate interpretation. Output values may in practice be real numbers that approximate to the values presented in the training output.

Several techniques have been developed for training neural nets. One that is commonly used and is applicable to satellite image classification is *back-propagation*. It is an iterative procedure which repeatedly generates outputs, at the output layer, compares these outputs with the desired outputs and changes all weights in a manner that is proportional to the difference between the actual and desired outputs. It is called back-propagation because the changes to the weights in each layer depend on the previous output produced at the following layer. The process for each iteration starts by calculating weights at the output layer and works back successively to the input layer. Each iteration is accompanied by a measure of performance of the net in terms of the difference between the actual outputs and the required output. The learning process terminates when the performance is regarded as adequate.

Summary

In this chapter we have made a distinction between ground-based surveys and remote-sensing surveys. In ground-based surveys we are often concerned with direct human observation of environmental phenomena that may be recorded on maps or documents. Socio-economic surveys may employ questionnaires. Ground-based surveys often entail the use of hand-held computers or data loggers to record the information in digital form. Surveyed locations can be determined by means of electronic total stations measuring distances and angles, while in recent years there has been increasing usage of Global Positioning System (GPS) receivers which use satellite transmissions to determine position to what can be very high accuracy, depending on the mode of operation.

In the context of remote sensing, we have focused on airborne technology on aircraft and satellites, though it should be remembered that there are other types of remote sensing, of the ocean floor and the subsurface, that are conducted from ships and on land. Major classes of remote-sensing technology are aerial photography, multispectral scanners and microwave scanners. Aerial photography can result in very high-resolution imagery which, when produced in overlapping stereo pairs, can be processed photogrammetrically to construct 3D models of the earth's surface. Such models may be used to create and update large-scale topographic maps. They may also be used for environmental surveys of, for example, geology and vegetation.

Multispectral scanners, mounted on satellites or aircraft, record in several visible and invisible wavebands to detect various characteristics of the earth's surface and atmosphere. Ground resolution of these devices is rarely better than about 10 m and thus they are used for medium- and smaller-scale mapping. Stereo coverage at the higher resolutions (as in SPOT panchromatic imagery) can, however, be used to create 3D models of the ground surface. The range of wavebands of multispectral imagers provides data that can be used to differentiate between different classes of phenomena including vegetation, crops, soils and geology, as well as human-made structures. Low spatial-resolution multispectral scanners can provide synoptic views of large parts of the earth's surface, which may be particularly useful in meteorological studies. Microwave imagery, in the longer electromagnetic wavebands, has the advantage of penetrating cloud cover. It is particularly sensitive to differences in surface texture, which may be used to differentiate, for example, between types of vegetation and geology. Processing and interpretation of remote-sensed imagery is a major overhead and we have briefly reviewed here some of the main techniques, including those for making geometric corrections, enhancing images through thresholding and contrast stretch, ratioing wavebands and spatial filtering. Conventional image-classification techniques study characteristic combinations of pixel values in different wavebands that correspond to known ground phenomena based on training sites. There is increasing interest in the application of neural net image-classification techniques which are well suited to some aspects of remote-sensed imagery.

Further reading

A short review article on the subject of photogrammetry and remote sensing has been written by Petrie (1995). For more details on surveying and photogrammetric techniques, see the collection of articles in Petrie and Kennie (1990). Lillesand and Kiefer (1994) provides a very good up-to-date text on the principles of remote sensing and image interpretation. For a recent set of articles on environmental applications of remote sensing, see Foody and Curran (1994). The journal *Photogrammetric Engineering and Remote Sensing* regularly publishes issues focusing on links between GIS and remote sensing.

CHAPTER 7

Data quality and data exchange standards

Introduction

The issue of data quality encompasses all aspects of spatial data processing. A brief definition of data quality, proposed by Chrisman (1984), is that it is a measure of the fitness for use of data for a particular task. It depends, therefore, upon what the user intends to do with a given set of data. Data quality is associated with a variety of interrelated concepts, (e.g. accuracy, error, precision and resolution) which can be defined in ways that help to quantify the quality of data.

In this chapter we link the subject of data quality with that of data transfer standards. Data transfer standards are partly concerned with formatting data in a standardised way that becomes widely recognised. However, effective data transfer depends upon providing enough information to ensure that data are as usable as possible and that the users are fully aware of the quality of all aspects of the data. Thus basic spatial data consisting of geometry and attributes needs to be accompanied by *metadata* that describe the data (Table 7.1). The quality of data depends not just upon their accuracy, but upon a knowledge of the feature coding schemes used to describe the attributes and upon a knowledge of when and how the data were acquired. The more information that is available regarding the origin of data, the better the user is able to assess its suitability and hence its quality for their own purposes.

In what follows, we start by attempting to clarify the terminology of data quality, with regard to accuracy, precision, resolution, scale and error. The subject of error models and the different types of error is then dealt with, before considering the reasons why error, or inaccuracy, may arise. In the second part of the chapter, spatial data transfer standards are introduced, with particular attention to SDTS and NTF.

Accuracy

In the context of spatial data, accuracy may be regarded as the difference between a recorded value (or measurement, or observation) and its true value. The concept is clearly a rather abstract one as it appears to assume that there is some truth which can be known. In practice the knowledge of a measurement may depend upon the type and scale of measurement. Categorical measures recorded on nominal and ordinal scales may in theory be determinable beyond reasonable doubt. Thus it is usually possible to agree that certain phenomena are described by a particular name and fall into a particular class. For example, it can usually be agreed that a particular person is of a particular species and has a particular name. Numerical measurements on interval and ratio scales may, however, be known only within some degree of certainty, referred to as the error in the observation. For example, the weight and the height of John O'Rahilly can only be measured to within some fraction of a kilogram and of a metre respectively.

In general, the assessment of the accuracy of an individual observation or measurement can only be

Table 7.1 Components of metadata.

Data exchange format	Specification of data storage format
Data summary	Source, classes of data, areal coverage, date, scale

Lineage	Agency of origin
	Method of data collection:
	   primary survey techniques
	   secondary data sources
	   digitising method
	Dates updated
	Processing history:
	   coordinate transformations
	   data model transformations
	   attribute transformation

Coordinate system	Type of coordinate system Map projection parameters
Spatial data model	Specification of primitive spatial objects Topological data stored
Feature coding system	Definition of feature codes and classification system
Classification completeness	Documentation on extent of usage of classification system
Geographical coverage	Overall extent Detailed specification of coverage if not complete
Positional accuracy	Statistics on coordinate error
Attribute accuracy	Statistics on attribute error
Topological accuracy	Methods of topology validation employed
Graphical representation	Graphical symbolism for each feature class Text fonts for annotation

made by comparison with the most accurate measurement or most commonly agreed classification that it would be possible to obtain. Although it can often be agreed that there are truths or at least approximate truths, the issue of data quality is concerned with the very real possibility of discrepancies between what is recorded in a database and the truth, so far as it can be known, of the phenomenon to which it refers. Accuracy in spatial data is often categorised according to whether it refers to position, topology or aspatial attributes. It is quantified by means of error, based on error models that we discuss in a subsequent section.

Precision

The term precision refers to the degree of detail with which a value is reported. For numerical data this should be the number of significant digits. The precision needs to be at least that for which a measurement can be recorded at its best known accuracy, but it is misleading to provide greater precision than this since any digits beyond the known accuracy of a measurement device are effectively redundant. Thus, if the coordinates returned by manual use of a digitising table cannot be relied upon to be more accurate than within 0.1 mm of a 'true' value, there is no point in reporting the value in millimetres with more than one decimal place.

Spatial resolution

Resolution is the smallest distinguishable difference between two measurable values. Spatial resolution may then be regarded as the smallest distance over which it is possible to record change. On a map to be read by human eye this would usually be determined by the minimum line width on the assumption that a line was used to record a boundary. Lines on maps are rarely drawn with a width less than about 0.1 mm. On a graphics plotting device the finest physical resolution is determined theoretically by the separation between the device's pixels. On a laser printer this is typically 1/300th of an inch (0.08 mm), and on high-quality laser plotters it is nearly ten times finer. The finest possible laser-plotted lines may not be easily discernable without magnification, though this depends on the contrast with background colour. Thus there is a difference between human visual resolution (acuity) and the device's physical resolution. A similar distinction can be made between the smallest distance which the human operator can consistently distinguish when operating the digitiser and the smallest distance that the digitiser hardware can report consistently.

Scale

Scale is the ratio between a recorded distance on a map and the 'real world' distance that it represents. The scale of a map will determine the distance on the ground represented by the width of a line on a map. For example, on a map of scale 1:10 000 a line of width 0.5 mm corresponds to a ground distance of 5 m. If this were a minimum line width, it would not be possible to represent phenomena smaller than 5 m in extent, without their size being exaggerated on the map. Although it is normal to quote a single scale value for a map, the nature of map projections is such that the actual scale will vary across a map. Whether this is significant or not depends on the type of projection and the extent of the earth's surface represented on the map. The idea of scale in the context of map projections is discussed in more detail in Chapter 4.

It may be remarked that knowledge of the scale of a map is a valuable parameter in the absence of other data on accuracy, since it carries with it an implication of the best accuracy that the data could possibly be as indicated above. The scale of a map also has implications for the level of generalisation of the data (see Chapter 16). It is sometimes regarded, therefore, as a surrogate for explicit data on accuracy.

Error

Having defined accuracy as the difference between a recorded measurement and its truth, it is clear that for most practical purposes its value will not be known exactly. However, some indication of the accuracy is certainly required and this can be provided by numerical error values, which are a measure of the expected accuracy of a measurement. In this context, error is a statistical concept based on some assumption of the nature of the measurement process. The measurement process can be understood by experiments involving repeated observations. If the statistical distribution of a set of measurements is Normal (or Gaussian), it can be used to provide a most likely value, represented by a mean, accompanied by a standard error which indicates that a certain proportion (68%) of observations are within a given distance of the mean. Ninety per cent of observations would be within 1.65 standard errors. Having established the nature of the measurement process by experiment, the associated standard error values can be used to indicate the expected accuracy of other single observations obtained by the same measurement instrument or method.

Positional error models

Positional error in points

In the case of a point recorded in terms of a pair of x and y coordinates, the error can be modelled as a function of the combination of the x and y measurements, on the assumption that they are normally distributed. Thus, if the values are normally (i.e. randomly) distributed, the probability that a point corresponds to the 'true' value (i.e. the mean of repeated measurements) is represented by a bell-shaped surface centred on the mean (Figure 7.1). The circle defined by the radius corresponding to the standard error of the x and y values is called the circular standard error (CSE), and the probability that a point lies inside this circle is 39.35% (Goodchild, 1991). A common way to define the accuracy of a map is to specify a circular map accuracy standard (CMAS), which states that 90% of points are within a given distance of their true position. Ninety per cent corresponds to 2.146 times the CSE. NATO defines four accuracy standards for points in which CMAS is 0.5 mm, 1.0, greater than 1.0 and undefined respectively. Note that 0.5 mm on a map of scale 1:25 000 corresponds to 12.5 m on the ground.

Positional error in lines

Although lines are usually specified in GIS terms as a sequence of points, the accuracy of a line is not well modelled simply by random error in the constituent points, since the points cannot be regarded as independent of each other. Due to the manual digitising process, it may be the case that sequences of points describing a line will all tend to be displaced, in part,

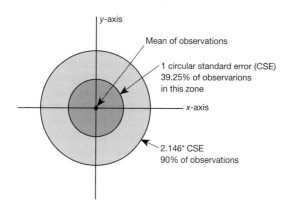

Figure 7.1 Probability distribution of point measurements. Error in the location of a point can be modelled by normally distributed error in the *x* and *y* directions resulting in a bell-shaped surface centred on the mean.

a normal distribution. A further possibility is to assume a bimodal shape, reflecting the expectation of systematic error to one side or the other. The box-shaped band is regarded as a deterministic model, while the others are probabilistic (Goodchild, 1991).

Positional error in polygonal maps

The error band model for lines has been extended to describe areas defined by linear boundaries. Blakemore (1984) used a box cross-section epsilon band model for lines to categorise points in the vicinity of a polygon as either definitely in, definitely out, possibly in, possibly out and ambiguous (Figure 7.2). Clearly the use of a probabilistic error model would enable a more precisely quantified measure of the likelihood of a point being inside or outside a polygon.

Attribute error

Methods of describing the error in non-locational attribute or 'feature' data vary according to the form of measurement. Numerical data, such as the concentration of a chemical at individual locations can be modelled with conventional parametric statistics that might assume a normal distribution of sampled values relative to an assumed 'true' or mean value. When such numeric data have been sampled at a set of locations it is possible to generate statistical

to one side or another of the 'true line'. This type of bias can sometimes arise from the line-following procedure and from the possibility of a map being misregistered in terms of its control points (i.e. all coordinates shifted in one direction), giving rise to a systematic error.

A possible error model for a line is a band centred on the true line. This is known as the epsilon band (Perkal, 1966). In its original conception this band was regarded as box-shaped in cross-section, but it could also be regarded as bell-shaped, in the form of

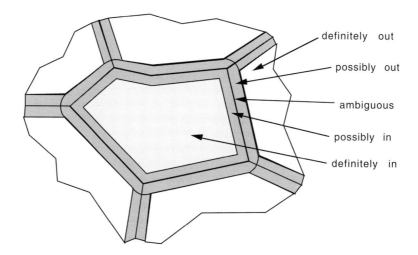

Figure 7.2 Zones of differing levels of probability of inclusion in a polygon. An epsilon band associated with the boundary of a polygon can be used to categorise the relationship between a point and the polygon (Blakemore, 1984). Thus a point may be regarded as definitely inside, possibly inside, ambiguous, possibly outside and definitely outside.

surfaces that interpolate the value at unknown locations and, using the geostatistical technique of kriging (see Chapter 12), provide estimates of the variance in the interpolated values. The reliability of individual variances in the interpolated values depends upon a prior analysis of the correlation between nearby samples as a function of the spatial separation.

Much of the attribute data in GIS are non-numerical categorical data in the form of nominal classifications of phenomena of interest, such as soil type or vegetation type. Error in such categorical data can be assessed using a *misclassification matrix*, also referred to as a *confusion matrix*, and as an *error matrix*, which is based on the results of testing the truth of a set of recorded data values (such as those derived from classification of a remote-sensed image). The horizontal axis of the matrix represents classes identified on the ground and the vertical axis classes resulting from interpretation (Table 7.2). Each cell of the matrix records the number of cells of a particular ground sampled class sc_i that were allocated to a particular interpreted class ic_j. Correctly classified cells will be recorded in the diagonal elements. Non-diagonal elements of the matrix represent cells interpreted with one class, but found on the ground to belong to another class. For each class it is then possible to study the pattern of misclassification and to calculate the percentage correctly classified. This percentage p can be used to derive a standard error SE in the classification

$$SE = \sqrt{\frac{p(100-p)}{N}}$$

where N is the number of samples in the true class (Hay, 1979). The nature of the error can be analysed in more detail, however. Within an individual column, the sum of the non-diagonal elements divided by the the total for the column gives the proportion of that category on the ground that was interpreted as something else. This is a measure of *errors of omission*. Conversely, within an individual row, the sum of the non-diagonal elements divided by the total for the row gives the proportion of that category that was classified as such, even though it was found on the ground to be something else. This is a measure of *errors of commission*. For other methods of analysing the matrix, see Rosenfield and Fitzpatrick-Lins (1986).

Some classification procedures, such as the maximum likelihood classification used in remote sensing, and the descriptions of soil units on maps, provide information about the probability of an areal unit (cell or polygon) belonging to each of a set of classes. However, images and maps are usually presented with each cell classified by the most likely class. Knowledge of a vector of probabilities of each areal unit can be used to generate a set of *realisations* of possible alternative interpretations. As Goodchild *et al.* (1992) point out, such realisations should take account of the correlation between adjacent cells in order to provide a 'realistic' representation of texture. If adjacent cells are highly correlated the map will tend to have relatively uniform regions made up from clusters of cells of the same class, whereas low or zero correlation gives rise to a speckled, relatively incoherent pattern (Figure 12.9). Generating a set of possible realisations of an area-class map can help the user to appreciate the degree of uncertainty associated with particular classifications. The generated realisation can also be used to perform calculations of area, and used in overlay operations, to create a variety of possible outcomes when combined with other map layers.

Table 7.2 A misclassification matrix (also called a confusion matrix and an error matrix) records the number of samples of a ground-observed class (horizontal axis) that have been interpreted as belonging to a particular class (vertical axis). Correctly interpreted samples lie along the diagonal. Correct classification of grassland = 300/600 = 50%.

Interpreted classes	Ground observations					Totals
	Buildings	Crops	Grassland	Woodland	Water	
Buildings	1850	150	100		200	2 300
Crops		3800	100	300		4 200
Grassland		700	300	200		1 200
Woodland		250	100	1000		1 350
Water	350				600	950
Totals	2200	4900	600	1500	800	10 000

Relationships between positional and attribute error

Although we have so far discussed positional and attribute error separately, it should be apparent from the previous considerations of area-class or categorical maps that if the classification of an areal unit is uncertain then a change in classification will introduce change in the position of the boundary between different class areas. Thus in area-class maps, position and attribute are intimately linked. In choropleth maps, however, the boundary between units is predetermined by some existing independent phenomenon, such as an administrative boundary. Thus uncertainty in classification within the boundaries cannot shift the position of the boundaries, except in so far as some boundaries may not appear on a map if adjacent units are of the same class. Conversely, there are numerous other cases where for most practical purposes the classification of a phenomenon such as a building or road is predetermined and accuracy of measurement of location does not depend upon the accuracy of the classification. Whenever the classification does fall into doubt, as for example in the case of knowledge of ownership of land, then this uncertainty may feed back into uncertainty of the position.

Sources of inaccuracy

Data may be subject to inaccuracy or error for a variety of reasons that may be related broadly to processes of

1. primary data acquisition,
2. secondary data acquisition,
3. data manipulation and analysis,
4. data transmission,
5. usage.

Most of these categories themselves include various types of error source.

Primary data acquisition errors

Errors due to primary data acquisition may be attributed to instrumentation and to human usage and interpretation. Individual instruments may be limited in the precision with which they can record information, dictated by the limits of resolution of the device. Errors may arise due to misuse of an instrument. It may not be set up correctly and, even if it is, the operator may make mistakes in transcribing results obtained from it. Primary data acquisition errors can also be due to mistakes or uncertainty on the part of an observer in identifying and classifying phenomena such as a type of rock mineral or a species of plant. It is also possible that the method of sampling was inappropriate in terms of a pattern or frequency of observations.

Secondary data acquisition errors

Secondary data acquisition errors may arise as a result of a digitising procedure or due to shortcomings of a source document or of a manual interpretation. A document such as a map will incorporate all the inaccuracy of primary data acquisition, but particular media such as paper may introduce their own problems due to distortion. Errors in the operation of a digitising device may be caused by failure to position a cursor correctly, insufficient sampling of a line, omission of features that should have been digitised, and mistakes in labelling or attributing features on a map. The latter may be due to misidentification and misclassification of the phenomena represented and incorrect transcription of text. Digitising errors are discussed in more detail in Chapter 5. An important cause of errors associated with secondary data acquisition is the failure to obtain adequate metadata, i.e. data describing the data. For a map this includes details of scale, map projection parameters, geodetic data, source organisation, methods of data acquisition, date of data acquisition and date of publication. Sometimes the information is not there to be acquired and sometimes it is there but not recorded as part of a digitising procedure.

Transmission errors

Data transmitted in digital form from one organisation to another or from one computer to another may be subject to numerical degradation if the transmission medium does not enable all significant digits of numerical data to be retained.

Data manipulation and analysis errors

GIS include many facilities for transforming and processing spatial data that may result in loss of accuracy of the processed data. Examples of these include changes in the coordinate system, such as from a grid system to a geographical coordinate system, and changes between spatial models. Conversion of a raster model to a vector model may result in parts of cells in boundary regions being forced into an adjacent category. Rasterisation (vector to raster conversion) results in formerly straight or curved boundaries being transformed to a jagged zig-zag shape (see Chapter 8). Some interpolation procedures, from irregular points (such as in a triangulated irregular network) to a regular grid, result in loss of the original data values and the creation of new data values that, due to the interpolation procedure, may or may not bear a close relationship to the original values. Interpolation between areal units is particularly prone to introducing error since inappropriate assumptions may be made about uniformity of distribution of recorded phenomena (see Chapter 12).

Considerable inaccuracies can arise when data from various sources are integrated or overlain to generate a new data set which is some function of the input data. Sometimes data integration requires that different classification schemes are homogenised, resulting in compromises and generalisations of the classes in one or more of the merged data sets. If several raster data sets are at different levels of spatial resolution, the use of a cell size that is the lowest common denominator may result in the loss of spatial resolution of some layers, while use of the highest resolution would result in a spurious sense of accuracy.

Polygon overlay processing with vector data is notorious for introducing sliver polygons in the vicinity of real-world boundaries that are common to two or more layers, but are recorded slightly differently in each. Often the sliver polygons will be removed automatically or manually but this could still leave some resulting boundaries less accurate than in certain of the source layers. Overlay processing and map algebra may combine several layers with a variety of logical and arithmetic operators. Arithmetic operators introduce errors in the product that are greater than in any individual layer and in some cases the final errors may be very large (see Burrough (1986) for a discussion of the accumulation of errors due to the application of the Universal Soil Loss Equation in overlay processing). By way of general guidelines, errors will be least if layers are added together. Subtraction can result in proportionally very large errors as can functions that include power terms. Errors will be increased if data values in different layers are correlated with each other.

Use errors

Errors in the course of data usage arise due to conscious or unwitting decisions made by the user of the data. For example, data chosen for a particular location may be out of date relative to the time of interest. Map data may be selected for usage, such as measurement or overlay processing, that is too generalised to produce results that are reliable or applicable. It is sometimes necessary to make use of data that are incompletely processed or are in measurement units that are related to, but not equivalent to, those actually required. Thus remote-sensed images that have undergone very little processing or classification may be used for approximate interpretation. Seismic times may be used to interpret subsurface geological structure before they have been converted to depths. Estimates of population may be based on surrogate phenomena such as housing type and density, which may result in usable approximations but will not be as accurate as some other forms of measurement.

Data transfer standards

A consequence of increasing usage of digital geographical data is the need to transfer data between organisations and between different computer systems. The need has resulted in pressure to use standard data transfer formats. If there was just one such standard, individual organisations would only need to acquire a facility for reading from and writing to the standard format in order to be able to communicate data. The alternative and in general less desirable option is to develop or acquire programs to translate between different formats specific to GIS systems. For organisations dealing with several data suppliers using different systems, the latter approach can create considerable overheads in maintaining appropriate software.

Recently, international standards for spatial data transfer have emerged, but for some time there have been a number of *de facto* standards developed by major data suppliers such as topographic survey organisations and major GIS software system developers. Widely used *de facto* standards include the Digital Line Graph (DLG) format developed by the United States Geological Survey (USGS), the DXF format for representing drawings and designs in 2D and 3D (associated with the Autocad graphics package), and the US Bureau of Census TIGER/Line files for topographic data.

In the UK, the Ordnance Survey has developed various formats for representing digital topographic data, culminating in the National Transfer Standard (NTF), which is becoming a British Standard and is now maintained by the Association for Geographical Information (AGI). Standardisation in the USA has focused on the Spatial Data Transfer Standard (SDTS) which has become Federal Information Processing Standard (FIPS) 173. It may be noted that there is a domain-independent international standard for transferring data, ISO 8211, and both SDTS and NTF have adopted this standard as a means of specifying data contents at a relatively low level.

The purpose of standards specifically designed for spatial and geographical data is to establish a common understanding and hence effective communication of spatial concepts (of metric geometry and of topology) and to ensure that data are accompanied by adequate metadata that provide a good understanding of the quality of data and its real-world interpretation. It is for this reason that interest surrounds NTF and SDTS despite the existence of very widely used standards such as DXF which, though effective for transfer of geometric data and of text, cannot communicate topological data effectively and lack adequate facilities for metadata. Spatial data standards may also be distinguished from more picture-oriented graphics communication standards such as Postscript and CGM (see Chapter 14).

The Spatial Data Transfer Standard

SDTS defines a conceptual model that is intended to be sufficiently descriptive of spatial phenomena, spatial objects and the relationships between them to enable users of the standard to define their own spatial information in terms of the SDTS concepts (Morrison and Wortmann, 1992). In the STDS model the world is regarded as being composed of *entities* which are characterised by *attributes* and represented digitally by *spatial objects*. The term *feature* is used to refer to the combination of a real-world entity and its corresponding digital, spatial object representation.

Spatial objects in SDTS include both vector and raster models. They are classified into two groups of geometry-only (G) spatial objects, including for example points, line segments and pixels, and of geometry-and-topology (GT) spatial objects, including nodes, links and so-called GT-polygons. The various types of SDTS spatial objects are illustrated in Fegeas *et al.* (1992).

SDTS allows features to be specified with respect to entities, attributes and included terms. Entities belong to a specified type (e.g. house), of which there are specific instances (e.g. 'Chez moi'). Attributes are of specified types (e.g. 'vegetation') and with specific values (e.g. 'heath'). There is a distinction between a *standard* term, which is the name given to an entity type or attribute, and an *included* term, which is a synonym or specialisation of a standard term. Thus 'dwelling' and 'semi-detached house' might be included terms relating to a 'house' which is a standard term. A set of 200 entity types, 244 defined attributes and some 1200 included terms are provided as part of the standard, but users can also define their own terms.

Information on data quality is a compulsory part of a SDTS transfer. The sender of the data must record what is known of data quality in five categories and it is the responsibility of the user of the data to determine on this basis whether the recorded data are fit for use. The five categories are lineage, positional accuracy, attribute accuracy, logical consistency and completeness. Lineage refers to the sources of the data, including the methods by which it was derived, transformations applied to it and the dates of the original recorded phenomena (or at least the date of publication of the data). Data-quality information may relate to individual spatial objects or to collections of data. If it is known, the lineage should include details of all mathematical transformations of coordinates with the values of the relevant parameters used. Information on positional accuracy distinguishes between deductive estimates, based on knowledge of errors at each transformation stage; internal estimates based on tests on the provided data; comparison to source based on the use of check plots, and an independent source of higher accuracy based on comparison of the data with a better quality representation.

Attribute accuracy may also be represented in SDTS by a variety of means, depending on whether the data are numerical (ratio or interval) or categorical (nominal) values. For categorical values this includes the possibility of using a misclassification matrix relating to counts of areas belonging to particular categories or to the sizes of areas found as a result of a polygon overlay operation involving a higher quality source.

Logical consistency is described with regard to tests performed and their results. The tests could include those for valid values (in an appropriate range) and those relating to digitising issues such as correct intersection of lines, undershoots and overshoots and duplicate data. Data may be classed as topologically clean, identifying the actual software used to establish this. The dates when tests were performed should also be provided.

Completeness refers to criteria used to select what information has been included, relating both to spatial properties such as minimum areas of objects recorded and to attribute properties such as the full use of geocodes for identifying phenomena.

Logical specification of SDTS

Much of the SDTS is concerned with defining how the conceptual level spatial objects and attributes and the data quality information are represented for data transfer. The logical specification is based on a set of 34 types of module, each of which consists of records consisting in turn of fields and subfields which themselves contain data items. There are five categories of modules which are *global* (metadata referring to the entire data transfer), *data quality*, *spatial object*, *attribute* and *graphic representation* (relating to cartographic information on text display and point, line and area symbology). In practice, only those modules that are relevant to the data being transferred are used. Which these are is defined by global modules. The global modules are also used for specifying coordinate systems and data dictionaries that carry information defining entity types and their attributes, including valid attribute values.

Physical level of SDTS

The conceptual and logical levels of SDTS may be regarded as independent of any particular stored format in the sense that they are concerned with specifying what information is transferred, not the details of the physical format of the computer files used to store the information. As indicated earlier, SDTS adopts the international standard ISO 8211 for the purposes of the physical implementation.

An ISO 8211 file is called a data descriptive file (DDP) and consists of data descriptive records (DDR) that contain data about the structure and description of data items that are stored in the data records (DR). Like the logical definition of SDTS, records contain fields and subfields, but they are defined in a way that enables a computer program to predict their presence. Thus actual numbers of bytes of data used are predefined in the DDR. Considerable flexibility is provided, however, by the use within a DDR and a DR of a directory which specifies a tag (i.e. a name), the length in bytes and the relative position in bytes of subsequent fields stored within user data areas of the records.

National Transfer Format (NTF)

NTF has been adopted as a British Standard for spatial data transfer (BS7567). It supports a conceptual model that, like SDTS, makes a distinction between real-world representation and spatial objects, though the terminology differs slightly, with the former being referred to as features while the latter consist of geometry and topology. The standard is characterised by the use of five levels of implementation. Level 4 can support all aspects of the conceptual model, including fully topologically structured space partitioned by faces that are bounded by edges, bounded in turn by nodes, that can also be isolated inside a face. Level 3 is described as partial topology, in which topological objects consist of nodes and chains associated with point and line geometry. Levels 1 and 2 are spaghetti models, differing in their degree of flexibility regarding multiple attributes but having no topological structure (i.e. no relations between geometry). Level 5 enables non-standard formats to be transferred. Thus it is possible to include 'foreign' file formats such as DXF as part of an NTF transfer. The standard takes a flexible approach to attribute coding, providing no pre-specified feature classes. The feature coding employed in any particular data transfer can be documented with the standard in a data dictionary.

At a logical level NTF defines various records for representing geometry, text, attributes and metadata. These definitions are specific to the various levels. At levels 3 and 4 this includes raster data. At present there are two physical implementations of NTF, one of which is called Plain and consists of the original detailed specification of field sizes and data types of individual records, while the other uses ISO 8211 to implement the logical records.

DIGEST

The Digital Geographic Information Exchange Standard, DIGEST, is the product of a collaboration between military agencies within NATO (Ley, 1992). The standard supports both raster and vector data models. The vector models are similar to those of NTF in that there is a gradation from a low-level spaghetti model with no topological data, to a chain-node, partial topological model, to a 'full topological' model consisting of faces, chains and nodes. One of the two implementations of the vector data model in DIGEST, called the Vector Relational Format (VRF), has been used for structuring the Digital Chart of the World data set, which is derived from mapping at a scale of 1:1 000 000.

DIGEST may be regarded as a more rigid standard than NTF in that the various levels correspond more precisely to the data model of the transferred information (in NTF it is possible to use a higher topological level to transmit data that could also be transmitted at a lower level). It also defines a feature attribute coding catalogue (FACC), which must be used to classify all transferred data. Thus it is not possible to include an application-specific data dictionary as is possible with both NTF and SDTS. DIGEST makes considerable use of ISO standards for the physical level of implementation, including ISO 8211, but also ISO 9660 for CD-ROM transfer and ISO 2022 for representation of raster data and names. DIGEST has been used as the basis of the French spatial data standard EdiGeo.

Summary

Data quality is an essential issue in GIS as the credibility of information obtained as a result of retrievals or analyses from GIS systems hinges entirely on the appropriate use of stored data. We have made use of the phrase 'fitness for use', which Chrisman (1984) coined and which highlights the need to be critical in the choice of data for a particular task. The appropriate choice of data depends on knowledge of metadata which describe the contents of a particular data set with regard to a wide range of parameters that include the accuracy, method of data acquisition, date of origin, map projection, and the types of transformation and processing that the data may have undergone. Accuracy refers to the difference between the measured value and the best possible knowledge of the actual value. It is described with respect to error models based on the statistics of the observation process. Positional error in points can be described in terms of the probability that the recorded coordinates are within a particular distance of the 'true' location. Error in lines can be modelled by deterministic or probabilistic bands that define ribbon-shaped zones of uncertainty. Error in categorical data can be described with regard to misclassification matrices that summarise experiments to determine the proportion of observations or interpreted values that correspond to the correct value. There are many causes of inaccuracy in data. These include inherent shortcomings of the instruments and of interpretation techniques, as well as shortcomings in the way instruments and data-processing techniques are used.

Proliferation in GIS has resulted in a demand for standardisation in the way in which data are stored for purposes of exchange between organisations. Standardisation in data formats helps to reduce effort in making data usable. Standard formats need to be flexible to ensure that all information and structure present in a particular data set can be retained in transmission. The subject of data exchange is inextricably linked with that of data quality, since in order for a data user to assess the quality of data, it must be accompanied by adequate metadata. Standards such the SDTS and the NTF in the UK make specific provision for including such metadata.

Further reading

Reviews of error and accuracy in spatial data are provided in the article by Goodchild (1991) and in Chapter 6 of Burrough (1986). The book edited by Goodchild and Gopal (1989) contains a set of articles covering many important topics related to the accuracy of data in GIS. There are numerous research articles that include results of practical studies of error in spatial data. See, for example, Chrisman (1987a) and Walsh et al. (1987) and, with regard to viewshed estimation, Fisher (1993). In the specific context of manually digitised data, see Bolstad et al. (1990) and Dunn et al. (1990). Heuvelink et al. (1989) and Heuvelink and Burrough (1993) consider the specific issue of propagation of error in spatial modelling. For studies that focus on error in boundaries defined by attribute data, see Mark and Csillag (1989), Goodchild et al. (1992) and Fisher (1991). For an overview of the SDTS standard, see Fegeas et al. (1992) which is part of a special issue of *Cartography and Geographic Information Systems* on implementing the standard (Morrison and Wortmann, 1992). For further details of NTF, see British Standards Institution (1992).

CHAPTER 8 Vector to raster and raster to vector conversions

Introduction

The history of GIS has been marked by a dichotomy between systems based on vector data models and those based on raster data models. In Chapter 2 we discussed the relative merits of the two models, from which it became apparent that each has its advantages and that neither can be expected to displace the other. This is reflected in the fact that many commercial GIS now provide facilities for both raster and vector processing.

Though there may be numerous application areas in which the requirement is primarily for either vector or raster data, there are relatively few in which, at some stage, a conversion from one representation to the other is not necessitated. The great majority of vector-oriented GIS employ graphics display devices which are based on raster technology. Thus, before a vector can be displayed on one of these screens or plotters, it must be rasterised using vector to raster algorithms which are built into the graphics system. There can also be a requirement for rasterisation as a preliminary operation within raster-based mapping packages which may need to use data that were originally in vector format. Equally, at the data acquisition stage of vector-based systems, there is an increasing usage of scanners generating raster images that must be converted to a vector format.

The objective of raster to vector processors operating on scanned maps is typically to produce vectorised versions of lines in the maps and to attach identifiers to them. Ideally, though not often in practice, the attachment of class identifiers would be an automatic process and some vectorising systems do provide a capability for the object recognition that the performance of this task requires. As well as distinguishing between different-shaped objects, such as buildings, it is highly desirable to be able to recognise text and to be able to identify the individual letters which compose it.

Applications which use remotely sensed data depend upon some preliminary process of image interpretation. This interpretation may consist of classifying the pixels in terms of known phenomena, which are by their nature continuously variable, in which case the data may conveniently remain in raster format. If, however, the images are to be used for recognising discrete objects, particularly human-made phenomena such as roads, buildings or field boundaries, it may be necessary to convert to a vector data structure to assist in the interpretation process.

The remainder of this chapter is divided into two parts which address first the problem of vector to raster conversion, also known as rasterisation and scan conversion, and secondly the problem of raster to vector conversion, which is also known as vectorisation.

Vector to raster conversion

The task of vector to raster conversion is one of finding a set of pixels in the raster space which coincide with the location of a point, line, curve or polygon in its vector representation. In general, this conversion process consists of an approximation, since for a given region of space, the raster model will only be able to address locations with a limited number of integer coordinates. In a vector model, the precision of terminal points of vectors is limited by the resolution of the mapping coordinate system, but the location of the intervening line is defined by a mathematical function which, on a computer, can be evaluated at numerous sample points limited in resolution only by the computer storage space allocated to individual numbers.

An ideal raster representation for a line may be regarded as one in which there are pixels set at fixed intervals along the length of the line. In practice, when we evaluate the equation of the line or curve, we will find that, apart from special cases, such as horizontal and vertical lines, most of the equidistant points along the true line do not coincide precisely with possible pixel locations (Figure 8.1a). A line rasterising algorithm must therefore find pixel locations which approximate as closely as possible to sample points generated by the mathematical definition. Rasterisation of a filled polygon differs

somewhat in that the location of all pixels which lie inside the polygon must be found. At the boundaries of the polygon, however, there may be a similar process of approximation to that required in line rasterisation (Figure 8.1b). If rasterised data are to be used for any form of spatial analysis, these boundary approximations inevitably result in errors; for example, in the estimation of the area of a polygon. Rasterisation in the context of GIS may differ somewhat from that in the context of computer graphics, since in the former case there is usually more need to ensure that rasterisation minimises the introduction of measurement errors.

The approximations involved in sampling vectors at a limited set of pixel locations give rise to a stepped representation of curves and edges. The failure to find pixels at equal sampling distances in any direction also causes variations in the density of pixels along rasterised lines. On computer graphics display devices, these errors in sampling may be visually apparent both in the sense of jaggedness and in the unwanted variation in line intensities. These effects are referred to as *aliasing*, following the

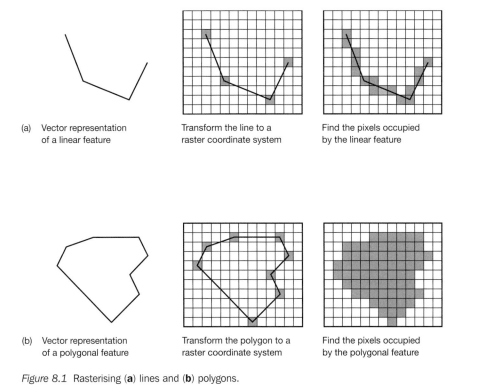

(a) Vector representation of a linear feature

Transform the line to a raster coordinate system

Find the pixels occupied by the linear feature

(b) Vector representation of a polygonal feature

Transform the polygon to a raster coordinate system

Find the pixels occupied by the polygonal feature

Figure 8.1 Rasterising (**a**) lines and (**b**) polygons.

terminology of signal processing. There are *anti-aliasing* procedures for counteracting these effects. They depend upon subtle changes in the intensity of pixels along lines and edges and introduce a significant computing overhead.

Rasterising straight lines

Straight-line rasterising algorithms usually work in an incremental manner starting at one end of the line and stepping along it, setting one pixel at each increment. We will start by describing a conceptually simple algorithm which performs line scan conversion, but does so in a computationally inefficient manner, in that it uses more floating point arithmetic than is actually necessary.

The minimum number of pixels that can be set along the line, while retaining the continuity between pixels, is determined by the maximum of the number of pixel locations in the x and y directions, between the beginning and end of the line (Figure 8.2). If the start and end of the line in the raster coordinate system are $x1,y1$ and $x2,y2$, the actual x and y displacements are given by

$$Dx = \text{abs}(x2 - x1)$$
$$Dy = \text{abs}(y2 - y1)$$

where abs refers to the absolute value function, available in many programming languages, to convert a number to a positive value. The total number of pixels to be set (n) will be the larger of these two values (D_{max}), plus one, i.e.

$$D_{max} = \text{max}(Dx, Dy)$$
$$n = D_{max} + 1$$

The x and y components, incx and incy, of distance to shift along the mathematical representation of the line are given by dividing each displacement, Dx and Dy, by the larger of the two:

$$\text{inc}x = Dx/D_{max}$$
$$\text{inc}y = Dy/D_{max}$$

Clearly, the resulting value will be unity in the direction of greatest displacement, but will be a value less than one in the other direction. In order to find the nearest pixel location at each increment, we can round the current sampled coordinates on the line to the nearest integer values. This can be done using a *round* function. Thus, after setting the first pixel in the line, the algorithm will include a for loop in which the integer pixel coordinates ix and iy are found with expressions of the form:

```
for i = 1 to (n–1) do
    x = x + incx
    y = y + incy
    ix = round(x)
    iy = round(y)
    setpixel(ix,iy)
end for
```

On raster scan graphics display devices, whenever a line is drawn on the screen the process of rasterisation must be performed. It is clearly desirable, therefore, to use an algorithm which performs this process as fast as possible, in order to minimise response times in an interactive system. The speed

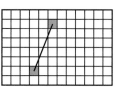

Figure 8.2 Straight-line rasterisation. Computer graphics algorithms usually determine the number of pixels to be set by the maximum of the displacements in the vertical and horizontal directions.

of numerical calculations on computers can be improved in general if they can employ integer rather than real arithmetic. This is what is done in Bresenham's line scan conversion algorithm, which is one of the most widely used such algorithms.

Bresenham's line algorithm works on the assumption that for a given slope of line, once one pixel has been set, the next pixel will always be one of a choice of two which have a fixed relationship to the previous pixel. Thus for lines of slope 0 to 1, i.e. 0° to 45°, the pixels to choose from are always the ones immediately to the right and diagonally to the above right respectively (Figure 8.3a). The choice between them is made on the basis of the value of a decision variable d, which takes on positive or negative values according to which of the two pixels is nearer the line. For lines of slope 0 to 1, d is zero if it is half-way between the two, positive if it is nearer to the upper right pixel, and negative if it is nearer the rightward pixel. The value of d is incremented each time a pixel is set. The way in which it is incremented is dependent upon which pixel was chosen. The initial value of d can be an integer and the two possible increments are also integer values. Hence all arithmetic required to execute the algorithm is integer.

Mid-point method of scan conversion

The technique of evaluating a decision variable, which is employed in Bresenham's algorithm to determine which of two possible pixels is to be set, can also be found in the mid-point method of scan conversion (VanAken, 1985). The method is based on the principle of evaluating a function, representing the curve to be drawn, at the mid-point between the two alternative pixels. If the curve passes through the

mid-point, it will evaluate to zero; otherwise it will be positive or negative depending on which side of the mid-point it passes. The curve function evaluated at the mid-point, which serves as the decision variable, can be calculated incrementally and, when applied to straight-line scan conversion, the derived algorithm is identical to Bresenham's. The mid-point method can also be applied to circles and to ellipses.

Thick lines

We have assumed, up to now, that lines and curves can be scan converted adequately by placing single pixels along their length. The graphic thickness of the resulting line will then depend upon the diameter of the pixels on the output device. To obtain a required line style, it will therefore often be necessary to set more than one pixel at each increment along the line. A simple approach to line thickening is to define a rectangular or circular pen which is centred on a given pixel location. A pen shape which approximates to a circle is desirable in that the resulting line thickness will be independent of the slope of the line. This and other techniques of creating thick scan-converted lines are discussed in Foley *et al.* (1990).

Polygon rasterisation

The process of polygon rasterisation entails finding those pixels that lie within a polygon which is defined by a sequence of vertices. The standard techniques for polygon rasterisation are based on those developed in the context of computer graphics, where they are also referred to as polygon scan conversion algorithms. We will now describe these techniques, but it is

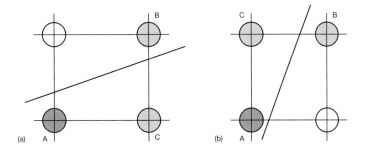

Figure 8.3 Bresenham's algorithm and mid-point method algorithms for straight-line rasterisation use a decision variable to choose between two alternative pixels at each step of the algorithm. In the two situations illustrated, pixel A has been set and the choice is between pixel B and pixel C depending upon whether the line slope is between 0 and 1(a) or 1 and infinity (b).

important to appreciate that, though they work well for determining interior pixels, there is an inherent error in the rasterisation at the boundaries of polygons, due to the finite size of the pixels. The decision for inclusion of a pixel within a polygon in the computer graphics procedures is a somewhat arbitrary one, in that it is not directly related to the proportion of the pixel occupied by the polygon. Thus at the boundary a pixel will be classified as belonging to the polygon if, in general, its centre is within half a pixel diameter of the boundary. Assuming that the boundary is shared with an adjacent polygon, part of the area of such a pixel will be 'stolen' from the neighbouring polygon. Alternative techniques include allocating a pixel to a polygon if the centre of the pixel is inside the polygon, called *central point rasterising*; or allocating it to the polygon that has the largest areal share of the pixel, called *dominant unit rasterising*. The problem is another aspect of aliasing, referred to above in the context of line rasterisation.

When using the scan line algorithms described below, the boundary pixels on a computer graphics device will often be determined by rounding the x coordinate of the edge/scan line intersection. Applica-tion of methods such as central point and dominant unit rasterising would require a modification to this procedure. Central point rasterising is relatively simple in that a decision to switch on the boundary pixel (found by rounding the x coordinate) could be taken on the basis of whether the centre of the boundary pixel lay within the range of the pair of unrounded edge intersections that were about to be filled.

The dominant unit rasterising technique is more difficult to implement since it requires calculating the area of intersection between the pixel and the vector representation of the polygon. This requires clipping the polygon against the rectangular cell of the pixel before measuring the area of the clipped piece of polygon. This would necessitate a polygon overlay operation, as described in Chapter 11. Note that these two methods of rasterisation can sometimes produce slightly different results (Figure 8.4).

In a study of errors introduced in polygon rasterising, Bregt *et al.* (1991) found that there is very little difference between the errors introduced by central point as opposed to dominant unit rasterisation. They also found a positive linear relationship, for three different raster cell sizes, between the error

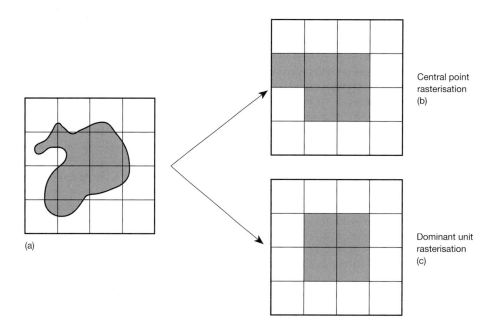

Central point rasterisation (b)

Dominant unit rasterisation (c)

(a)

Figure 8.4 Rasterisation of the polygon in (**a**) produces different results according to whether (**b**) central point rasterisation or (**c**) dominant unit rasterisation is used.

introduced and the complexity of the polygon boundary. This is in combination, as may be expected, with a general increase in error with raster cell size. Their research showed that for a given raster cell size, the associated error could be predicted on the basis of a boundary index which is a measure of the boundary complexity (BI). This BI is based on the total boundary length of polygon boundaries on a map, divided by the area of the map.

Scan line coherence

It is possible to approach the problem of polygon rasterisation by treating all pixels in the vicinity of the polygon independently, to determine whether they are inside or not. It is more efficient, however, to find continuous rows of pixels which lie inside. This latter approach exploits the properties of spatial coherence exhibited by typical polygons, whereby their interior can be defined by relatively uniform, continuous regions of pixels, as opposed to the pixels being randomly distributed. Thus instead of dealing with one pixel at a time, the algorithm operates by searching for the boundaries of continuous sequences of pixels within a row or scan line of the image and setting these as a block (Figure 8.5).

The scan-line coherence polygon scan-conversion algorithm works by examining each scan line that crosses the polygon to find the intersections between it and the edges of the polygon. Having calculated the intersection coordinates, they are sorted in increasing order along the scan line before filling in between each successive pair of intersections. When searching for intersections, horizontal polygon edges,

i.e. those parallel to the scan lines, can be ignored since they will be represented by the rows of pixels lying between the intersections of the vertices of their adjacent edges with the relevant scan line.

Care must be taken, when applying the technique of filling between successive pairs of intersections, to avoid counting the intersection point twice, due to the fact that adjacent edges terminate at the same shared vertex. If such points are counted twice, there is a possibility of creating a row of one pixel length and leaving a gap across the interior of the polygon.

One way to avoid such aberrations is to count only those intersections with a vertex if they are at the lower (minimum y) end of an edge. A side-effect of this strategy is that vertices which are maxima of the polygon boundary will not be represented at all, since the vertices of both of the adjacent edges are at the upper ends and so neither will be counted. Such situations can, however, be detected and treated as special cases.

An important element of the scan-line algorithm is that of finding intersections between edges and scan lines. The amount of computation for this part of the algorithm can be reduced if the intersections are calculated in an incremental manner, rather than from scratch for each scan line. For any given edge, once the intersection with the initially intersected scan line is known, being given in fact by the x-coordinate of one of the vertices of the edge, the intersection with the other successive scan lines can be found by repeatedly adding the reciprocal of the slope. This is based on the assumption that the y interval between successive scan lines is unity, which it is in an integer raster coordinate system.

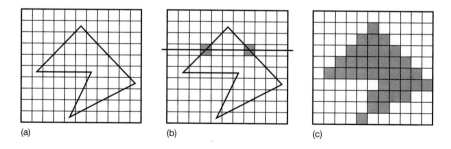

(a)   (b)   (c)

Figure 8.5 Polygon rasterisation using scan-line coherence. (**a**) Convert the polygon to a raster coordinate system. (**b**) Find the intersections between scan lines (horizontal lines passing through centres of pixels) and polygon edges. (**c**) Fill between pairs of intersections for all scan lines.

Raster to vector conversion

In the remaining part of the chapter we will consider the problem of transforming a raster image to a vector representation of the objects within the image. The processes involved form part of the broader subject of *computer vision*, which concerns the automatic recognition of objects in an image consisting normally of a 2D array of pixels. The creation of a vector representation of the objects is an important preliminary stage in many computer vision systems. In general, computer vision systems search for the shape of objects using *edge detectors*. Although later in this section we will indicate briefly what is involved in edge detection and in some of the object-recognition procedures that may follow it, the emphasis here is on the techniques employed to vectorise linear features. As was pointed out at the beginning of this chapter, these techniques are essential to the process of interpreting and structuring digital images obtained by scanning existing map documents.

In addition to digitising and recognising linework, the techniques of raster to vector conversion are also useful in generating linear representations of objects that are represented originally with significant areal extent. This can be useful in cartographic generalisation, when features such as rivers need to be transformed from a two bank form to a single line, which is referred to as a collapse operation. The linear expression of a polygon, also called the *skeleton*, or the *medial axis transformation* can be of use in producing generalised polygonal representations of originally more detailed polygons.

Bilevel images

The initial output from a document scanner usually consists of either grey level or colour pixel values recording the changes in brightness and colour across the document. If the document contains linework which contrasts clearly with the background, as can be expected in printed maps, then it is desirable to

Figure 8.6 Processing steps for scanned topographic linework (from Greenlee, 1987). (**a**) Portion of scan-digitised topographic map. (**b**) After edge enhancement. (**c**) Linework after thresholding to a binary image. (**d**) Linework after filling and thinning. (**e**) Linework after vectorising. (**f**) Linework after splining (smoothing) and deletion of short arcs.

convert the pixels to bilevel values which distinguish simply between background, say with value 0, and foreground linework with value 1. This can be achieved by the process of thresholding in which a pixel value in the original range of values is chosen as representing the division between foreground and background. All pixels above or equal to the value are then set to 1, while all those below are set to 0. Figure 8.6(c) illustrates the result of thresholding a grey scale image (Figure 8.6a). Before thresholding was applied in this case, an edge-enhancement operator was applied to improve the contrast between the linework and the background (Fig 8.6b).

A threshold value may be selected by examining a histogram of all intensity levels (pixel values) in the image. Provided the image has uniform background brightness, there should be a bimodal distribution of levels representing the distinction between pixels which are either foreground or background. If this is the case, a value at the minimum between the two peaks can be chosen for thresholding. For documents in which the general brightness level changes between different parts, there may be no bimodal distribution for the whole document. This could necessitate a localised approach to thresholding in which, for example, the image was subdivided into regions, each of which was thresholded individually.

Linear features in bilevel images

Once a bilevel image has been produced, linear features will be defined by connected sets of foreground pixels (Figure 8.7). Depending upon the resolution of the scanner, each line may be several pixels wide. For an initially 'clean' document the line width in pixels will be fairly constant, but on a poorer quality document it may vary considerably. There is also the possibility that if linework was locally indistinct the thresholding process may have introduced minor gaps, in what should be continuous lines, due to low-intensity pixels being misclassified as background. The presence of gaps in the scanned linework can be compensated for, prior to generating the vectorised lines, by a process of *filling*, in which minor discontinuities are removed by changing the value of those pixels which cause the gaps from the background to the foreground value.

The main phase of converting to vectors then consists of thinning the multi-pixel wide linework to the width of a single pixel (Figure 8.6d). This involves

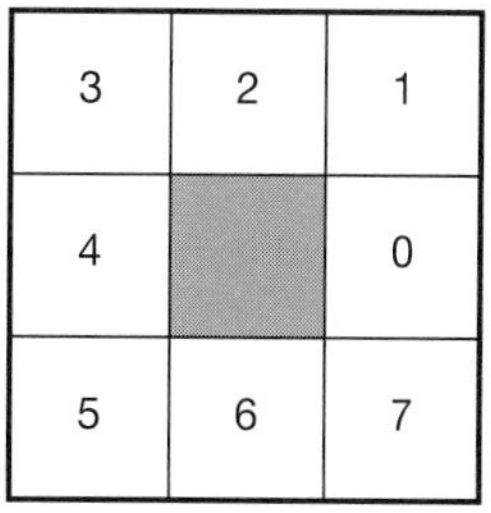

Figure 8.7 Pixel neighbours and connectivity. The shaded pixel has eight neighbours. In the numbering system illustrated (after Pavlidis, 1982), the four neighbours in the horizontal and vertical directions are referred to as N-even neighbours, while the diagonal neighbours are referred to as N-odd neighbours. In eight-connectivity space, all neighbours are regarded as connected to the central pixel. In four-connectivity space, only the N-even neighbours (0,2,4,6) would be regarded as connected to the central pixel.

changing pixels on the boundary of the line from the foreground to the background value. The thinning process is also known as skeletonisation, as it results in the skeleton of the original object. Once the rasterised linework has been reduced to one pixel in width, the start and end (i.e. the nodes) of each line are found. Nodes will correspond either to junctions between lines or to the terminal pixel of lines that are not connected to other lines at one or both ends. A vector representation of each line may now be found quite easily by recording the coordinates of the sequence of pixels which leads from one of its nodes to the other (Figure 8.6e). The resulting vector lines may have an unnecessarily high number of coordinates along straight sections and they may also have locally sharp angularities due to the fact that the coordinates can only lie at the coordinates of pixel locations. These shortcomings can be overcome by using a line simplification algorithm to remove redundant points (see Chapter 16), and by means of filtering procedures to fit smooth curves through the original sample points (Fig 8.6f).

Local image processing using masks

The processes of filling, thinning (skeletonisation) and node detection can all be performed using algorithms that classify and if necessary modify individual pixels of the raster image, after examining their immediate neighbourhoods. Each pixel has eight neighbours which we will label from 0 to 7, as in Figure 8.7. The horizontal and vertical neighbours may be called the N-even neighbours (with labels 0, 2, 4, 6), while the diagonal ones are the N-odd neighbours (1, 3, 5, 7). In a bilevel image each neighbour may have the value of either 0 or 1 which provides for the possibility of 2^8 (256) different combinations of these values. The different combinations, or possible surround conditions, are treated as *masks* which are compared with the actual surround conditions of each pixel. The comparison is used to decide how a pixel is to be classified for the purposes of filling, thinning and node detection and what, if any, action is to be taken (Greenlee, 1987). Masks and the actual surround conditions for a pixel can be stored in single bytes of computer storage, since a byte consists of eight bits, each of which can be used to represent one of the eight neighbour pixels (Figure 8.8).

The comparison of the actual surround conditions with the possible surround condition masks can be performed by carrying out logical AND operations between the one byte containing the actual surround condition and the bytes containing the possible surround conditions. If the actual surround condition byte has bits set to one at all of the same positions as those in the mask with which it is ANDed, the result will be a byte with the same value as the mask.

Filling

The process of filling is concerned with identifying and removing small gaps and voids due to imperfections in the original linework. Pixels of the background value are examined to determine whether they are surrounded sufficiently by foreground pixels to be likely to constitute a gap or a void. Such background pixels are then set to the foreground value. The decision to change a pixel from the background to the foreground value is based on the fact that the pixel under consideration has at least three of its N-even neighbours set to the foreground value.

Thinning

Several methods of line thinning have been developed. A notable distinction is between those that 'peel' pixels off the boundary of the line in an iterative fashion, until only the skeleton is left, and those that attempt to determine skeletal pixels explicitly on the basis of their being those pixels which are furthest from the boundary. We may refer to the former as

Figure 8.8 Surround conditions. If the eight neighbours of a pixel are allocated values which are powers of two from 0 to 7 respectively, an individual pixel's surround condition can be characterised by a value between 0 and 255, which is given by the sum of the numeric values of those neighbours which are foreground pixels.

peeling algorithms (Peuquet, 1981) and to the latter as distance skeleton algorithms (Rosenfeld and Pfaltz, 1966). The removal of pixels from the boundary of an object in an image is also referred to as *erosion*, being complementary to the addition of pixels which is referred to as *dilation*. Erosion and dilation are also described as mathematical morphology operators (Serra, 1982).

Peeling algorithms examine foreground pixels on the boundaries of lines to determine whether or not their removal would introduce a gap in an otherwise continuous line. If no gap would be created the pixel is set to the background value. Pixels which must not be modified because they would result in a broken line are called skeletal pixels and the objective is to remove all non-skeletal pixels. Pixels which are skeletal are characterised by the fact that, when considering their surround conditions, there are two neighbours which are linked by the central pixel, such that if the central pixel were to be deleted there would not be a path between them (Figure 8.9).

As pixels are removed from the boundaries of lines, it is desirable to ensure that both sides of a line are peeled off at the same rate, so that the boundary moves in toward the original centre of the line, rather than migrating toward one side or another. Boundary pixels, which are also known as contour pixels, are distinguished by the fact that at least one of their N-even neighbours is a background pixel (Pavlidis, 1982). A balanced peeling procedure can be implemented by examining first all boundary pixels which have a 0-neighbour which is the background, then those which have background pixels as their 2-neighbour, 4-neighbour and 6-neighbour in turn (using the numbering scheme in Figure 8.7).

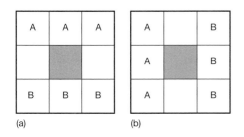

(a)   (b)

Figure 8.9 Identification of skeletal pixels. The central pixel in (**a**) and (**b**) is a skeletal pixel if it is a foreground pixel and at least one of the A pixels and one of the B pixels are also foreground pixels. The unidentified (white) pixels in (**a**) and (**b**) are background pixels. Note that the central pixel is skeletal because it provides the only path from an A pixel to a B pixel.

In a similar way to the filling algorithm described above, if a pixel is found to be a candidate for deletion, it is first flagged temporarily to a value such as 2, which in this case will be treated as a foreground pixel, before being set to 0 when all of the particular-direction N-even neighbour boundary pixels have been examined. This four-phase procedure is itself iterative in that the result of a single four-phase pass is to reduce the width of a line by a maximum of two pixels (one from either side). Thus the total number of iterations required before the line is reduced to a single pixel depends on the original width of the lines in the scanned image (Figure 8.10).

The technique just described follows approximately that which Pavlidis (1982) calls the Classical Thinning Algorithm. In practice there are various other possible strategies. For example, rather than considering four types of boundary pixel separately, it is possible to consider those with N-even neighbours either above or to the left in one phase before considering those with N-even neighbours below or to the right (Rosenfeld and Kak, 1982). Figure 8.6(d) illustrates the result of the thinning procedures referred to by Greenlee (1987).

Distance skeleton thinning

The distance skeleton algorithm differs from the peeling algorithms in that the number of passes through the image is fixed, not being a function of the width of lines in the original image. The initial objective of labelling those pixels which are furthest from the boundaries of the original lines is achieved in two passes through the image, the result of which is called

the *distance transform*. In the first pass we regard the starting point as the top left and proceed in a left to right fashion along each scan line. The value of each pixel as it is encountered is recalculated as the sum of one plus the minimum of the previous value on the scan line and the value immediately above the current pixel. The outside of the image is regarded as having value 0 (background). As a result of this pass many pixels may be expected to have values higher than 1, which was the original maximum value.

In the second pass we start at the lower right and proceed in a right to left direction along each scan line. This time each pixel is recalculated as the minimum of its own value, the value of the pixel to the right (previous pixel) plus 1 and the value of the pixel immediately below plus 1. Again the outside has value 0. After this pass those pixels with the highest value are the pixels that are equidistant from the boundaries. In order to create the distance skeleton by identifying these skeletal pixels as 1, while all others are set to 0, a further two passes are required, corresponding to the reverse of the previous two passes (Rosenfeld and Pfaltz, 1966).

Node detection

Having reduced linework to the skeletal representation, the nodes constituting the ends of the lines and junctions between lines can be identified by examining the surround conditions of each skeletal pixel. An unconnected end of a line can be identified by patterns such as those of groups in which the central pixel has either only one neighbouring pixel or two neighbours which are adjacent to each other.

Vectorised line extraction

Transformation of the line skeleton in the raster image to a vectorised line requires tracing all pixels from the start and end node pixels of each line (Figure 8.11). The sequence of coordinates of each pixel found in the tracing process then constitutes the vectorised line. As Peuquet (1981) points out, there are different possible approaches to line extraction. One possibility is to use a list of the node pixels already found to identify start points for each line and trace each line in its entirety before going on to the next. Since each node may have several lines connected to it, the line pixels (other than the node) can

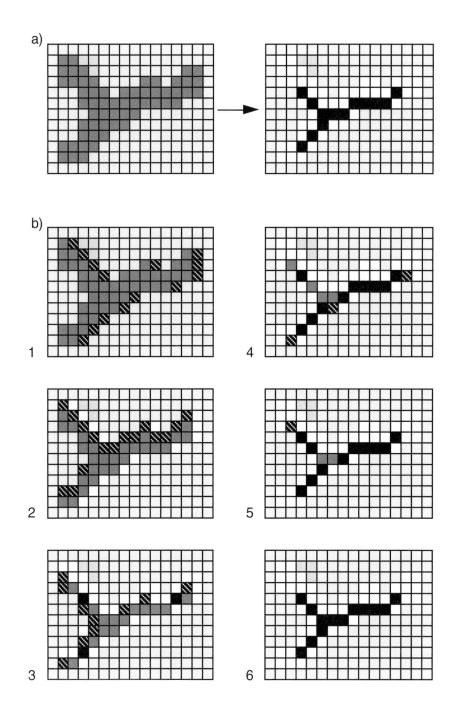

Figure 8.10 (**a**) Thinning raster data. (**b**) Stages in the process of applying a simple thinning algorithm. The striped pixels are those identified as contour pixels at each stage of the procedure, while the black pixels are those identified as skeleton pixels.

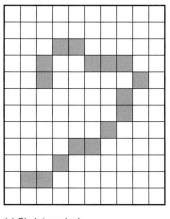
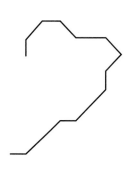

(a) Skeleton pixels   (b) Result of vectorising

Figure 8.11 Vectorising a skeleton.

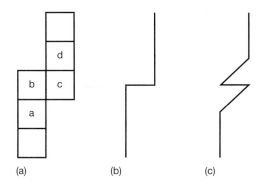

(a)   (b)   (c)

Figure 8.12 Ambiguity in line vectorisation. The pixels in (**a**) could be vectorised as in (**b**) or as in (**c**). The line (**b**) is the most likely in this case, but the vectorising algorithm would need to look ahead to detect this.

be flagged as having been examined in the course of tracing them, to avoid repetition of the same line.

The procedure for tracing the pixels along a line skeleton will be dependent upon the properties of the skeleton produced by the line thinning algorithm. Provided that all non-skeletal pixels (as defined earlier) between the two nodes of the line have been removed, there will be a unique path from one node to the other. Thus each pixel other than the nodes will have only two neighbours. Therefore, in the course of tracing the line, having arrived at a pixel via one of its neighbours, there will only be one possible neighbouring pixel to move onto. The situation would become complicated if the skeleton did not obey the rules for skeletal pixels referred to earlier. If, for example, the skeleton could include the sequence of pixels in Figure 8.12, then ambiguity would arise in determining the path from pixel a, since it has two more neighbours, pixels b and c. The correct path would probably be a–b–c–d. An alternative path is available given by a–c–b–d, but this introduces a twitch in the line. The former path could be found (assuming it was the correct one) by looking ahead at the surround conditions of b and c, or by choosing the N-even neighbour when given a choice.

Simplifying and smoothing vectorised lines

The output of the vectorising procedure can, in general, be expected to contain redundant coordinates arising from vertical, horizontal and diagonal sequences of pixels, in which several successive pixels lie on the same straight line. The redundant co-ordinates can be removed by means of a line simplification algorithm such as the Douglas–Peucker algorithm (see Chapter 16). This discards coordinates which are less than a pre-specified perpendicular distance (tolerance) away from a straight line constructed between its previous and successive coordinates in the vectorised line definition. By setting the tolerance value to a small distance in excess of zero, it would be possible to use the algorithm to remove minor steps in approximately straight sequences of coordinates.

Because all coordinates of the vectorised line are constrained to lie on the fixed set of possible locations dictated by the resolution of the raster image, it is likely that the line may have a somewhat angular appearance which may not reflect the shape of the original line represented by the image. This angularity can be alleviated by smoothing the line using spline-fitting techniques in which the vectorised line is described by a sequence of low-order, typically cubic, polynomials which are forced to be continuous with each other and to pass nearby, but not necessarily exactly through, the line coordinates (see e.g. Foley *et al.* 1990; Rogers and Adams, 1990). Figure 8.6(f) illustrates the result of a splining operation.

It may be noticed in Figure 8.6 that the result of the procedures described is flawed owing to unwanted gaps in contour lines and to inappropriate connections between contours. Clearly these vectorisation tasks have not been fully automated, though recent commercial software may produce better results using more intelligent methods for bridging gaps. The need remains, however, for manual validation and interactive editing to correct errors.

Edge detection

The raster to vector conversion techniques that have been described so far are intended to operate on essentially linear objects, or on objects which are to be reduced to linear form. It should be pointed out, however, that in the field of image processing, considerable research efforts have been devoted to the related problem of detecting and vectorising edges which constitute the boundaries of objects of any form. The topic is of increasing importance in GIS in that it represents one aspect of the problem of *automatic feature recognition* as applied to remotely sensed images. Some aspects of interpretation of remotely sensed imagery, relating in particular to the classification of regions of the image, were described in Chapter 6. Here we will introduce some low-level image processing operators for edge detection and indicate how boundaries can then be constructed.

The initial stage of edge detection is concerned with searching for sharp changes in the intensity levels of pixels in a grey scale image. Clearly these can be expected at the boundaries of objects, since the presence of any discernible object in the image is in

general only apparent by virtue of such changes. If we regard the image as a 2D mathematical function, the changes in intensity will be equivalent to high values in the gradient, or first derivative, of the image. The gradient in the x and y direction can be estimated at each pixel location by means of simple arithmetic operators consisting of a small array of values, also called a mask or a kernel, which are multiplied with the pixel values in the immediate neighbourhood of the pixel to be evaluated.

The Sobel gradient operators are given by the following two masks for the x and y gradients respectively:

$$\text{mask}_x = \begin{matrix} -1 & 0 & 1 \\ -2 & 0 & 2 \\ -1 & 0 & 1 \end{matrix} \qquad \text{mask}_y = \begin{matrix} -1 & -2 & -1 \\ 0 & 0 & 0 \\ 1 & 2 & 1 \end{matrix}$$

If the masks are superimposed with their centres on a pixel with coordinates $p(i,j)$ then, assuming that the origin is at the upper left, the results, $g_x(i,j)$ and $g_y(i,j)$, of the operators are given by

$$g_x(i,j) = -p(i-1,j-1) + p(i+1,j-1) -2p(i-1,j) + 2p(i+1,j) - p(i-1,j+1) + p(i+1,j+1)$$

$$g_y(i,j) = -p(i-1,j-1) - 2p(i,j-1) - p(i+1,j-1) + p(i-1,j+1) + 2p(i,j+1) + p(i+1,j+1)$$

The effect of these operators is to generate high (positive or negative) values if the pixel values to the left and the right, or above and below, are significantly different from each other. Note that if the values of pixels on either side of the central pixel are all the same then the operator will return a value of zero. To obtain a single value indicating the magnitude of the gradient, whatever the direction, the two values for x and y can be combined, thus

$$g_m(i,j) = \text{sqrt}\,(g_x(i,j)^2 + g_y(i,j)^2)$$

or as a computationally simpler approximation (in which the vertical line symbol means take the positive value, or magnitude, of the enclosed parameter):

$$g_m(i,j) = |g_x(i,j)| + |g_y(i,j)|$$

The Sobel operator introduces an element of averaging in its estimation of the gradient. Higher degrees of averaging, or smoothing, can be obtained using larger masks, which examine a bigger neighbourhood. Conversely, a smaller neighbourhood providing very little smoothing is provided by the Roberts operator which does separate the x and y components of the gradient. It consists of the two masks

mask1 = 1 0   mask2 = 0 1
         0 –1   –1 0

These give the two values

$$g1 = p(i,j) - p(i+1,j+1)$$

$$g2 = p(i+1,j) - p(i,j+1)$$

which can be combined, as previously, either as the square root of the sum of squares, or as the sum of their magnitudes.

Summary

Conversions between vector and raster representations occur in several contexts of GIS and cartography. Vector to raster conversion is referred to as rasterisation or, in the context of computer graphics, as scan conversion. It is a necessary pre-processing step, for originally vector-structured data, in the use of raster-based GIS systems. The vast majority of graphics display devices operate in raster mode, and thus require scan conversion of all vector data. The characteristics of widely used algorithms have been reviewed and we have seen that line-scan conversion algorithms are characterised by an incremental approach, which steps along the direction of the line in increments determined in advance by the gradient. Area-scan conversion techniques often exploit properties of spatial coherence, by identifying discontinuities (in practice the boundaries) and filling with continuous runs of pixels between these boundaries.

Raster to vector conversion is a more challenging task than rasterisation as it is effectively a process of reconstruction of vectors from what may be fragmented or imprecisely defined images. It involves several image-processing techniques, the use of which varies according to the type of data being vectorised. A major application is in digitising images that are first scanned to a raster format before being transformed to a structured vector format. Scanned images are often converted, by the process of thresholding, to bilevel images in which the pixels are classified as either foreground or background. The locations of the boundaries of linear features of this image are identified using edge-detection procedures. The linear features are then reduced to single pixel width by thinning or skeletonisation procedures, which may be regarded as eroding the boundaries of the features. Filling procedures are used to ensure continuity of what are assumed to be complete features. Having produced relatively 'clean' images of single pixel width lines, they can be transformed to vector format by procedures that identify nodes and follow the paths of the lines to create linear sequences of pixel coordinates. These may then subsequently be filtered to remove redundant coordinates.

Further reading

The subject of vector to raster conversion is dealt with in most computer graphics textbooks, notably Foley *et al.* (1990) and Rogers (1985). The papers by Bregt *et al.* (1991) and Knaap (1992) deal with aspects of the problem specifically relevant to GIS. For a text on image processing which addresses the problem of vectorisation see Gonzalez and Wintz (1987), and for a book on computer vision, see Ballard and Brown (1982). As indicated in the chapter, procedures for transforming image objects by eroding or expanding their boundaries are part a field of image processing called mathematical morphology. We will see in Chapter 16 that similar techniques have been applied in the context of map generalisation. See Serra (1982) for an extensive review of the subject. For an illustration of advanced research efforts aimed at solving problems concerned specifically with automatic recognition of map features, see McKeown (1990).

PART 3 Data Storage and Retrieval

CHAPTER 9   Computer data storage

Introduction

In our introductory treatment of the representation of geographical information (in Chapter 2), a distinction was made between higher level conceptual views of information and the lower level representations that are nearer to the computer's facilities for data storage. In this chapter we take a closer look at these lower level representations. Many of the concepts and associated terminology introduced here, particularly those introduced early on in the chapter, are quite important for users of GIS to understand, since they are integral to the facilities typically provided for storing and managing data.

The early sections of the chapter introduce the units of data storage that are used to organise and quantify the storage space of data in files and within computer programs. This is illustrated with respect to some very simple file formats. The subject of raster data storage is addressed and some of the commonly used techniques for compressing raster and image data are described. In the remainder of the chapter some of the fundamental techniques for organising data in conventional (non-spatial) database files are described as these are relevant to, though not in general sufficient, for GIS databases. We focus in particular on indexing methods that enable rapid access to data of interest that may be located within potentially very large files. An appreciation of these techniques will provide appropriate background for the following two chapters which address issues of database management and of specialised techniques for accessing spatial data respectively.

Units of stored Information

When a computer program or data management system needs to store individual items of geographical data, such as a coordinate or a feature attribute code, it assigns them to logical units called *fields*, which are grouped together to form *records* (Figure 9.1). The contents of a record are usually logically associated, in that each field stores a property of an individual *entity*, such as a particular state or city. Records are collected together, often in a strictly ordered sequence to form *files*, which are named units of storage on long-term memory media such as disks and tapes. The term *database* refers to a higher level category of storage and normally consists of a set of related files, each containing its own characteristic type of information. The contents of a file in a spatial database could be distinguished by the type of geometric object, or class of feature, that it contained. Thus there could be files of point-referenced information, such as town and village locations, files of linear information, such as the coordinates of rivers, roads and railways, or files devoted purely to roads of a particular class. Other possible types of file include dictionaries of feature descriptor codes and files which are indexes to the contents of other files.

In programming languages, data items are stored temporarily in *variables*. Certain languages, such as Pascal and C, allow the programmer to define records explicitly, in which case individually named variables are allocated to specific fields within the records. In this way there may be little or no difference between data structures stored in permanent files and data structures manipulated within the

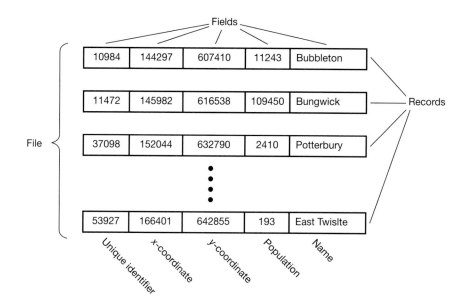

Figure 9.1 Files are usually subdivided into records, each of which may be further subdivided into fields that store logically distinct items of data.

program. Thus the program can create records and it can write them to a file, or it can read their contents from pre-existing files.

The physical storage space occupied by fields and variables can be measured in *bytes*, which are units of data storage consisting of eight binary digits, or *bits*. A bit is an element of the storage medium capable of being set to one or other of two states which may be regarded as symbolising the digits 0 and 1. A byte can store a numerical value between 0 and 255, i.e. ($2^8 - 1$) or, taking account of negative numbers, values between −127 and +127. For the purposes of storing numerical data values, bytes are grouped together into *words*, which are typically 2, 4, 8 or more bytes in length. Bytes and words may themselves be grouped into higher order units called *blocks*. On a disk, the blocks might be 512 bytes long, but other, often larger, sizes may also be used.

A four-byte word can store *integers* ranging in size between -2^{31} and 2^{31}. By using part of the word to represent an exponent, much larger magnitude, *floating point*, or *real*, numbers can be stored, though to fewer significant figures. When it is required to store text, such as the names of towns and rivers, it is usual to allocate each letter or character to a separate byte. This is done by assigning numeric values to alphanumeric characters. The ASCII and the EBCDIC systems are commonly used character coding schemes. A field consisting of a collection of bytes making up an item of text is referred to as a *character string*. We can see then that there are several types of field, of which integer, floating point and character are three of the most important. Other examples of field types are *Boolean*, for storing the value true or false, and *double precision*, for storing floating point numbers with a high number of significant figures.

Two simple file structures for storing vector data

Let us consider a file concerned with the towns in a given area. It could consist of a list of identically structured records, each of which contained the following fields: a unique *feature serial number*, an *easting* coordinate, a *northing* coordinate, the *population*, the town *name* and the *county*. The first four, numerical, fields could each be represented by four-byte integers, while the town name and the county could be stored in 30 and 20 byte text strings respectively (Table 9.1a).

The structure of all records within a file need not always be identical. Thus a file of data about rivers

Table 9.1 Example record descriptions for two simple file structures. (a) Record description for a simple file of data on towns. (b) Record descriptions for a file of digitised sections of rivers, for which there are two record types, a header record type which is followed by a sequence of coordinate data records storing latitude/longitude coordinate values.

(a) *Field name*	*Description*	*Type*	*Length (bytes)*
FEATNO	Unique feature serial number	Integer	4
EAST	Easting in metres	Integer	4
NORTH	Northing in metres	Integer	4
POP	Population of town	Integer	4
NAME	Town name	Character	30
COUNTY	County	Character	20

(b) *Field name*	*Description*	*Type*	*Length (bytes)*
Header record			
RFEATNO	Unique feature serial number	Integer	4
RCLASS	River feature class	Character	2
RNAME	River name	Character	30
NCOORDS	Number of coordinates	Integer	2
Coordinates record			
LAT (1)	Latitude in decimal degrees	Real	4
LONG (1)	Longitude in decimal degrees	Real	4
.	.	.	.
.	.	.	.
.	.	.	.
LAT (10)	Latitude in decimal degrees	Real	4
LONG (10)	Longitude in decimal degrees	Real	4

might consist of a mixture of two types of record. The first type could contain a unique *feature identifier*, a *feature code* for the class of river, the *name* of the river, and the *number of coordinates* used to define the centre line of this section of river. The second record type might then consist of a predetermined number of latitude and longitude *coordinates*, in decimal degrees, for storing the digitised representation of the river. The coordinates would be ordered according to their original digitised sequence. Each record of the first type would be followed by a variable number of records of the second, depending upon how many points were used to define the line of the given river section (Table 9.1b).

Compressing vector-encoded linear feature data

Given a map-grid coordinate system, such as a national grid or UTM zone, each of the eastings and northings could be seven significant digits in length. If the data in the file covered a relatively small part of the area of the grid system, the coordinate values may be regarded as being stored with considerable redundancy, in that the first few digits of all eastings and all northings respectively would be the same throughout the file. Some saving in storage space can be achieved by defining a local origin, such as the lower left corner of the region to which the file relates, and storing all coordinates relative to this origin. These *relative coordinates* will be smaller numbers than the original coordinates and will therefore be capable of being stored in a smaller amount of computer storage space. Provided the absolute value of the origin is also stored, all of the original values can subsequently be reinstated by adding the value of the local origin to the relative values.

When a sequence of points define a linear feature, we find that there is further redundancy in that the differences between successive coordinates in the file are likely to be relatively small, compared with the absolute values of the coordinates, being the distance between the respective digitised points. Further compression can therefore be achieved by storing the 'complete', possibly local, coordinate values of only the first points of each linear feature, all remaining coordinates of each line being incremental distances between successive points.

A vector data compression technique which is suitable for lines that have been rasterised onto a regular grid of possible coordinate locations is that of *Freeman coding* or *chain coding*. As with relative coordinates, the initial point of each line must be stored to full resolution. Subsequent points of the line are assumed to be located on adjacent grid cell locations, which are horizontally, vertically or diagonally connected (Figure 9.2). Each increment of the chain then consists of a vector defined by a directional code, which can take any value from 0 to 7, representing the eight possible directions from one cell to its eight neighbours. Three bits of storage would therefore be required for each incremental element of the chain code. The technique is suited to lines derived from raster images, but is less appropriate for digitised line data that consist of discrete points which may need to be plotted at different scales and hence if subsequently rasterised, on different raster resolutions.

Some lines on maps, such as meridians, may be defined by a mathematical function, rather than a list of digitised points, in which case the data describing the line, or arc, will consist of the relevant parameters of the function.

Raster data storage

One of the most interesting properties of a raster, from the information-retrieval point of view, is that the coordinates of grid cells, or pixels, are implicit within the row and column ordering of the matrix. This differs from the situation with vector-encoded information, in which location is stored explicitly as *x* and *y* coordinate values.

When stored in a computer's internal memory, the most natural data structure for a raster is a 2D *array*. This is an area of memory that is treated as a table of rows and columns, each element of which can be addressed, or referred to, directly in terms of the row number and the column number. A raster that consisted of a remotely sensed image of intensity values within a particular spectral waveband could be stored in an array of one-byte integers. Each element of the array could then store a grey level value in the range between 0 and 255. Scanning digitisers of the sort described in Chapter 5 may, after preliminary processing, produce rasters of binary values in which each element signals the pres-

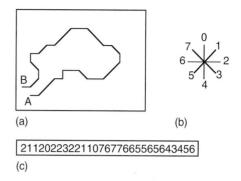

(a)   (b)

211202232211076776655565643456

(c)

Figure 9.2 Chain codes (or Freeman codes) represent a line by (**a**) a sequence of vectors or direction codes that are on a grid of fixed resolution and can only be oriented in a limited number of directions, typically eight. The eight directions are represented by the digits from 0 to 7 (**b**). The chain code for the line in (**a**) from A to B is described by the sequence of digits illustrated (**c**). Each digit can be stored in three bits of memory. To locate the line correctly, the coordinates of the first point must also be stored.

ence or absence of the digitised feature. This could be stored by a 2D array of single bits.

A file can be structured for storing the equivalent of a 2D array by allocating a record for each row of the array. Each record would then contain a field for every column in the array. If a raster element is to be stored in a single byte or, for larger values, within a two- or four-byte word, then this is generally a simple matter. However, if each element is to be stored in a single bit, or say four bits, then, in order to use only this much space in the file, a small programming overhead may be incurred, since few database management systems or programming languages provide the facility for manipulating single bits as variables in their own right. It might be necessary, for example, to regard a two-byte word as being composed of 16 separate bits, or four sets of four bits, each corresponding to a pixel requiring one or four bits respectively. Each bit could then be manipulated by integer arithmetic based on powers of two. Thus if the bits were numbered 0 to 15, the *n*th bit could be set to 1 by adding 2^n to the 16-bit integer.

From the point of view of data storage there are two notable shortcomings of rasters. The first is that, unless the pixels are very small in areal extent, they will constitute only an approximation to the actual position of the entities they represent. This is particularly the case for point and line features, in which the representative pixels will be those that lie nearest to the recorded location, within the resolution of the raster. For areal objects, these positional errors are confined to the boundaries. The second issue regarding storage is that a straight line represented in vector format by just two end points must, in raster form, be represented additionally by all the intervening pixels, while an area, which can be defined economically in vector form by a polygon, may be represented by a potentially enormous number of pixels in a raster. Related to this is the fact that in a raster dedicated to the storage of point or linear features, it is likely that a large proportion of pixels may constitute the background, or non-information. This can result in a very great storage space overhead.

Compression of raster data

The data storage overheads of raster encoding can be overcome to some extent by the use of some data compression techniques which take account of redundancy in the storage of the original values. The tendency for adjacent pixels to have the same or similar values is a measure of the *spatial coherence* and *spatial autocorrelation* of the raster image. If adjacent values often do have the same or relatively similar values, compression can be achieved by storing the differences between successive values, these being smaller numbers and hence requiring fewer bits of storage space than the original values. Reconstruction of the original values requires having kept the initial value, say at the lower left of the raster, to which the differences are successively added to build up the raster in a repeated alternating right to left and left to right row order.

Another widely used compression technique that exploits spatial coherence within rows of pixels is that of *run-length encoding* (Figure 9.3b). Its effectiveness derives from the fact that in raster data sets which are digitised from, or which represent, highly structured maps composed of discrete points, lines and areas, there will usually be many consecutive sequences of pixels of the same value. This will tend to be the case if regions between linear boundaries are blank or are occupied by solid colour. The primary data elements in run-length encoding are pairs of values or *tuples*, consisting of a pixel value and a repetition count which specifies the number of pixels in the run. The run-length data are built up by working successively row by row through the raster, creating a new tuple every time the pixel value changes, or the end of the row is reached.

For a 256×256 raster, each pixel of which could have a value between 0 and 255, the run-length tuples would each require two bytes of storage. One byte would record the length of the run, the maximum run length being 256 pixels (indicated by a value of 255), while the other would record the pixel value. A file structure for storing run-length encoded data could consist of records containing a fixed number of tuples, each composed of two fields. The file could include a single header record which specified how many tuples were in the file.

Run-length encoding only takes advantage of 1D spatial coherence within rows of pixels. Greater benefits in data compression may be achieved by exploiting the 2D nature of spatial coherence, whereby, within a raster, pixels are in general equally likely to be of the same value in a vertical direction as they are horizontally. Two-dimensional spatial coherence can be used by aggregating pixels of the same value into cells. There are a number of cell shapes, in particular triangles, squares and hexagons, which tessellate in a regular manner and could therefore be used for subdividing a raster.

The most common tessellations applied to raster data are based on rectangular blocks or squares. The square has the useful property that individual cells can be subdivided or aggregated into cells of the same shape, thus enabling a simple combination of variable-sized cells. This is exploited in the *quadtree* data structure, as applied to 2D images (Figure 9.3c). A quadtree structure can be built from a square raster by taking a quadrant, starting with the entire raster, and testing whether it is homogeneous in so far as all pixels are of the same value. If a quadrant is homogeneous, then by noting its position, size and pixel value, it can be used to represent all the contained pixels. Otherwise the quadrant is subdivided into four equal subquadrants, each of which is treated in the same fashion, with a quadrant only being subdivided if it is found to be non-uniform.

The principle of quadtrees has been applied to partitioning vector data and to creating spatial indices for the resulting cells. A discussion of this subject is deferred to Chapter 11.

High levels of compression of raster data depend upon the use of a number of general-purpose data compression techniques, some of which are not specific to raster data or image data. An important principle in data compression is the substitution of frequently occurring data items, or symbols, with short codes that require fewer bits of storage than the original symbol. The techniques of Huffman

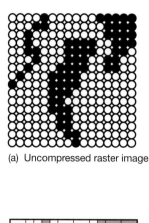

(a) Uncompressed raster image

(b) Run-length image encoding

(c) Quadtree image encoding

Figure 9.3 Compression of raster data image (**a**) using run-length encoding (**b**) which looks for rows of pixels of the same value, and quadtrees (**c**) which identify square blocks of pixels in a regular recursive pattern.

coding and arithmetic coding both involve preliminary analyis of the frequency of occurrence of symbols. Huffman coding creates, for each symbol, a binary data code, the length of which is inversely related to the frequency of occurrence. Decoding requires translating the codes back to original data items. Arithmetic coding is potentially more efficient than Huffman coding and entails constructing a single, extremely large, number that encodes the original data.

The symbols to be encoded might themselves result from transformation of the original data, for example to create differences between successive data values. In the context of gridded digital elevation models, Kidner and Smith (1992) used Huffman coding to represent the differences between actual and predicted elevations of successive grid points, employing a three data points prediction formula used both when encoding and decompressing the data.

Some image-compression methods, such as PKZIP, use the techniques of Ziv and Lempel, often referred to as LZ and its variants such as LZW (Welch, 1984). These methods identify repeating sequences of data items (pixels or characters), which are entered (conceptually at least) into a dictionary and subsequently referenced by tokens. The JPEG (Joint Photographic Experts Group) image compression scheme is based on the Discrete Cosine Transform which is comparable to Fourier analysis and involves filtering of spatial frequences (see Chapter 6). JPEG in fact uses a mix of compression techniques that can provide controlled levels of loss of detail (Wallace, 1991; Lammi and Sarjakoski, 1995). It is therefore referred to as a lossy compression technique, as opposed to lossless methods, like Huffman coding and LZW, in which no loss of original detail need take place.

Character files and binary files

It is apparent from the earlier discussion, that a record may be built of a mixture of numeric and non-numeric data fields. Numeric data are not, however, always stored in the form of integer or floating point types of field. It is possible for a character string to be used for storing numerical data, in which case each byte of the string could contain the ASCII code of a decimal digit. This is not a very efficient way of storing numerical data, since the text string representing the number 15600550, for example, requires eight bytes, while the same number can be stored equally accurately as a single binary number within a four-byte word.

Database management systems designed for space efficiency and speed of access generally use the binary method of storing numbers, and the resulting files are termed *binary files*. It is very common, however, to store independent files of map data which are wholly in character format, such files being referred to as *character files* or *text files*. The main reason for this is that the standard editing and file printing utilities of most general-purpose computing systems (as opposed to database management systems) are oriented to inputting and outputting strings of characters represented in a character format. Since the characters are often in ASCII format, text files are also frequently referred to as ASCII files.

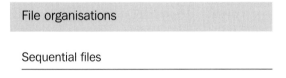

File organisations

Sequential files

The file for storing town locations and populations and the file of river names and their line geometry,

referred to previously, were described as lists of records. The simplest way for a computer to read the contents of such files is to access each record in the file in turn, in the order in which they were added to the file. So, to read the hundredth record it would be necessary to access each of the previous 99 records, even if their contents were of no interest. This way of reading files is called *sequential*, or *serial*, access and files intended to be accessed in this way are called *sequential files* (Figure 9.4a).

Direct access files

If a file is very large, and it is not uncommon for geographic data files to contain many thousands of records, then sequential access can be a very slow mode of retrieval if only a small subset of the total number of records is actually required. To obviate this problem, other types of file have been created in which it is possible to read, or address, individual records directly (Figure 9.4b), provided that the identity of the record, i.e. its address, is known. There are two notable and interrelated ways in which records can be uniquely identified. One is to associate explicitly with each record a number, allocated in a purely serial manner, based on its position within the file if it were to be read sequentially. Thus if a file could contain 10 000 records, they would be numbered 1 to 10 000 from the first to the last. Programs which access them need to specify the number of a record before it can be read.

An alternative way in which records can be identified is in terms of the contents of their constituent fields. A field treated in this way is called a *key field*.

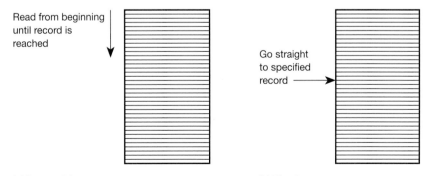

(a) Sequential access   (b) Direct access

Figure 9.4 File access methods based on sequential access and direct access.

The field which is expected to be used most commonly for identifying the record is designated as the *primary key* field, while other fields which are to be used in this way are *secondary keys*. There are a number of ways in which a file access mechanism may use the contents of a key field to find a record number and hence gain access to the keyed record. We will consider now the methods of *binary search*, *hash addressing* and *indexing*. The subject of *indexed files* will also serve to lead us into the use of *pointers* and their role in implementing *linked-list* and *tree* data structures.

Binary search

A binary search is one of the simplest methods of accessing a file on the basis of its primary key. It depends upon the records in the file being sorted into a sequence dependent upon the contents of the primary key field. Given a primary key field value of interest, i.e. a *search key*, the method starts by reading the record mid-way within the sequence of allocated records. Its record number is deduced by halving the highest record number. If this record's key value is not equal to the search key, then the record mid-way within either the upper or lower half of the file is read, depending upon whether the search key is less than or greater than the primary key of the accessed record. Having confined the search to one-half of the file, the procedure is repeated, continually halving the range of search until the required record is found.

Hash addressing

Hashing entails attempting to transform a key field value to the number of the record in which it is stored in a single calculation. It uses a *hashing algorithm*, or *hash function*, to apply a sequence of numeric operations to the numeric equivalent of the contents of the key field, resulting in a number within the range of possible record addresses. Ideally, when building the file initially, this process of hashing results in a unique record number for each unique primary key field value. In this case, the record will be allocated to the resulting record number. If the hashing algorithm produces a record number that is already in use, causing a 'collision', then some course of action will be

taken to find an unused record. In order to retrieve a record, the search key value is transformed by the hashing algorithm to generate the record number in which the data reside.

Indexed files

Another approach to generating a record number from a key field value is to create an index, ordered by the key field values. If records in the indexed file are also stored in order of the primary key, then this is referred to as an *index-sequential* (or *indexed-sequential*) file. In a dense index, there may be an entry for every record, with a unique primary key in the main, indexed file. Each entry in the index might consist of the key field and the record number of the corresponding record in the main file (Table 9.2a). This record number, stored in the index, is called a *pointer*. Note that because the only user-supplied data contained in the index is the key field, the index itself may be much smaller than the main file, and hence much quicker to read.

For a large file, however, the index could become unacceptably slow to read. One solution to this problem is to record in the index only a subset of the key fields, say every tenth one. This results in a *sparse index* which, because both it and the indexed file are sorted in key field order, provides direct access to a record in the indexed file which is never more than a limited number of records away from the record containing the key field of interest. Use of the index involves reading it sequentially until reaching a key field value that is greater than or equal to the search key. The associated record number is then used to begin a short sequential search of the main file (Table 9.2b). In the event of a sparse index becoming excessively long, the logic of sparse indexing can be applied to produce an index to the index, resulting in a hierarchical structure. This is done by creating an outer index consisting of an ordered subset of the key fields in the first, inner index, along with the numbers of the associated records in the inner index.

An index that is based on the primary key is called a *primary index*. If it is expected that the search criteria may frequently depend upon the contents of a field other than the primary key, i.e. a secondary key, then it may be considered worthwhile to create a *secondary index*, to gain direct access to records identified by the secondary key. In our file of towns, the

Table 9.2 Index sequential files. Each pointer directs the search to a set of records with key field values less than or equal to that in the index entry.

(a) Dense index

Key field	Pointer	Record no.	Town name	County	Population	x	y
Nancekuke	1	1	Nancekuke	......	......	......	......
Nancledra	2	2	Nancledra	......	......	......	......
Nanhoron	3	3	Nanhoron	......	......	......	......
Nannau	4	4	Nannau	......	......	......	......
Nannerch	5	5	Nannerch	......	......	......	......
Nanpanton	6	6	Nanpanton	......	......	......	......
Nanquidno	7	7	Nanquidno	......	......	......	......
Nantddu	8	8	Nantddu	......	......	......	......

(b) Sparse index

Key field	Pointer	Record no.	Town name	County	Population	x	y
Nannau	1	1	Nancekuke	......	......	......	......
Nantddu	5	2	Nancledra	......	......	......	......
Nantgwyn	9	3	Nanhoron	......	......	......	......
Nantmor	13	4	Nannau	......	......	......	......
Nant-y-gollen	17	5	Nannerch	......	......	......	......
		6	Nanpanton	......	......	......	......
		7	Nanquidno	......	......	......	......
		8	Nantddu	......	......	......	......
		9	Nantellan	......	......	......	......
		10	Nant Ffrancon	......	......	......	......
		11	Nant-glas	......	......	......	......
		12	Nantgwyn	......	......	......	......
		13	Nantlle	......	......	......	......
		14	Nantmawr	......	......	......	......
		15	Nantmel	......	......	......	......
		16	Nantmor	......	......	......	......
		17	Nant Peris	......	......	......	......
		18	Nantwich	......	......	......	......
		19	Nant-y-dugoed	......	......	......	......
		20	Nant-y-gollen	......	......	......	......

county field might be used for creating such a secondary index (Table 9.3).

Linked lists and chains

A secondary index of the towns file could be ordered by the county name and could contain an entry for every record in the file. Since there would be many towns for each county, there would be considerable repetition of the county field in the index. If the primary key is assumed to be the town name, it would not be possible to create a sparse index in quite the same way as referred to above, as towns belonging to the same county would not be adjacent to each other, and hence direct access to one record belonging to a particular county would not help in locating other records relating to that county. It is however possible to create an index with just one record per county if, within the town records, an additional field is added containing the record number of another town belonging to the same county (Table 9.3).

In this way all records sharing a common property, in this case county, can be chained together, while the address of the first record in each chain is stored in

Table 9.3 A secondary index may consist of a sorted list of values of the secondary key, accompanied by the numbers of associated records in the file. All records containing the same secondary key may be chained by pointers.

Key	Pointer		Record no.	Town name	County	Other data	Pointer
Cheshire	18		1	Nancekuke	Cornwall		2
Clwyd	5		2	Nancledra	Cornwall		7
Cornwall	1		3	Nanhoron	Gwynedd		4
Gwynedd	3		4	Nannau	Gwynedd		10
Leicestershire	6		5	Nannerch	Clwyd		
Powys	8		6	Nanpanton	Leicestershire		
Shropshire	14		7	Nanquidno	Cornwall		9
			8	Nantddu	Powys		11
			9	Nantellan	Cornwall		
			10	Nant Ffrancon	Gwynedd		13
			11	Nant-glas	Powys		12
			12	Nantgwyn	Powys		15
			13	Nantlle	Gwynedd		16
			14	Nantmawr	Shropshire		20
			15	Nantmel	Powys		19
			16	Nantmor	Gwynedd		17
			17	Nant Peris	Gwynedd		
			18	Nantwich	Cheshire		
			19	Nant-y-dugoed	Powys		
			20	Nant-y-gollen	Shropshire		

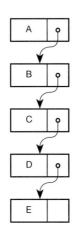

Initial list

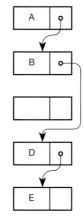

List after
deleting C

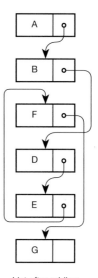

List after adding
F and G

Figure 9.5 Editing a linked list.

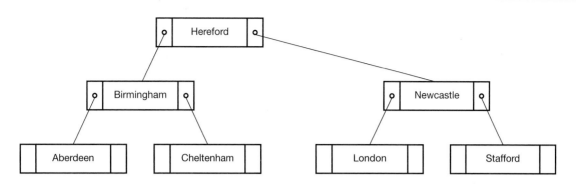

Figure 9.6 Town names ordered alphabetically in a binary tree. The Hereford record is the root of the tree. Aberdeen, Cheltenham, London and Stafford are leaf nodes.

the index. Chains connected together by means of record addresses are called *linked lists* and the 'next record' field is a pointer. If the file contained a continuous chain of all records sequenced on the basis of the secondary key, it would then be possible to create a sparse secondary index. Note that in Table 9.3 this would require the end of each individual county chain to point to the beginning of the next.

In addition to connecting together records that have a common property, linked lists can be used for accessing records in a predetermined order, such as alphabetical or ascending numerical order. When new records are to be added to a linked list, the pointers enable them to be inserted at the correct logical place to maintain the ordering of the list, even though they may be stored in records which are physically distant from their neighbours in the linked list (Figure 9.5). Because of this facility for updating an ordered sequence, it is often desirable to chain all records in an index-sequential file into a linked list ordered by the primary key, in addition to the chains connecting records related by the secondary key. In this event there would be pointer fields for both the primary and secondary key chains.

used by a computer program. We have seen that linked lists provide a versatile means of storing data in a predetermined order, but when we are dealing with large quantities of information, such as all the towns in a state, then, when sorted in alphabetical order, it may be necessary to read through many items in the list before reaching the ones of interest. This problem has led to the use of tree data structures, which provide for the possibility of alternative paths through the structure depending upon the object of the search.

In a *binary tree*, each record may contain some data in a key field, and two pointers which lead to records containing data items which, relative to that of the key field, are respectively higher and lower in the sorting sequence (Figure 9.6). Traversing a binary tree consists of starting at the root record and descending either left or right at each record (node) until the data item of interest is reached. For a large number of records this approach will, on average, result in the need to access far fewer records than in an equivalent linked list search. The number of records that must be examined to find any specified record in a binary tree is proportional to the logarithm to the base 2 of the total number of records.

Trees

Pointers, which are in practice record numbers or addresses of records related in some way to the record storing the pointer, provide a very powerful way of selectively accessing data within individual files, within sets of related files and within records stored temporarily in the main computer memory

B-trees

The principle of the binary tree, in providing quick access to data organised hierarchically according to a sorting order, has been extended to create complete database index mechanisms, in which the nodes of the tree store key field values and pointers to records of corresponding database files. The B-tree is the classic database indexing structure and is a multiway tree in that, unlike a binary tree, each node can refer to more than two descendents. It is a highly adaptable structure, in that by enabling a variable number of pointers per node, the tree remains balanced, when data are inserted and deleted from the database. This means that the height of the tree (number of levels) is the same for all leaf nodes. Since the height of the tree affects the maximum number of pointers that it is necessary to follow before the target data are found, the performance of the access mechanism is maintained to a consistently good level.

Summary

This chapter has provided an introduction to the principle of data storage in computer files and databases. At the level of a computer's storage facilities, we have seen how spatial objects and their attributes can be allocated to logical units of storage such as fields, records and files. These in turn are represented by physical units of storage, such as bytes, words and blocks. The organisation of the computer's data storage can be made to reflect the logical structure of the spatial models such as the sequence of points in a line or the membership of a feature classification. This may be done by controlling the order in which fields and records are stored, or by linking associated records and files explicitly by means of pointers, in linked lists, or logically by means of unique key fields. The subject of file indexing has been introduced here but our treatment of the subject has been confined to conventional indexing methods intended originally for non-spatial data. An important such method is that of hashing, whereby a key is transformed to a record address. Many databases make use of hierarchical indexing schemes which involve traversal of a conceptually tree-shaped index which organises the key fields according to a sorting order, and records with each each key field the address where the associated record is stored in the main file.

The fact that spatial data are referenced locationally in two or three dimensions means that some of the techniques described here are not adequate for all aspects of geographical data storage. We will pursue this topic in more depth in Chapter 11, in which we will examine multidimensional indexing methods and techniques for partitioning the contents of a database in a way that corresponds to the spatial distribution of the stored objects.

Further reading

For an understanding of the way different types of variable are used for storing data, it helps to gain some basic familiarity with a programming language such as Pascal, C or Fortran, which provides facilities to manipulate each of the different types of numerical and character variables. To find out more about particular types of data structure, see for example Tenenbaum and Augenstein (1981), which assumes a knowledge of Pascal. For further reading on the basics of file and database indexing methods see one of the standard texts such as Date (1995). The subject of spatial data handling is pursued in more detail in Chapter 11, which provides further references.

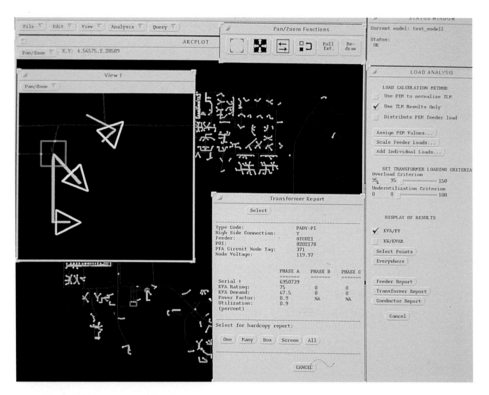

Plate 1 GIS for electricity supply maintenance. Map shows connections of supply to individual households. Associated non-spatial data is also displayed. *Courtesy of ESRI.*

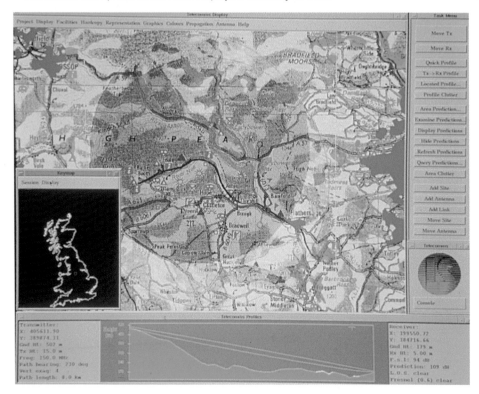

Plate 2 Digital terrain models can be used to plan the location of radio transmitters and receivers, taking account of the effect of terrain on radio transmissions. *Courtesy of Laser-Scan Ltd.*

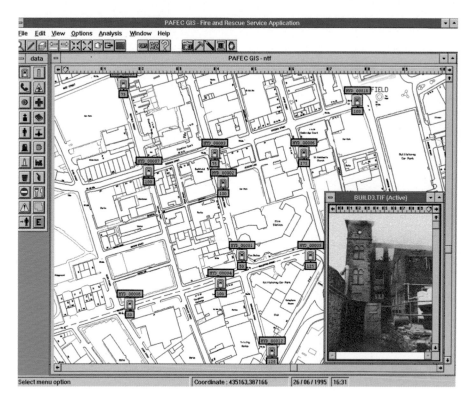

Plate 3   GIS for fire and rescue services. The GIS maintains the locations of fire hydrants and stores images of buildings. Base map from Ordnance Survey data, Crown Copyright. *Courtesy of PAFEC.*

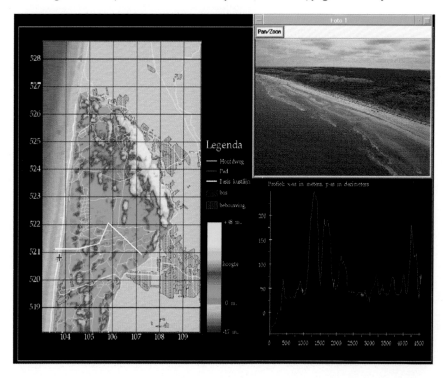

Plate 4   Digital terrain models used in predicting coastal locations liable to flooding. The areas in blue would be affected by a 3.5m flood. The graph represents the profile of terrain along the cross-section marked in white. *Courtesy of Coastal Zone Management Centre, The Netherlands.*

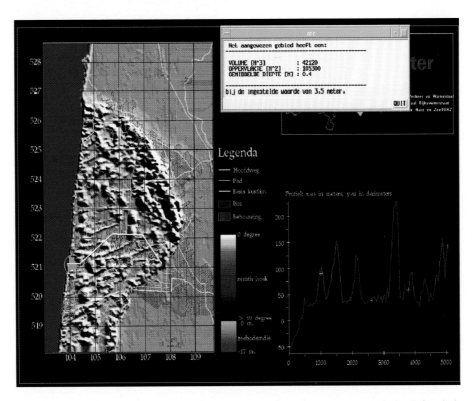

Plate 5   Digital terrain models in coastal zone management. The image uses a shaded relief technique to represent the same area as in Plate 4. Areas liable to a 3.5m flood are marked in green. The graph represents the profile of terrain along the cross-section marked in white. *Courtesy of Coastal Zone Management Centre, The Netherlands.*

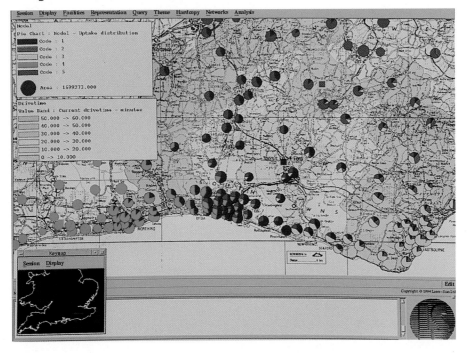

Plate 6   GIS in market analysis. The pie charts show predicted uptake per postal sector by store, where the size of the pie chart is proportional to the estimated total money available. Base map from Ordnance Survey data, Crown Copyright. *Courtesy of Laser-Scan.*

Plate 7 Simulated 3D scene for use in planning new buildings and street developments. *Courtesy of Grintec.*

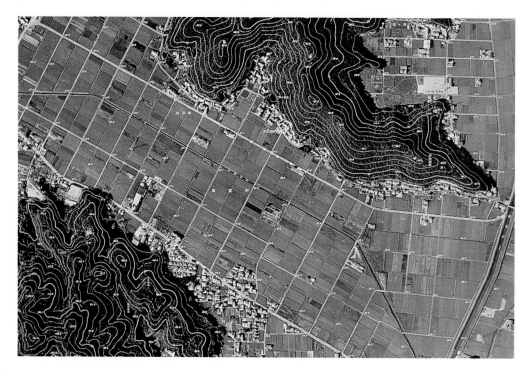

Plate 8 Aerial photograph with contours and other map features plotted following photogrammetric processing. *Courtesy of Leica.*

Plate 9

Landsat multispectral scanner false colour composite image of an area of 90 x 135 km over southern California. Los Angeles is visible in the lower right. The red tones represent vegetation in parts of the city and on the lower slopes of the mountain ranges. The Mojave Desert is in the north-east with irrigated fields to the south of the desert. The San Andreas Fault crosses the centre of the scene. *Copyright National Remote Sensing Centre.*

Plate 10 A Coastal Zone Colour Scanner (CZCS) image from the Nimbus satellite, covering 1300 x 900 km over the east coast of North America. Several narrowly defined wavelengths are combined to highlight nutrient-poor eddies of the Gulf Stream in blue and, towards the inshore, increasing levels of phytoplankton represented by green, yellow, orange and brown. *Copyright National Remote Sensing Centre.*

Plate 11   Landsat Thematic Mapper mosaic covering the south-west of England. *Copyright National Remote Sensing Centre.*

Plate 12   SPOT image over Edinburgh created by merging 10 m resolution panchromatic data with 20 m resolution multispectral High Resolution Visible (HRV) data. *Copyright National Remote Sensing Centre.*

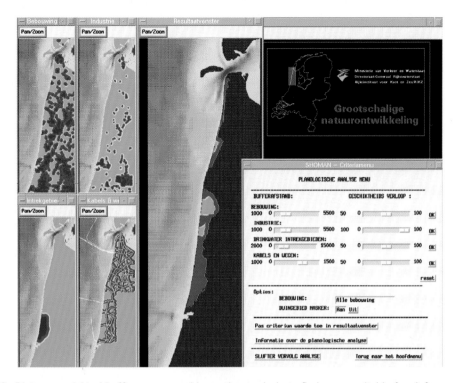

Plate 13 Distance-weighted buffer zones used in overlay analysis to find areas suitable for slufter development in the Netherlands. The four buffer maps on the left, representing distances from (clockwise) housing, industry, roads and drinking water supply are overlaid to create the larger map, in which green areas are most suitable. The slider bars are used for setting weights. *Courtesy of Coastal Zone Management Centre, The Netherlands.*

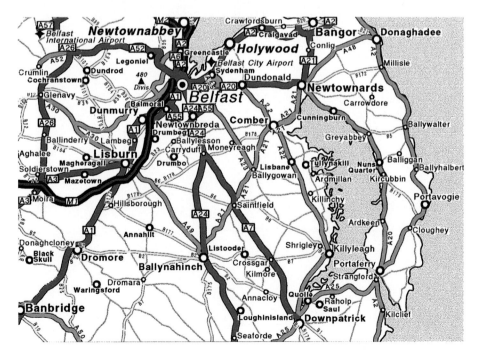

Plate 14 Significant differences in hue are usually used to distinguish between major categories, such as classes of road. Map produced using Maplex with digital map data. *Courtesy of Automobile Association.*

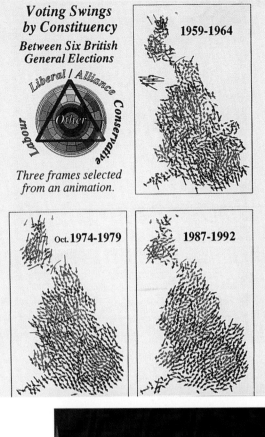

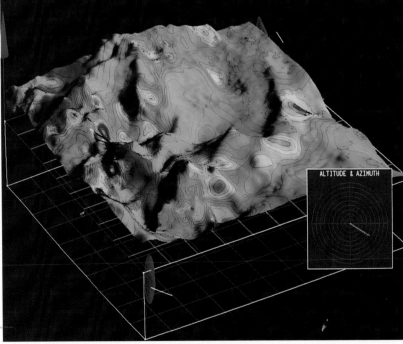

Plate 15

The graphic variable of orientation is used in the figure to indicate voting swing from one electoral party to another. The information is reinforced by the use of hue and lightness to distinguish the major parties (Dorling, 1992).

Plate 16   Visualisation of geophysical data. A Bouguer anomaly map (based on gravity) is represented by a shaded relief surface, on which is superimposed (draped) the corresponding magnetic field represented by a coloured contour map. The display helps in studying the correlations between gravity and magnetics. *Courtesy of British Geological Survey.*

CHAPTER 10 Database management systems

Introduction

For many organisations using GIS technology, the data they store will be one of their most valuable assets. Having invested so much, whether in time or money, it becomes imperative that the organisation manages the resulting database in a way that both exploits and preserves the asset.

The importance of structuring spatial data for the purposes of effective and efficient access has already been stressed in Chapter 9. In Chapter 2 we introduced the idea of a spatial model which is implemented by lower level data structures. In considering the data management aspects of a GIS in this chapter, we will take a broader view of data modelling, examining real-world conceptual models intended for database design and computational or logical models oriented to particular types of database management system (dbms). It will also be necessary to appreciate the importance of a number of issues that are vital to the successful operation of a database management system. Many, but by no means all, of the problems which arise in maintaining large quantities of geographical data are no different in principle from those encountered by numerous commercial organisations. For this reason we will consider in this chapter some of the objectives in commercial database management systems in so far as they are relevant to GIS. It must also be appreciated that GIS introduce data processing problems that are different from those of typical commercial data processing applications.

In the remainder of this chapter, we start by considering objectives in database management and the components of database systems in general, before looking at issues of conceptual modelling of data in order to create a design suitable for a database. Following this is a review of the way in which database systems are organised for the purposes of implementing the conceptual design, focusing on the relational model that is the most widely used. The potential of the newer technologies of object-oriented and deductive database models are then reviewed. Finally, the subject of spatial query languages for GIS applications is introduced.

Desirable characteristics of database management systems

Because the cost of maintaining a database increases with its size, care must be taken when working with large quantities of data to ensure that the same information is not duplicated unnecessarily. Avoiding such data redundancy helps to serve another important database function, which is to achieve *consistency*, whereby data values referring to the same entity are not contradictory, as could happen if a value was updated in one part of the database but not in another.

Consistency may itself be regarded as part of a larger problem of *data integrity*. This concerns the correctness of the database contents. In general, a database cannot 'know' whether its data values are correct, but it can perform certain checks, such as for unrealistic values and for data items that have been corrupted by the computer hardware. Checks on the initial validity of data can be performed when data are entered. Database transactions concerned with data entry, or loading, may also include automatic update of associated data to maintain consistency if particular data items are known to be duplicated and if certain data items are a function of the updated item. The latter could occur in the case of data items which record measurements or statistics that are based on the stored contents of the database.

Updating of GIS databases can be complicated by the need to maintain several representations or versions of the same geographical area, reflecting either real-world changes, or proposed changes, that may be part of development plans. This leads to the need for *version control* to keep track of associated versions and to record when changes were made in the database and when they relate to in the real world.

The fact that there will not in general be any automatic mechanism to prove that the contents of the database are correct means that it may be important to impose restrictions upon personnel entitled to make changes to the database contents. This includes the addition of new data and the deletion of existing data. This is one aspect of database *security*. Another aspect of security concerns control over who can retrieve information from the database and from particular parts of the database. It may be appropriate to give individuals authorisation to read only certain types of information to which they are entitled to have access.

In addition to rights of update and access, the problem of security extends to protecting against loss of data, whether accidental or intentional. *Roll-back* and *recovery* procedures are necessary to ensure that it is possible to return to earlier (perhaps uncorrupted) states of the database, and to reinstate a database that has been damaged by computer failure. As such, they may be regarded as a further aspect of maintaining consistency. To implement recovery procedures it is necessary to perform regular *back-up* of the database. This involves making copies of files onto separate back-up disks or tapes and storing them in a safe place.

In the course of development of a database it may prove necessary to add new types of data which had not been foreseen at the design stage, and it may be appropriate to make changes to the database structure to improve efficiency. If there were application programs which were closely tied to the structure and organisation of the database, such changes could introduce a major overhead in modifying the programs to work with the updated database. In a well-designed database, however, the query programs should not be so closely coupled to the physical structure and contents. The query programs or application programs should request the retrieval of particular data items, while a lower level of the database is concerned with servicing the request and translating into the physical form of the database. This approach is known as program data independence, or just *data independence*.

In a large organisation there may be many data users with diverse interests in particular aspects of the database contents and with varying responsibilities for accessing and for updating the contents of the database. This is especially true in many GIS environments for which data may be derived from numerous sources and applied to numerous activities. Such diversity of use will often be accompanied by diversity of user location and may result in the development and operation of *distributed databases* in which the contents of the database may be subdivided between several user sites.

If a database is to be distributed rather than centralised then it places additional burdens on the dbms to maintain the integrity of the database. It also raises problems of communication between the component parts when database queries demand access to files located at different sites. On the other hand, a distributed database, if well organised, may improve the performance at the local sites with distinct needs, since in general it should reduce the overheads of data transmission which characterise remote access to a central database.

Whether a database is centralised or distributed, the need commonly arises for more than one user to access the same part of the database at once. Database management systems must therefore support *concurrency* of user requests to access the same data. If the request is to read data, the problem can be handled fairly easily, but if one or more users attempt to update the same item of data the situation becomes complicated since it could be possible to read incomplete data. Maintenance of concurrency therefore requires strict controls which result in part or all of a file of data being locked while updates are taking place. This means that for certain operations only one user will be allowed access to a part of the database until the transaction (of writing or deleting records) is fully complete.

Components of a database management system

There are at present a considerable number of commercially available packages for building and operating database management systems. Although they may differ in several important respects, regarding the way they are implemented, there are several common components that can usually be recognised.

The desirability of achieving data independence, referred to above, generally results in a clear distinction between the logical and physical levels at which the database is defined. It is in fact possible to identify several levels of information representation in a database. These include the conceptual, logical (or internal) and physical models (Figure 10.1). A dbms can be regarded as a tool for representing, in a computer, a real-world oriented model of a set of data. This relatively high-level representation, or abstraction, is referred to as a *conceptual model* and its specific description in the database is often referred to as a *schema* or a *conceptual schema*. The database tool for defining the conceptual schema is a *data definition language* or DDL. The syntax of this language differs somewhat between different database systems.

The conceptual model and the database-specific schema that records it must refer to all items of data which are to be stored. In a database of any considerable size, there will normally be subsets of data which are required for particular applications. In fact, the model will probably have been designed by first identifying all specific, application-oriented data storage requirements, before integrating them into the complete model. This latter process is called *schema integration*. Thus each application may have its own local view of the database, referring only to the relevant subset of data. This application-oriented representation is called an *external model* and is implemented in the database by a *sub-schema* or *external schema*. Again the DDL, or a part of it, is used for describing the sub-schema.

External models and the corresponding sub-schema are an important aspect of the dbms because they provide a convenient means of restricting the operations that particular application programs can perform on the database. Thus associated with an individual sub-schema will be specific rights controlling whether each of the component data items can be read, written to (updated) or deleted.

Both the external and conceptual models are independent of the data structures and retrieval mechanisms on which the efficient operation of the database depends. Most commercial database management systems are implemented by one of three widely recognised data models, also referred to as *logical models* and *internal models*. These are the hierarchical, the network and the relational model. More recently, object-oriented and logic-based, or deductive, models have been introduced but are not yet widely used, although some of the associated techniques have been introduced in 'post-relational'

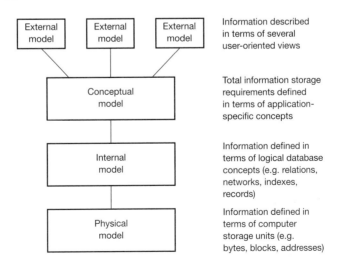

Figure 10.1 Levels of information representation in a database.

database management systems. This level of the database architecture conventionally specifies logical record structures and access paths corresponding to all data referred to in the conceptual and external models. We have seen in Chapter 9 that there are several ways in which data files may be organised for the purpose of retrieving specific items of data. When we refer here to access paths we are specifying which particular mechanism to use, in terms of key fields, indexes and pointer chains, for example. The characteristics of the various logical data models will be described later in this chapter.

The logical model, unlike the conceptual model, is certainly dependent upon the workings of a specific dbms, but it is still independent of the physical organisation of the data on the storage device. This lowest level of the database architecture is the *physical model* and it is at this level that data items are defined in terms of parameters such as the number of bytes of storage they occupy and the address at which they are to be found in physical memory.

We have already mentioned a means for building a database, namely the data definition language, or DDL. There must, of course, also be a mechanism for retrieving data and for adding, changing and deleting data once the database has been initialised. This is provided by the *data manipulation language*, or DML. Typically this will be accessible from an application program in the form of special-purpose procedures or subroutines, and from a query language, such as SQL, which allow an interactive dialogue with the database management system. Commands in the DML will allow the user to retrieve specific items of data or collections of data which meet specified retrieval criteria, relating for example to a spatial window or a class of objects of interest.

Queries to the database are expressed in terms of the data entities and attributes that constitute the conceptual model and its local subsets which are the external models. Execution of a query results in a request to the database manager software to find the named data items or classes of data items. This might require, in theory, that there be an internal database mechanism for translating from the external model to the conceptual model, from the conceptual model to the internal model, and from the internal model to the physical model (Howe, 1989). In practice, for the purposes of efficiency, there may be short cuts to the translation between the various models. However, a common factor in many dbms is the use of a *data dictionary*. This contains an entry for all data entities and their attributes, which may be accompanied by a translation to their lower level representation in the internal and perhaps physical models.

Understanding the data: conceptual modelling

In our explanation of the components of a database management system we have identified several levels of abstraction, the highest of which are the conceptual model and its application-specific views or external models. To create a database, therefore, we need to formulate the conceptual and external models. This process is known as conceptual modelling. It consists of defining both the entities or objects for which data are to be stored, and the relationships between them which together support the processes and applications for which the database is required.

A widely used technique for building conceptual models is called *entity relationship (E/R) modelling*. A characteristic of the resulting E/R models is that they lend themselves to translation into the lower levels of database abstraction, of which hierarchical, network and relational models are the prime examples. Entity relationship models are usually represented by entity relationship diagrams (ERD), in which the main components of entities, relationships and attributes are displayed by means of rectangles, diamonds and ovals respectively (Figure 10.2a). The connections between them are indicated by straight lines. Figure 10.2(b) illustrates a very simple entity relationship diagram in which Highway and County entities are associated by the relationship maintained_by.

The idea of an *entity* was introduced in Chapter 9, when we described the purpose of records in storing data pertaining to entities and their attributes. It should be remembered that an entity is a loose term that refers to any thing, whether physical or abstract, which can be distinguished from another thing. It must therefore be possible to allocate to an entity a unique identifier which consists of one or a combination of its attributes. An example of an entity type is a County (Figure 10.2b). Attributes of a County could include its county_name, county_town and county_population. Another example is a Highway. Attributes could include road_class and road_name.

A relationship between two or more entities may be said to exist if the combination of entities is required to perform an application process or func-

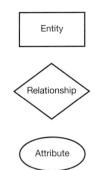

(a) Symbols in entity relationships diagrams

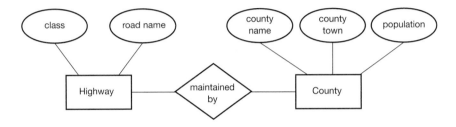

(b) Simple entity relationship diagram

Figure 10.2 Entity relationship diagrams.

tion. For example, the determination of paths through a road network could require determining those Highway_links connected to a particular road Junction. To service such a requirement we could create a relationship called Junction/Highway_link which associates highway links to junctions. This is

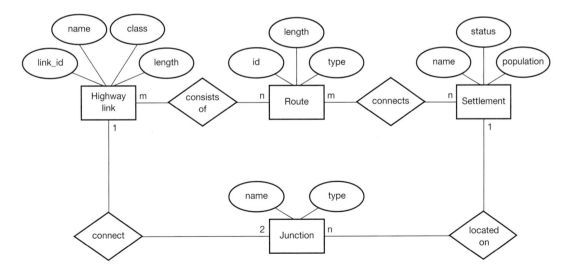

Figure 10.3 Example of an entity relationship diagram representing the components of a road network.

illustrated in Figure 10.3 (adapted and modified from similar types of diagram in Calkins and Marble (1987)). Route entities are associated with the Highway_link entities that compose them and Settlement entities that that lie along them. Settlement entities are associated with Junction entities which may be located within them.

Relationships between entities fall into one of three main categories of *cardinality*, which are one-to-one, one-to-many and many-to-many. The least commonly occurring of these is the one-to-one. It is distinguished by the fact that any entity in one of the entity sets is only ever associated with a single entity in the other entity set. For example, in the allocation of British members of parliament (MPs) to electoral constituencies, each constituency can have only one MP and each MP is only allowed to represent one constituency. Thus an entity set MPs would have a one-to-one relationship (called, say, represents) with the entity set Constituencies. In a one-to-many relationship, each entity in one of the entity sets may be associated with more than one entity in the other. This is a very common type of relationship. An entity set Counties would have a one-to-many relationship with the entity set MPs. It would also have such relationships with entities Towns, Highways, Parks, Swimming_pools, etc. Many-to-many relationships occur when one entity in either of the entity sets can be associated with more than one entity in the other entity set. For example, considering highways and counties, individual highways could pass through several counties while each county could include several highways.

Referring to the E/R diagram in Figure 10.3, the Settlement/Junction relationship between the entity sets Settlement and Junction is a one-to-many relationship, allowing for the general case of a settlement containing more than one junction, while a junction could not be in more than one settlement. The relationships Highway_link/Route and Route/Settlement are both examples of many-to-many relationships. In the case of the Route/Settlement relationship a route may link up several settlements, while a settlement may lie on several different routes. The Highway_Link/Junction relationship is a special example of a one-to-many relationship that is constrained to two (allowing for the case of a terminal junction).

Entity relationship models are subject to *existence dependencies*. This means that certain entity sets depend for their existence on the presence of another entity set to which they are related, while other entity sets exist independently of any other, whatever the relationships. Thus entity sets County and Settlement would usually have independent existences and be called *regular entity sets*. If, however, a UK administrative District is subdivided into small administrative units, such as the Ward, these smaller units would depend for their existence on the District and a corresponding Ward entity set would be called a *weak entity set*. A further sort of dependency in E/R modelling is *ID dependency*, in which the unique identifier of an entity depends on its association with another entity. In the above example a ward might have an ID dependency if it was identified by a name which might be used for wards in other districts, being only unique within the parent district.

We have noted that E/R models include attributes in addition to entities and relationships, and gave an example of some attributes that might be associated with a county. It is also the case that relationships can have attributes. These attributes distinguish the relationship, but might have little meaning when associated with only one of the associated entities. In the road-routeing example, the Highway/Route relationship could have the attribute sequence number, indicating its order in the sequence of links composing the route. The Route/Settlement relationship could have an attribute to indicate whether the settlement was terminal (beginning or end) or internal to the route. It might also have a sequence number within the route. Another example could be in monitoring the maintenance of highways with respect to the various counties through which they pass. A relationship 'County/Highway_maintenance' could then include attributes concerning dates of maintenance work and the identities of the maintenance contractors involved.

Semantic modelling

The entity relationship model was originally developed as a tool for constructing conceptual database models in a way that could fairly easily be translated into any of the main internal database models which underlie the implementation of dbms. It has been followed by a number of other conceptual modelling schemes which have come to be known as semantic models and of which the E/R model is now seen as an important example (Hull and King, 1987). The

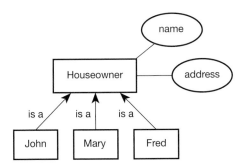

Figure 10.4 Classification. The individuals John, Mary and Fred may all be classed as Houseowners. The attributes of the class Houseowner are Name and Address.

objects or entities, semantic modelling schemes have introduced several abstraction methods which help to distinguish between different sorts of relationships. These abstractions include *classification, generalisation, aggregation* and *association.*

Individual instances, or occurrences, of objects which are of the same type may be categorised as such, the process being one of classification (Figure 10.4). In the E/R model an entity corresponds to the type of entities that belong to it. In most real-world situations, classifications are hierarchical and this is reflected in semantic modelling by the process of generalisation which places a number of types of object that share some property or characteristic into higher level types (Figure 10.5). The process of classification may be regarded as an aspect of generalisation, but for the purposes of data modelling the process of classification is only applied to the lowest level objects in a hierarchy. These lowest level objects are called *tokens*, a set of which is classified into a *type*. When types are collected together to form more complex types then the term *generalisation* is used. Thus classification in this usage is the first step in generalisation. Individual generalisation relationships are sometimes also called *is_a* relationships. They may also be called *a_kind_of* relationships. The opposite process to generalisation is that of *specialisation* and involves the identification of the component members of an object which is a complex type (Smith and Smith, 1977).

purpose of introducing other conceptual modelling schemes was to provide further levels of abstraction which could better represent the real-world functions of database management systems. The expression 'semantic model' is intended to suggest the greater level of meaning that such models can represent. By raising the level of abstraction, semantic models become more detached from the computer record-based structures that characterise the lower level hierarchical, network and relational models.

Semantic models commonly use the term 'object' rather than, or interchangeably with, the term 'entity' to refer to the real-world things that they model. Instead of simply referring to relationships between

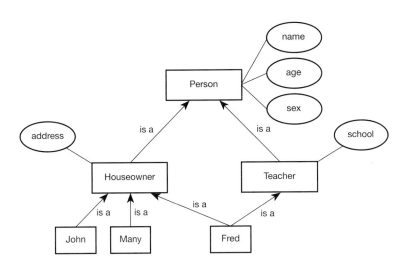

Figure 10.5 Generalisation. Individuals John, Mary and Fred can be classed as Houseowners and Teachers. The classes Houseowner and Teacher belong to the superclass Person and inherit the attributes of Person, which are Name, Age and Sex.

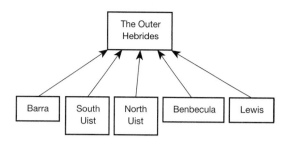

Figure 10.6 Association. Several objects of the same type are associated in a group. In this example, the objects belong to the class 'Island'. The association relationship refers here to a natural group of islands.

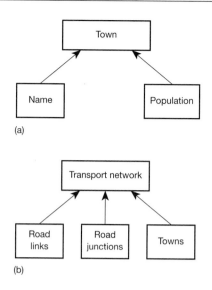

Figure 10.7 Aggregation. The aggregation relationship defines an object in terms of a collection of parts that may simply be regarded as attributes of the object (**a**) or they may be composite or aggregated objects in their own right (**b**). The parts of an aggregation do not in general belong to the same class.

An example of classification would be that an individual domestic dwelling, such as 'number 8 Baileywick Crescent', is classified as a 'semi-detached house'. A semi-detached house could then be generalised to being a kind of house, which could in turn be generalised to being a kind of building.

The abstraction mechanism of *association* provides a means of referring to a group of objects of the same type sharing some property or condition. For example, associations could be formed from all towns that were located inside a particular county, or the islands belonging to a natural cluster (Figure 10.6). Association serves the purpose of collecting together a subset of objects for the purpose of a specific analysis or operation. The collection of objects are all of the same type, but the collection is not of sufficient generic significance to form a database class in their own right, other than for the purpose of the operation to be performed on them. We should note here that the term association is also used in a broader context to refer to any relationship between two distinct entity types (not related by generalisation).

In the context of the E/R model we have referred to the fact that an entity is usually defined by a set of attributes. The process of combining a set of properties or a set of objects to form a single entity or object is called *aggregation*. This abstraction method differs fundamentally from generalisation in that the objects which are aggregated are of types that are different from each other, rather than being similar in some way. The result of the aggregation should, however, be a coherent and semantically higher level object. The object 'town' could be an aggregation of its name, population and the coordinates of a representative point. A transport network could be a higher level aggregation of towns, interconnecting roads and road junctions, each of which with their own lower level components (Figure 10.7). For further discussion of aggregation in the context of GIS, see Egenhofer and Frank (1987).

An important feature of the generalisation and aggregation abstractions is that they allow a data model to represent explicitly the hierarchical relationships which frequently arise when attempting to represent complex situations. Though the original E/R model only included aggregation by allowing the specification of the attributes which constitute an entity, the model has subsequently been extended to include generalisation (Teory *et al.*, 1986).

Logical models for database implementation

The result of the conceptual modelling process should be a clear representation of all entities and their attributes and all relationships between entities that are required to meet the foreseeable information storage and retrieval requirements. The purpose of the logical data model is to represent the conceptual model components in terms of the computational concepts of a particular type of database.

Commercial database management systems are conventionally categorised as either hierarchical, network, relational or object-oriented. The hierarchical data model is currently the least widely represented (IBM's IMS is an example). It can be regarded to some extent as a specialisation of the network model, for which several commercial systems are currently available. The relational model is very much the most widely implemented for commercial applications and it is characterised by quite simple concepts of data organisation and related query languages, which we will consider below.

The ideas of semantic data modelling discussed in the previous section have resulted in the development of object-oriented databases, which have attracted a great deal of attention in the GIS community, but at the time of writing they are still in the minority in a commercial context. Another category of database which has been the subject of considerable research activity is the deductive database or logic database. Deductive databases include inference mechanisms and are sometimes seen as extensions of the relational model, though they have also been developed in object-oriented environments.

The relational model

The main data storage concept in the relational model is a table of records, referred to as a *relation*, or simply a *table*. The records in a table contain a fixed number of fields, which must all be different from each other, and all records are of identical format. There is, therefore, a simple row and column structure. In relational database terminology the rows, or records, are also referred to as *tuples*, while the columns of fields are sometimes referred to as *domains*. Each record of a table stores an entity or a relationship and is uniquely identified by means of a *primary key* which consists of one field, or a combination of two or more fields in the record. The need for *composite keys*, consisting of more than one field, arises if no one field can be guaranteed unique. The fields of an entity table store attributes of the entity to which the table corresponds. Table 10.1 illustrates two example tables for Settlement and County.

Relationships between entities in a relational database are maintained by including the key of one table within the record of another to which it is related. A

field that stores the key of another table is called a *foreign key*. It is important to realise that the primary key of a table and any foreign keys that it may store consist of logical data items which may be attributes such as names or some allocated numerical identifier. They do not consist of physical addresses in the database. They will, however, be used as the basis of indexing mechanisms which the database management system uses to provide efficient query processing.

In Chapter 9, we gave an example of a data file that contained mixed record types and in which records could store multiple data items of the same type. In a relational database this approach would require significant re-organisation to conform to the simple tabular view of information storage. There are well-established procedures for carrying out this re-organisation, referred to as *normalisation*. One of the tasks of a database designer would be to reduce all information to normalised form. The process can be regarded as an aspect of the entity relationship modelling techniques referred to earlier. Thus a relational database table can be regarded as representing a set of entities, each of which is stored in a record of the table. Alternatively, a table can represent a relationship which links key fields of associated entities. There are several degrees of normalisation ranging between first normal form and fifth normal form. They differ in various respects, including the extent to which data items within a record are dependent upon each other, as opposed to having an independent

Table 10.1 Examples of relational database tables for settlements and counties.

(a) Settlement

settlement name	settlement status	settlement population	county name
Gittings	village	243	Downshire
Bogton	town	31 520	Downshire
Puffings	village	412	Binglia
Pondside	city	112 510	Mereshire
Craddock	town	21 940	Mereshire
Bonnet	town	28 266	Binglia
Drain	village	940	Mereshire

(b) County

county name	county population	county area	county capital
Downshire	632 511	142	Forage
Binglia	1 520 388	205	Bunge
Mereshire	490 265	170	Pondside

identity and therefore being stored in separate relations that could subsequently be linked (or joined) to other relations by means of common fields.

For purposes of handling spatial data there is a problem concerning the definition of records in a relational database. The records are intended to store a set of data fields of different type. Several important entities in spatial data consist of sets of data items of the same type, such as the coordinates making up a line or the arcs making up a polygon. In a standard relational database such data items of the same type must be stored in separate records, the consequence of which can be overheads in storage space and poor performance in accessing all the data items, such as coordinates, that constitute a logical entity such as a line. Relational databases are sometimes used in GIS in combination with special-purpose file organisations (as in ESRI's ARC/INFO system), and in their own right in systems which compromise some of the tenets of relational database theory. We return to this issue again in a later section of the chapter.

Relational operators

Retrieval from a relational database involves creating, perhaps temporarily, new relations which are subsets or combinations of the permanently stored relations. There are several relational algebra operators that can be used to search and manipulate relations in order to perform such retrievals. Some of these (*selection*, *project*, *union* and *join*) are illustrated in Figure 10.8. Other standard operators include *product*, *divide* and *intersection* (see, for example, Date (1995) for further details). From the user's point of view, the operators are not named as such but are implemented by means of the standard Structured Query Language (SQL) using a number of commands and key words. We can illustrate the commands with reference to Table 10.1. The projection operator consists of retrieving only a subset of the fields of a relation. For example, the command

SELECT settlement_name, county_name
  FROM Settlement

will create a new table which consists only of the settlement name and county fields of the Settlement table.

The selection (or restrict) operation is concerned with retrieving a subset of the records of a table on the basis of retrieval criteria expressed in terms of the contents of one or more of the fields in each record. For example, to retrieve all settlements in the county of Mereshire with a population greater than 20000, the SQL command would be

SELECT
  FROM Settlement
  WHERE county_name = Mereshire AND
    settlement_population > 20000

Note that the WHERE condition consists of a logical expression. This query could have been combined with a projection operation by specifying field names after the SELECT command.

The join operator is more complicated than projection and selection in that its purpose is to combine fields from two or more tables. The operator depends on the tables being related to each other by means of a common field. Considering the examples in Table 10.1, a retrieval requirement could be to create a table consisting of the names of towns of population greater than 30000 along with the name and population of the county they occupy. The value of the county_population attribute will have to come from the County relation. The desired join operation will have the effect of creating a new table which appends to selected Settlement records the county_population field of the corresponding county record, identified by the county_name, which is the common field. Clearly the relevant county records can be expected to match with many settlement records and the retrieved county_population fields could therefore be repeated accordingly in the new relation. The SQL command for this operation would be

SELECT Settlement.settlement_name,
  Settlement.county_name,
  County.county_population
FROM Settlement, County
WHERE Settlement.settlement_population > 30000
  AND Settlement.county_name =
    County.county_name

Storing coordinate chains and composite objects in a relational database

The frequent occurrence, within geographical data, of variable length chains of coordinates presents special problems in a relational database. According to the theory of relational dbms, each field or domain of a relation represents a distinct attribute such that the order of the fields is of no consequence. Similarly the order of records or tuples within a relation is arbitrary since each one is uniquely identified. The geometry of linear features,

Selection (restrict)

  Retrieve a subset of the
  records in a table

e.g. settlements with
       populations greater than 30000

| Bogton | town | 31 520 | Downshire |
| Pondside | city | 112 510 | Mereshire |

Project

  Retrieve specified fields
  from a relation

e.g. settlement name and
       population

Gittings	243
Bogton	31 520
Puffings	412
Pondside	112 510
Craddock	21 940
Bonnet	28 266
Drain	940

Union

  Combine the records of
  two tables without duplication

e.g. union of names of settlements
       with population greater than
       30000 and county capitals.
       Note that the two tables to
       be unioned in this example
       must first have been created by
       selection and projection operations

Join

  Make a table which combines
  fields from two specific tables
  using a common field as a link

e.g. retrieve the names of settlements
       with populations greater than 30000,
       their county and population of the
       county.
       Note that settlement name comes from the
       settlement table, county population from
       the county table and county name is
       common to both tables.

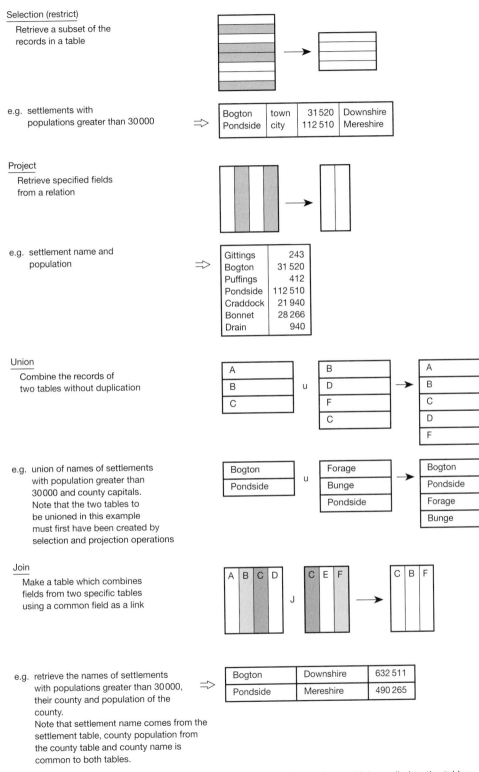

Figure 10.8 The relational operators of selection, project, union and join applied to the tables as described in Table 10.1. Other relational operators, not illustrated here, include product, intersection, difference and divide.

Table 10.2 Storage of geometric data in relational tables.

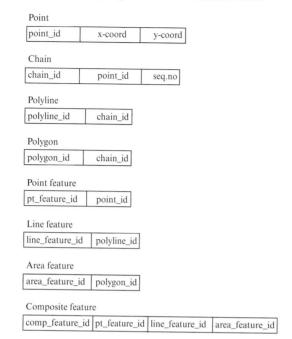

Point

point_id	x-coord	y-coord

Chain

chain_id	point_id	seq.no

Polyline

polyline_id	chain_id

Polygon

polygon_id	chain_id

Point feature

pt_feature_id	point_id

Line feature

line_feature_id	polyline_id

Area feature

area_feature_id	polygon_id

Composite feature

comp_feature_id	pt_feature_id	line_feature_id	area_feature_id

however, is normally represented by a chain or string of coordinate values, the sequence of which is essential to the representation. To store coordinate chains in a relational database without compromising the theory, it is necessary to identify each point with a sequence number indicating its position within the chain. This sequence number might also serve as a unique point identifier (point_id). One approach to relational storage of geometry would then involve the creation of a Point table consisting of point_id and the coordinates of the point (Table 10.2). A relation for storing a Chain of coordinates would need to include fields for the chain identifier (chain_id), the point_id and a point_sequence_number. It would then be possible to retrieve, by means of a join operation between Point and Chain, all coordinates belonging to a named chain, or those coordinates lying within specified ranges of coordinate values. Before plotting, or analysing, such coordinates it would be necessary to ensure that they were sorted by the point_id, within the chain_id.

Note that in Table 10.2 in the case of the chains, polygons and polylines relations, the chain_id, polygon_id and polyline_id would be repeated through several records, since each geometric object type is made of multiple components. The primary key of these relations is therefore composite, consisting in each case of the only two fields which make up the relations. The Point feature, Linear feature and Polygon feature relations attach attributes to the various geometric objects. Records of these feature relations are uniquely identified by their respective feature_id fields. A fourth Composite feature relation has been defined to enable the specification of complex objects consisting of a combination of polygon, polyline and point features.

Although this approach to the storage of spatial data is possible, it would usually be regarded as undesirable in a database due to the storage overheads incurred by repetition of the chain identifier and by the need to store a sequence number for every point. In addition, the creation of 'long thin' files results in the internal creation of large indexes to maintain them in sorted order. The tabular file structure has, however, been proposed as a standardised approach to data transfer since it is explicit and hence amenable to translation to and from more efficient, though less explicit, storage schemes (van Roessel, 1987).

There have been several attempts to overcome the shortcomings of using relational database management systems, but they diverge from standard relational theory. Typically these non-standard schemes use a special 'bulk' or 'long' data field type, in which a sequence of coordinates or identifiers may be treated as a single unit, identified by a unique record identifier (Waugh and Healey, 1987). A single chain may require several such records if the special data type is of fixed length (though some systems provide the non-standard facility for variable length records). Using this method it is then necessary to interpret the contents of the bulk field using special-purpose software, since the standard query languages will not be able to interpret it. This approach is employed in several GIS products.

Spatial indexing in a relational database

If large quantities of geometric data are to be stored in a relational database, the problem arises of how to provide efficient access to data within an arbitrary spatial window. Assuming that coordinates are blocked up in special record types, as described above, each record could be identified not only by the chain identifier, but also by a spatial key such as a linear quadtree address referring to a quadtree cell containing the chain coordinates (see Chapter 11). This is an attractive approach since it can then exploit the standard indexing facilities present in commercial relational databases, by designating the quadtree address as a primary key. Assuming that the dbms succeeds in storing records with adjacent keys at similar physical locations on the computer disk, then records relating to a particular spatial window should be retrieved in relatively few disk accesses.

Another method of spatial indexing is to store in each record the coordinates of the minimum bounding rectangle of the coordinate data (Charlwood *et al.*, 1987). This approach has the shortcoming that there are four defining values for a rectangle and thus there is no natural sorting order as there is for a quadtree address. It may be noted that R-trees (see Chapter 11) provide a means of efficient access to rectangles but they are not present as a facility of standard relational databases.

The shortcomings of the standard facilities of a relational database have led to the use of special-purpose spatial indexing software used in parallel with the standard facilities. Charlwood *et al.* (1987), in their account of the use of a commercial relational database, point out that they developed their own spatial indexing scheme, in association with the storage of geometric data in variable length bulk fields provided by the commercial system. The ARC/INFO system also combines special-purpose spatial data handling software with a commercial relational database for handling the non-spatial data.

Object-oriented databases

A recent trend in both software engineering and in database design is towards the use of object-oriented techniques. For the purposes of geographical databases these techniques are of great interest since they hold the promise to overcome significant shortcomings, from the point of view of GIS, of the widely used relational database methods. We saw in the previous section that attempts to use relational databases for operational GIS are almost invariably accompanied either by the need to add special-

purpose data handling facilities for the geometric data, as in the case of ARC/INFO, or to modify the way in which data are stored within relational tables, with the consequence that the standard SQL query language cannot be used without additional software for interpreting the spatial data. When SQL was originally designed it was intended to be adequate for formulating and answering the majority of types of query which might be made upon a commercial business-oriented database. Normal queries to a GIS require spatial data processing operations which such standard query languages cannot currently handle.

Object-oriented techniques provide the tools for building databases which, unlike relational databases, model complex spatial objects. The database representations of objects include, in addition to stored data, specialised procedures for spatial searching and for executing queries which may require geometric and topological data processing. Objects in an object-oriented database are intended to correspond to classes of real-world object and are implemented by combining data, which describe the object attributes, with the procedures, or *methods*, which operate on them. Accessing an object involves sending a *message* to it, which results in the addressed object using its internal methods to respond to the message. A variety of types of message may be sent to an individual object, depending upon its properties and the methods that it has implemented. Examples of the types of message that might be sent to a polygon class of object would be to return its coordinates, to return the result of a measurement, such as area or perimeter calculation, or to display the polygon on a graphics device.

An individual object is an *instantiation*, or a particular example, of a class of objects, and as such it is uniquely identified within the database with an object identifier. An object class may inherit the properties, data attributes and methods of one or more other object classes. Thus having defined typical object classes, new ones may be created which are combinations of or subclasses of existing ones. The stored instances of a class of object implement the semantic modelling concept of classification. The relationship of generalisation is implemented when one object class is the superclass of another class which is derived directly from it, but is in some way a specialisation of it. The use of different classes of object in the definition of a new class implements the concept of aggregation.

From a programmer's point of view, the object-oriented approach can be advantageous when developing a variety of systems based on similar classes of real-world object, since having identified and defined the commonly used classes, new classes which are modifications of them can easily be defined by referring to the existing object classes. This leads to the important benefit in software engineering of re-usable code. The attraction of object-oriented methods of software development is further enhanced by encapsulation, whereby the way in which the methods of an object are implemented is independent of the interface to the object, by which messages are passed. Thus changes to an object's methods should not have any side-effects on any other objects with which it might communicate. Chance *et al.* (1990) claim that though object-oriented programming does not necessarily result in any improvement in system performance over other methods, significant benefits can be reaped when making changes and enhancements to existing software.

An early example of the commercial use of object-oriented techniques for storing and manipulating spatial data is to be found in the Intergraph TIGRIS system (Herring, 1991), which combined a fully topologically structured spatial data model with the R-tree multi-dimensional indexing scheme, which is well adapted to spatial data.

It is not uncommon for database systems which are described as object-oriented to be built using a combination of object-oriented language techniques and a relational database for storing data. If this is done, then the domains of a table may be used for storing complex data types consisting of multiple data items of differing type. They may also be used for storing procedures which operate on these complex data types. This hybrid approach is typified by the POSTGRES dbms (Stonebraker and Kemnitz, 1991) which is a development of the earlier, more conventional, INGRES relational database.

Deductive databases

An important development in database design is the inclusion within a database of rules and of an infer-ence mechanism that can use the rules to deduce additional facts or relationships from the stored data. Databases of this type are called deductive databases or logic databases. They represent an effort to integrate in a database management system, facilities normally found within expert systems and knowledge-based systems.

The approach is relevant to GIS since geographical data are often encoded with multiple or hierarchical classification schemes and with spatial (topological) relationships which, for a human, imply the presence of classifications and spatial relationships, but which may not be explicitly encoded (Egenhofer and Frank, 1990; Abdelmoty *et al.*, 1993; Jones and Luo, 1994). For example, knowledge of the fact that a building was located within a particular electoral ward would imply that it was also in a particular administrative district and a particular county. A deductive database could make this implication if it contained facts, about the inclusion of wards in districts and of districts in counties, and a rule specifying that if object A is contained in object B and B is contained in C, then object A is contained in object C. This is an example of a transitive relationship, of which there are several other important examples that are applicable to GIS. Thus in addition to the topological relationships of inclusion and containment, the temporal relationships of 'before' and 'after' are also transitive, and so are the semantic   relationships of class generalisation and class specialisation. The potential for deduction in spatial databases has motivated some recent research into formalising spatial reasoning, particularly with regard to topological relationships (e.g. Egenhofer and Franzosa, 1991; Cui *et al.*, 1993).

Deductive or logic databases are sometimes implemented as extensions of the logic programming language Prolog, such that the facts and rules which constitute a Prolog program are stored in a permanent database environment with efficient data access facilities. Some deductive databases have been constructed by using a loose coupling approach in which a relational database, for storing facts, is linked with the Prolog language. This can be used to demonstrate the principle of deductive databases, but it is not very efficient, since the internal search mechanisms of Prolog may not work in unison with the database data access mechanisms. The alternative, more efficient, approach is one of close coupling, in which the inference mechanism is integrated with the database access mechanism (Bocca,

1991). Deductive databases can also be constructed in the context of object-oriented databases, in which case they are described as deductive object-oriented databases or DOODS .

Spatial query languages

In the earlier section on relational databases we saw some examples of the use of SQL. SQL is based on the assumption that data are organised in the form of tables which characterise relational databases. Retrievals based on an SQL query use algebraic comparisons, such as equal, greater than and less than, between values stored in fields and values specified in the query. Some GIS queries can be executed with such operations, but in general it is necessary to specify spatial relationships which require geometric processing and cannot therefore be expressed in standard SQL.

Several researchers have extended conventional SQL to include spatial relationships and hence to render the language useful for expressing typical GIS queries. Egenhofer's Spatial SQL (Egenhofer, 1991) is very similar to standard SQL, in that it is based on the SELECT FROM WHERE construct. It differs, however, in that the WHERE clause can include spatial 'constraints' on objects, using the relationships inside, overlap, contains, equal, neighbour and disjoint. The following is an example (from Egenhofer) illustrating the use of Spatial SQL.

```
SELECT   city.name, city.geometry
FROM     city, state
WHERE    state.name = "Maine" AND
         city.type = "County Capital" AND
         city.geometry INSIDE state.geometry AND
         city.population > 5000;
```

This query retrieves the names and geometric description of cities inside the state of Maine which are county capitals and which have a population greater than 5000. Spatial SQL also enables the user to employ spatial functions such as distance, direction and measurements of area, length and volume. For example, to find the geometry of cities within 50 km of the city of Bangor, the following query could be used:

```
SELECT   city.geometry
FROM     city, city cityBangor
WHERE    cityBangor.name = "Bangor" AND
         distance ( cityBangor.geometry,
         city.geometry ) =< 50;
```

Note that cityBangor is used as a duplicate name for the city relation.

Summary

GIS databases have several distinctive characteristics concerning the requirements of handling the complex structures of spatial information. They also have much in common with other commercial database management systems in needing to deal with several fundamental issues in data management. These include database integrity, consistency, the avoidance of redundancy, the need for security and the provision of concurrency.

Effective organisation of data within a database requires a clear conceptual view of the data, which is then mapped to the lower level logical data models provided by commercial database systems. We have seen that entity relationship modelling is an effective way of defining the inherent relationships within information. E/R modelling and its semantic modelling extensions take account of several major types of relationships, including those of aggregation of lower level components of data into higher level entities; classification of information of related types into classes; generalisation of classes to higher level classes; and association of entities of the same type into sets.

In commercial data processing the relational approach to database management is currently the dominant one. It is based on a simple and powerful idea whereby information is represented by tables of data. These tables are subjected to various relational operators which transform them to the subsets of data required by the database users. In the context of geographical information processing, the relational paradigm works well for non-spatial data but has limitations in handling spatial data. The geometric components of spatial data are described by collections of geometric objects such as points, lines and polygons, which must be treated as a whole rather than in terms of the individual data items which compose them. The result has often been that manufacturers of GIS technology have adopted a hybrid approach to database management which combines specialised spatial data handling facilities with the conventional relational database tools.

A recent trend in database management is towards the use of object-oriented techniques. This approach includes the concept of complex objects which can treat collections of data in combination with procedures specifically adapted to operating on the object's data. Since some relatively simple GIS queries can require geometric procedures, such as clipping of data against a spatial window, the ability to integrate such specialised procedures within a database is clearly relevant to GIS. Object orientation also brings with it a capacity for greater flexibility in the types of data stored within a single database. Object-oriented databases are thus well adapted to handling multimedia, whereby images, sound and video can be combined with textual and numerical data. For these reasons, object-oriented databases appear to be of particular relevance to geographical information systems. The technology is, at the time of writing, still very much in a state of flux and there are very few commercial object-oriented database systems. However, in the GIS sphere object-oriented ideas have been applied in some commercial products.

Another trend in database technology that has considerable potential for improving GIS databases is that of deductive databases. Their chief characteristic is that it is possible to store implicit data, in the form of rules, for deducing the answers to queries on the basis of explicit stored facts, that may be related logically to the required data. Since any one set of spatial data carries with it an enormous number of implicit spatial relationships between the stored objects, and semantic relationships between classifications, such a capacity for deduction is clearly applicable to a GIS database. Deductive database technology has not yet reached the commercial arena.

Further reading

For one of the many textbooks on commercial database management systems, with an emphasis upon relational database technology, see Date (1995). A straightforward explanation of the way in which relational database technology can be linked with specialised storage systems for handling geometric data is provided by Morehouse (1985) in a description of ARC/INFO. Waugh and Healey (1987) have described an early research effort to create a closer integration of geometric data with relational databases (reflected in some current commercial GIS). Practical examples of problems that arise in GIS database design and specifically in integrating multiple databases are provided by Nyerges (1989). A notable and extensive article on sematic data modelling is that of Hull and King (1987). Worboys et al. (1990) have described the application of object-oriented data modelling to GIS, while an overview of object-oriented database concepts is given in the article by Worboys (1994). Herring (1991) has described the data modelling concepts underlying the commercial TIGRIS system. David et al. (1993) describe an application of the O_2 object-oriented database system to create a GIS database. For a textbook on object-oriented databases, see Khoshafian (1993). A review of the ways in which deductive database technology is applicable to GIS is given in Abdelmoty et al. (1993). For an appreciation of recent developments in spatial database technology, see the proceedings of the Symposia on Spatial Databases (Buchmann et al., 1989; Gunther and Schek, 1991; Abel and Ooi, 1993; Egenhofer and Herring, 1995).

CHAPTER 11

Spatial data access methods for points, lines and polygons

Introduction

Fundamental to all information systems, whether geographical or otherwise, is the need to search through a quantity of data that is often very large, in order to find a subset which satisfies the user's query. The distinguishing characteristic of the typical geographical data retrieval is that it is expressed in terms of spatial locations and spatial relationships. In general, spatial queries may be either location-based or phenomenon-based, or a combination of the two. A location-based query specifies a location and asks what phenomena are to be found there. A phenomenon-based query specifies particular phenomena and asks where they are to be found.

Locations can be defined in terms of named places, coordinate-based geometric objects and spatial relationships. Access to data purely on the basis of a name can be achieved using conventional data-indexing methods, which were introduced in Chapter 9. Access to data specified in terms of coordinate-based geometry, of points, lines and areas, and of spatial relationships between them, has introduced the requirement for specialised storage and data-search procedures. This need arises because there is not, as a rule, an exact match between locations or spatial relationships, as they are expressed in a database query, and the data items that are stored in the database.

In a relatively simple case, called a *range search*, the query may request all data of particular classes that are inside a rectangular spatial window defined by ranges of coordinates in two dimensions. Stored geometric objects may actually lie within the ranges, i.e. be entirely inside, in which case they can be retrieved as a whole. Alternatively they may overlap the range, in which case the overlapping objects may need to be clipped at the boundary of the ranges of the search region to find the parts that are inside. This need for the calculation of geometric intersection, in order to answer the query, may also arise in more complex queries that specify required locations in terms of spatial containment regions, or windows, that are irregularly shaped.

In more complex, phenomenon-based queries the required result may be generated from the intersection of several layers of geometry corresponding to particular thematically specific phenomena or to buffers created around these phenomena. When working with vector-defined geometry, the processing required to find the subset of data that constitutes the solution to the query again involves determining intersections between boundaries, as well as procedures to maintain the structure of any new polygons that may be formed.

Queries that include topological relationships between phenomena may make use of stored topological relations. They may also require geometric data processing to determine the nature of topological relationships from the coordinate-based geometry. Commonly used procedures are those to test whether a point, a line or a polygon is located inside a specified polygon. Other related procedures test whether geometric objects are coincident or adjacent with each other.

In what follows of this chapter we start by summarising some general principles of spatial data organisation for purposes of efficient search and retrieval. This leads into a major section devoted to a variety of spatial data access, or *spatial indexing*, methods that retrieve data from within specified rectangular windows. This is followed by a short section summarising the principles of multi-resolution spatial data access. The following section then describes some of the techniques associated with geometric intersection and the determination of topological relationships from metric geometry.

General principles of spatial data access and search

Although there are numerous specific data structures and algorithms for performing locational searches, they are mostly governed by only a few important principles. One of these principles is the *partitioning* of the search space into regions that are usually, but not necessarily, rectangular in shape. Considered simply, this consists of placing data into uniquely identifiable boxes or cells. In *regular decomposition* methods, the data space is partitioned in a regular or semi-regular manner that is only indirectly related to the objects in the space. The geometry of the contained objects may be subdivided and hence distributed between several adjacent cells. More commonly, the object descriptions are kept intact, while the spatial index cells store references to the database locations of the complete objects that intersect them.

In *object-directed decomposition* methods, partitioning of the index space is determined directly by the objects. In one such approach the cells consist of minimum bounding rectangles, or extents, of the objects. In another approach to object-directed decomposition, partitioning is achieved by applying *divide and conquer* strategies, whereby individual data points or lines may be selected to subdivide the data space into successively smaller half-spaces. Such strategies generate hierarchical or tree data structures, in which the descent down each branch of the tree should result in reducing the relevant volume of data by at least half at each stage. This is the principle of the binary tree described in Chapter 9 and we may recognise it thus as the application of *sorting* to the spatial search problem. The subdivision of space into regular cells with predictable locations, in regular decomposition, may equally be regarded as a process of sorting into discrete intervals along the spatial dimensional axes.

Regular decomposition

The idea of superimposing a regular pattern of cells over the geometric data to be stored has much in common with the raster model of data storage discussed in Chapter 3. It has already been pointed out that there are three clear possibilities for the shape of the cells, namely triangular, rectangular and hexagonal. The rectangle is generally found to be most convenient because its edges can be aligned with the axes of a 2D Cartesian coordinate system, thereby simplifying tests for inclusion within rectangular search windows. Triangles and hexagons, however, have other properties which make them more attractive than squares in certain applications. Both triangles and hexagons are suitable for representing approximately spherical (global) surfaces (Dutton, 1984), while hexagons are useful for mapping statistical properties since their neighbouring centres are equidistant in all six directions. Triangles, like squares, but unlike hexagons, have the advantage that they can be regularly subdivided any number of times, in the manner of a quadtree.

Regular grids

Just as with a raster, once the size of a regular grid has been specified in terms of the length of the sides of the constituent cells, a point in a 2D coordinate system can readily be translated into the address of a cell which contains it. The main difference between a regular grid and a raster is that rather than the cells being uniform, equivalent to a pixel, they are compartments capable of storing geometric objects (Figure 11.1). The data associated with each cell will normally be stored in one or more records, the address of which is given either directly or indirectly in terms of the coordinates of the lower left corner of the cell.

Assuming that the *x* and *y* coordinates were made into a single 'composite' number, this could be used as the key for a hashed index, for example, or it might be translated directly into a relative record address of a direct access file. If there was more than one record for each cell then the remaining records could be referred to by a continuation (or next record) pointer within the first and subsequently addressed records. As a simple example of cell addressing, if the entire grid extends from 0 to 100 000 units in each direction, and each cell was 10 units

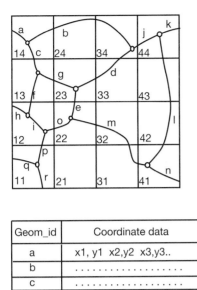

Index File

Cell_Id	Geometry_ids
11	q, p, r
12	o, p, i, h
13	c, f, g
14	a, b, c
21	
22	e, m, o
23	g, e, d
24	b
31	m
32	m
33	d
34	b, d, j
41	m, n, l
42	l
43	l
44	j, k, l

Geom_id	Coordinate data
a	x1, y1 x2,y2 x3,y3..
b	. . . .
c	. . . .
d	. . . .
e	. . . .
f	. . . .

Figure 11.1 Regular grid spatial indexing.

square, the key K for a cell with coordinates $x = 45000$ and $y = 68480$ could be $K = 45006848$. Note that the last digit of each value is redundant since cells are always 10 units apart. To retrieve data from a rectangular spatial window, it is only necessary to derive the address of all cells covering the window. For the window $x_{MIN} = 36210$, $y_{MIN} = 21488$, $x_{MAX} = 41000$, $y_{MAX} = 27555$, the corresponding range of cell addresses would be all those keys whose x components lay between 3621 and 4100 and whose y components were between 2148 and 2755 (inclusive), the most south-westerly key being 36212148.

To retrieve data in order to process a proximity query to find the nearest neighbours of a given object, the regular grid provides a simple strategy. Thus, knowing a 'source' cell containing the given object, its neighbouring objects can be found by examining the contents of the immediately adjacent cells, the addresses of which can be deduced by incrementing or decrementing the x and y components of the address of the source cell. If necessary the search may proceed in a spiral pattern emanating from the source cell.

An advantage of using an index, provided by hashing or B-trees, for a regular grid is that records need only be allocated for cells containing data, whereas a

scheme in which the cell addresses were equivalent to relative record numbers would require that records were present for all cells, whether occupied or not.

If a regular grid is used for storing independent points, there will usually be a one-to-one relationship between a point and the cell it occupies. When allocating a point to a cell, the only ambiguity arises when it lies on the boundary. This can easily be resolved by adopting a rule that places border points in the cell immediately above or to the right of the border. This can then be taken into account when retrieving. Thus if the rightmost edge of a spatial window coincided with a cell boundary, it would be necessary to access the next cell to the right.

In the case of allocating linear geometry to grid cells, the situation is not quite so simple in that line segments can frequently be expected to cross cell boundaries. One solution to this problem is to cut the line at the cell boundary and to store the resulting boundary point twice, in both of the cells which share the boundary. This is not in general a satisfactory solution however, as it will tend to degrade the quality of the data by introducing points that ought to be collinear but are not, due to the numerical imprecision of the computer. If linear and polygonal data are not cut at cell boundaries in regular grids, the data stored in the cells may be references (or pointers) to the storage location of the complete geometric objects, as indicated in Figure 11.1.

When implementing a regular grid scheme, an important issue is the choice of cell size. Because the contents of each cell are stored in one or more records of a file, it would clearly be desirable to match the quantity of data in the cell to the size of the record (or vice versa), in an attempt to avoid wasted space within the record (assuming fixed length records). Since spatial data are rarely uniformly distributed, it will rarely be possible to ensure that the records are always fully occupied since different cells will inevitably contain different amounts of data.

Quadtrees

The problem of selecting a suitable cell size for regular grids can be reduced by adopting a grid structure in which the cell size can be varied to adapt to changes in data density. Quadtrees provide this facility by subdividing a square region into quadrants, each of which may be further subdivided in a recursive manner until the contents of the cells meet some criterion of data occupancy.

The original quadtree data structure has major shortcomings from the point of view of database implementation in that it requires the maintenance of a large number of pointers which may occupy considerable amounts of storage space. The storage overhead due to pointers can, however, be reduced by organising the quadtree node data as a list of records, the order of which is determined by their position in the quadtree (Gargantini, 1982). In the resulting *linear quadtree*, cells can be numbered according to their ancestry, as illustrated in Figure 11.2 (which is explained below).

Once a quadtree is represented in linear form, it becomes well suited to conventional database storage schemes which are oriented to working with 1D indexes of sorted lists of numbers. The linear quadtree numbering sequence has a useful characteristic for the purposes of spatial search in that objects which are regionally adjacent will tend to have similar cell addresses and hence be stored close together in a file ordered by the quadtree addresses. Note that the regular grid addresses described in the previous section create a row-dominated spatial ordering in which horizontally adjacent cells have similar addresses but vertically adjacent cells have addresses that are separate numerically by a quantity related to the length of a row of cells.

The pattern of the linear quadtree numbering sequence is that of a *Peano curve*, which is one of a variety of space-filling curves that may be of interest for indexing spatial data (Figure 11.3). The first use of this particular numbering sequence for spatial indexing is usually attributed to Morton (1966), and it is sometimes referred to variously as Morton sequence, Morton matrix or Morton numbering, while individual addresses may be called *Morton numbers*.

An important property of Morton numbers is that they can be generated by alternating successive bits of each of the binary representations of the x and y coordinates of the lower left corner of the cell to which they refer (Figure 11.4). The process is called *bit interleaving*. When expressed in base four, the Morton numbers correspond exactly to the quadtree addresses illustrated in Figure 11.2(a). As indicated above, these numbers do not by themselves indicate the size of the cells to which they refer. This can be done variously by recording the cell size explicitly (which is rather wasteful of space), by recording the quadtree level, which implies cell size, or by modifying the numbering system to the base five scheme of Figure 11.2(b) in which cell size is apparent from the number of trailing zeros. With this last scheme, the smallest size quadtree cell must be decided in advance, since all addresses will be of the same length. For a general-purpose scheme, the finest resolution is given by the number of significant digits in the grid coordinate system. Since cell dimension reduces by half at each subdivision, it is convenient to regard the entire data space as being square with a maximum dimension which is a power of two. The smallest cell division will then be unity.

It may be noted that bit interleaving can be used for creating the addresses of the fixed-sized cells in the regular grid. To gain the full advantage of doing so it would be necessary to ensure that the grid data records were ordered according to the address sequence. The approach may be regarded as a type of hashing.

The homogeneity criterion, or what is in a cell

When working with raster data, as in the earliest applications of quadtrees, the criterion for stopping subdivision when building the quadtree is that the pixels within a quadrant are all of the same value. This is a criterion of homogeneity. When used for storing points, as in the PR quadtree (Samet, 1990a), the criterion may simply be the number of points that are to be stored in the record corresponding to an individual cell. For storing lines defined by vertices,

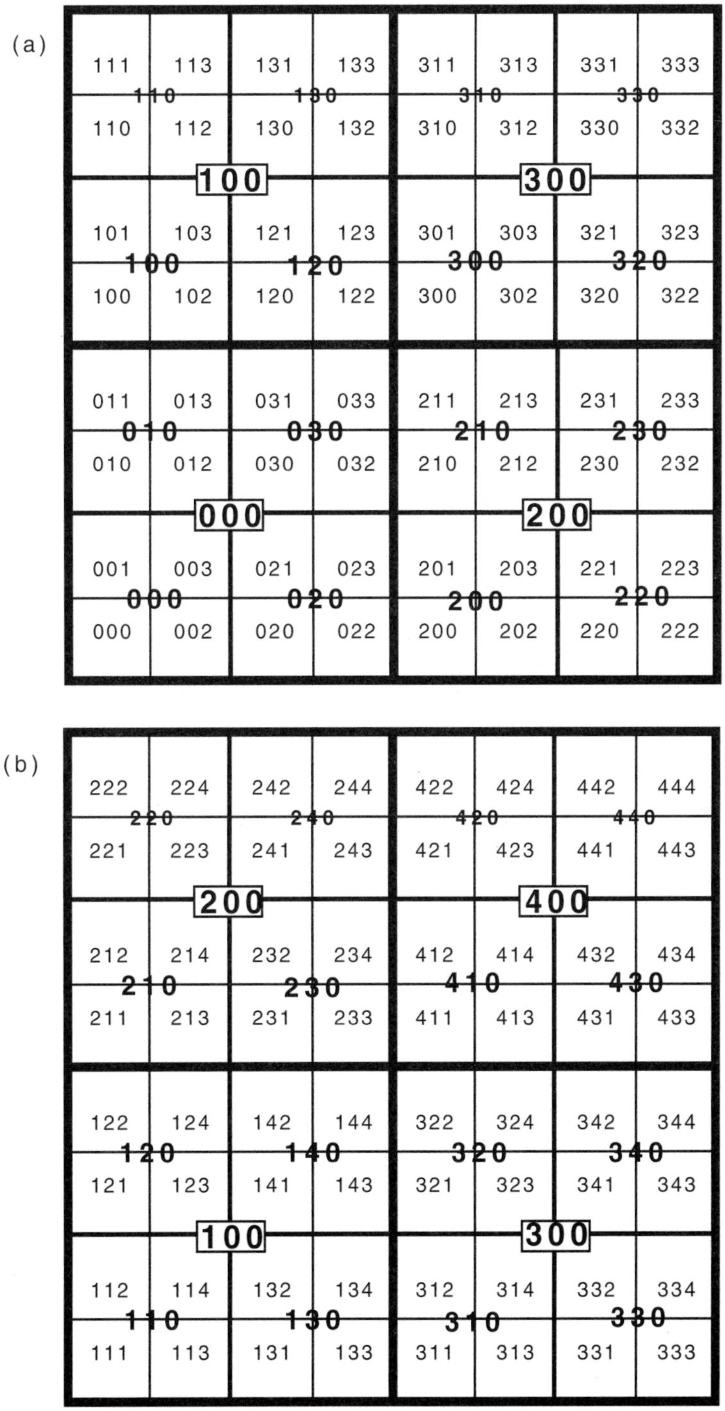

(a)

111	113	131	133	311	313	331	333
110	112	130	132	310	312	330	332

110 · · 130 · · 310 · · 330

100 **300**

101	103	121	123	301	303	321	323
100	102	120	122	300	302	320	322

100 · 120 · 300 · 320

011	013	031	033	211	213	231	233
010	012	030	032	210	212	230	232

010 · 030 · 210 · 230

000 **200**

001	003	021	023	201	203	221	223
000	002	020	022	200	202	220	222

000 · 020 · 200 · 220

(b)

222	224	242	244	422	424	442	444
221	223	241	243	421	423	441	443

220 · 240 · 420 · 440

200 **400**

212	214	232	234	412	414	432	434
211	213	231	233	411	413	431	433

210 · 230 · 410 · 430

122	124	142	144	322	324	342	344
121	123	141	143	321	323	341	343

120 · 140 · 320 · 340

100 **300**

112	114	132	134	312	314	332	334
111	113	131	133	311	313	331	333

110 · 130 · 310 · 330

Figure 11.2 Linear quadtrees. The cell numbering scheme may use base four digits (**a**), in which case some means of recording cell size must be provided, or base five (**b**), in which case the trailing digits (zeros) imply cell size.

(a)

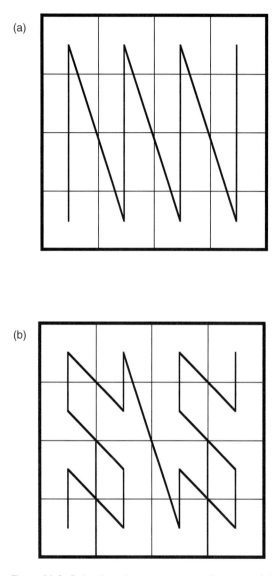

(b)

Figure 11.3 Paths through space corresponding to spatial indexing systems. (**a**) A column-ordering pattern corresponding to the regular-grid indexing scheme illustrated in Figure 11.1. (**b**) The Peano space-filling curve of quadtree indexing (Figure 11.2), whereby cells that are adjacent in space are more likely to have similar spatial index addresses than in column (or row) ordering schemes. Hence, data that are close in space are close in the storage system.

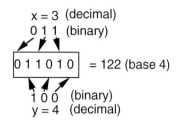

Figure 11.4 Bit interleaving with quadtree addresses. The decimal coordinates (3,4) are converted to binary (011, 100) before being interleaved to produce (011010) which is the quadtree address 122 in base 4 (and 26 in decimal).

either one vertex and the edges which are connected to it, or by a single edge, the vertices of which are in other cells. The parts of edges which cross individual cells are referred to as q-edges. In practice it is convenient for the quadtree record that stores edges to contain pointers to a separate list data structure where the coordinates of the vertices are stored (Figure 11.5). This avoids the storage overheads (and numerical imprecision) which arise when each cell record stores the coordinates of the q-edge's intersections with the cell boundaries, or the coordinates of the q-edge vertices. If the edges belong to polygonal objects, as in the PM3 tree, then the attributes of the regions on either side of the polygonal edges may also be stored in the associated list data structure, rather than in the quadtree cell records themselves.

Searching quadtrees

To search a linear quadtree index, in order to find stored data inside a search window, the window itself may be described in the form of a list of quadtree cells that cover it. It is not necessary for this search list to correspond exactly with the window, provided it covers it entirely. Once stored data cells are found that overlap the search cells, precise comparison can be performed with an exact (vector format) geometric definition of the search window. The search process is driven by the contents of the search list. Initially the stored data must be searched, via its index, for the first record address which matches the first search address. Matching means that the stored data cell is either equal to, or an ancestor (bigger and overlapping), or a descendant (smaller and overlapping) of the search address. For a quadtree search strategy to find the nearest neighbour to an arbitrary point, using a circular search region, see Hoel and Samet (1991).

the criterion may be more complex in that it needs to consider both vertices and the edges which join them. In the PM3 quadtree of Samet and Webber (1985), the criterion is that a cell is occupied by

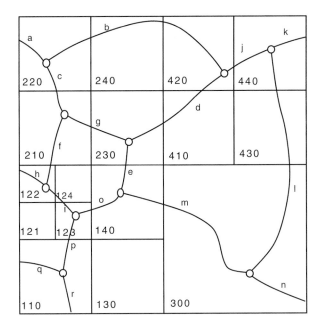

QUADTREE INDEX FILE

Quadtree address	Geometry ids
110	r,p,q
121	
122	h,f,i
123	i,p,o
124	i
130	
140	o,e,m
210	f,c,g
220	a,b,c
230	g,e,d
240	b
300	m,n,l
410	d
420	b,d,j
430	l
440	j,k,l

Figure 11.5 Quadtree spatial indexing for linear features. The geometry ids are regarded as addresses or references to the storage locations of the geometry describing the linear features.

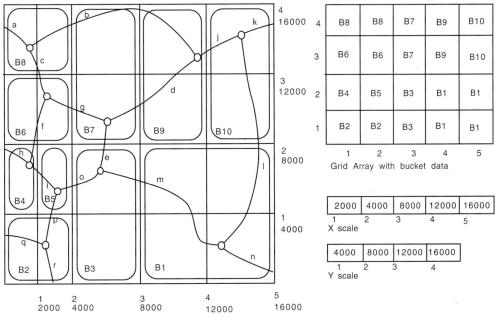

Grid File Directory with buckets indicated as B1 to B8

Figure 11.6 Grid file spatial indexing in which space is partitioned by an irregular grid which adapts to the data distribution.

Octrees

The principle of quadtree spatial indexing can be extended to 3D data with the use of octrees (Gargantini, 1989). The original octrees were based on the voxel spatial data model, just as quadtrees were originally based on rasters. The octree subdivision of space can, however, be used as an indexing method to reference 3D geometric objects of points, lines, polygons, surfaces and volumes, in which case they are sometimes referred to as vector octrees. The terms polytree and exact octree have also been used (Jones, 1989a).

Grid files

An alternative spatial access method to quadtrees is the grid file (Nievergelt *et al.*, 1984) in which space, of whatever dimension, is divided in a slightly less regular manner, but which, like a quadtree, adapts to the spatial variation in data density. The cells of a 2D grid are referenced by a 2D grid array, the elements of which store the address of other data records (called buckets) storing the geometry that is inside or intersects the cell (Figure 11.6).

The geographical dimensions of the grid (in 2D) are defined by a set of vertical and horizontal partition lines. The relationship between real-world grid coordinates of the cells and the grid array elements is maintained by 1D arrays called linear scales. The coordinate values of the x-direction grid lines are stored in one 1D array while those in the y-direction are stored in another. Given a pair of x,y coordinates, the grid directory cell in which they lie can be found by reading through the respective linear arrays to find the first element with a corresponding coordinate value that is greater than that of the coordinate of interest.

A characteristic of the grid file is that a bucket is assumed to be able to store several items of data (actual geometric data, or references to the storage of relevant geometric data) and that several directory cells may reference the same bucket. As data are added to or removed from the database the contents of buckets may need to be modified. It may also be necessary to refine the grid by adding new partitions, or by merging cells together following the removal of data. Changes to the form of the grid are accompanied by updates to the linear scales and to the grid array.

The object-directed decomposition search and data-structuring techniques partition space by means of the coordinates of individual data points or of the extents or bounding rectangles of geometric objects which are to be stored. When applied to linear and polygonal objects, these methods have the characteristic of referencing entire spatial objects, rather than their component parts as is done with some of the regular decomposition methods. There is a multiplicity of object-directed decomposition search methods. It will only be possible here to describe a few of them although an attempt will be made to identify the various characteristics of the methods as a whole.

Two-dimensional binary search for points

One of the best known object-directed decomposition techniques applies the principle of divide and conquer to the organisation and search for point data. This range search technique makes use of a binary tree to order the data with respect to their x and y coordinates (Preparata and Shamos, 1988). Initially a point located approximately centrally within

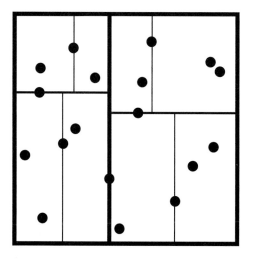

Figure 11.7 Binary search schemes, such as in the *k*-D tree, select individual points to split the data space recursively. Only the first few levels are shown in the diagram (see text).

the range of x coordinates is chosen to partition the data set vertically into two halves (Figure 11.7). In each half another point is chosen in a similar manner to partition the halves along horizontal lines passing through these points. Each new region is itself split into halves, alternately about horizontal and vertical lines. The process of splitting stops whenever a new subregion contains no other points.

The resulting tree can be searched to determine points that lie inside a search rectangle D. Starting at the root, a test is performed to determine whether D lies in one or other of the two regions separated by the point stored in the root node. If D does lie entirely within one side or the other, the corresponding branch of the tree is descended and a similar test performed with the point in that node. If, however, D is found to cross the partitioning line, a test is performed to find whether the point in the node lies inside the window. If it does, it is saved. The search then continues down both branches of the tree before applying the same logic to each of the two branch nodes. The search terminates at individual nodes when the node is a leaf, i.e. there are no branches to descend.

The range search approach can be extended into higher dimensionality by considering planes or hyperplanes which partition the k-dimensional space into two. As the tree is descended, splitting will take place for each dimension in turn. The general tree structure is called a k-D tree (Bentley, 1975; Knuth, 1973), standing for k-dimensional binary-search tree. The k-D tree and its k-D-B variant (Robinson, 1981) have been used for point search problems in cartography and GIS (Openshaw *et al.*, 1987; Doerschler and Freeman, 1989), but their performance with regard to speed of access is not very impressive when compared with that of standard binary tree methods applied to conventional 1D problems (see Preparata and Shamos (1988) for further discussion).

R-trees

The R-tree (Guttman, 1984) is intended for indexing two (and higher) dimensional objects in terms of their minimum bounding rectangles (MBR). Nodes of the tree store MBRs of objects or collections of objects. The leaf nodes of the R-tree store the exact MBRs or bounding boxes of the individual geometric objects, along with a pointer to the storage location of the contained geometry (Figure 11.8). All non-leaf nodes store references to several bounding boxes for each of which is a pointer to a lower level node. These non-leaf boxes record the extent, i.e. enclosing box, of all boxes stored in the lower nodes to which it refers. The tree is constructed hierarchically by grouping the leaf boxes into larger, higher level boxes which may themselves be grouped into even larger boxes at the next higher level. Since the original boxes are never subdivided, a consequence

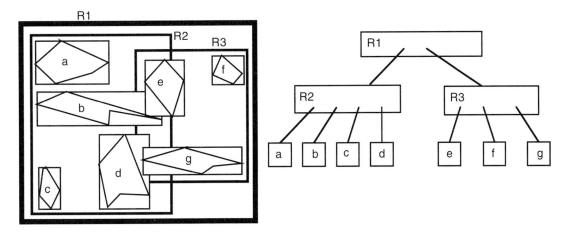

Figure 11.8 R-trees are an object-directed spatial indexing scheme based on a hierarchical representation of bounding rectangles of referenced objects.

of this approach is that the non-leaf node 'covering boxes' can be expected to overlap each other.

Searching the R-tree consists of comparing the search window with the boxes in each node, starting at the root, and following the child pointers of those boxes which are included in or overlap the ranges of the search window. The procedure is continued, possibly down several branches, until reaching the leaf nodes, the contents of which are then tested against the extent of the search window.

Roussopoulos and Liefker (1985) proposed a method of packing the boxes so as to reduce the degree of overlap and hence make retrieval more efficient by reducing the number of nodes that must be visited to retrieve the contents of a given search window. This is called the R*-tree. Overlapping of rectangles is avoided in R+-trees (Sellis et al., 1987); the rectangles are clipped against each other if they overlap, creating additional, smaller rectangles.

Several variants on the R-tree have been devised in which the enclosing boxes are replaced by more tightly fitting shapes which may be polygons or spheres. Examples of these data structures are referred to as polygon trees, cell trees (Gunther, 1989) and sphere trees (van Oosterom and Claasen, 1990).

Multi-resolution spatial data access methods

The accuracy and level of detail of stored geometric data can vary greatly, from large-scale plans of the boundaries of properties to small-scale, highly generalised representations of entire countries. For some purposes the ideal GIS database might be one that could store data at the highest level of accuracy available, but enable users to access the data at different levels of detail and generalisation according to the purpose. This objective is not at present achievable, since procedures do not yet exist with the capability to generalise over all possible scale ranges. The pragmatic solution most commonly adopted is to store several versions of data, derived from different-scale map sources. There is, however, another possibility, i.e. to use multi-resolution spatial data structures that provide access to subsets of the geometry of linear, polygonal and surface features which approximate or generalise the features over certain limited ranges of applicable scales. They cannot entirely remove the need for multiple representations but they may limit

the number required and make multi-scale access more efficient.

One of the earliest multi-resolution data structures is the strip tree (Ballard, 1981), for storing vector-digitised lines (Figure 11.9). This can be regarded as a binary tree, the root of which stores a rectangular strip enclosing the entire line. The rectangle is oriented parallel to the straight line joining the first and last points of the line. The two sons of the root each contain narrower strips that approximate two subsegments of the original line divided at a point where the line touches one side of the length of the root strip. Each node of the tree is further subdivided into two narrower strips. Leaf nodes are strips of a given limiting width, or of zero width, corresponding to the original straight segments of the digitised line.

The strip tree is a main memory resident data structure, which is not efficient in data storage. Several multi-resolution data access methods have been developed specifically for database usage. These include the Multi-Scale Line Tree (MSLT) of Jones and Abraham (1986; 1987), the Reactive Tree (van Oosterom, 1993), the Priority Rectangle (PR) File (Becker et al., 1991) and the Multiresolution Topographic Surface Database (MTSD) of Ware and Jones (1992).

The MSLT uses the Douglas algorithm (Chapter 16) to classify points according to their scale-related priority. Points are given unique identifiers, recording their original order, and are assigned to one of several predefined database levels, each of which corresponds to a certain limiting tolerance of the Douglas algorithm. Each level is spatially indexed with a quadtree. The Reactive Tree also classifies vertices with the Douglas algorithm but stores them in a binary tree (Binary Line Generalisation or BLG Tree), similar to the strip tree but much more space-efficient in that only an individual vertex is stored in each node (rather than a rectangle definition). The BLG trees are themselves spatially indexed by an R-tree related data structure. The PR File separates vertices into priority-related levels, like the MSLT, but creates a spatial index for each level based on the bounding rectangles of small sets of successive vertices in the level.

The MTSD combines a hierarchical triangulated surface data structure, based on the Delaunay Pyramid (Floriani and Puppo, 1988; Floriani, 1989; with linear and polygonal constraints that are also organised hierarchically in a scheme similar to the MSLT. A flexible and space-efficient modification of the MTSD is provided by the use of the Implicit TIN

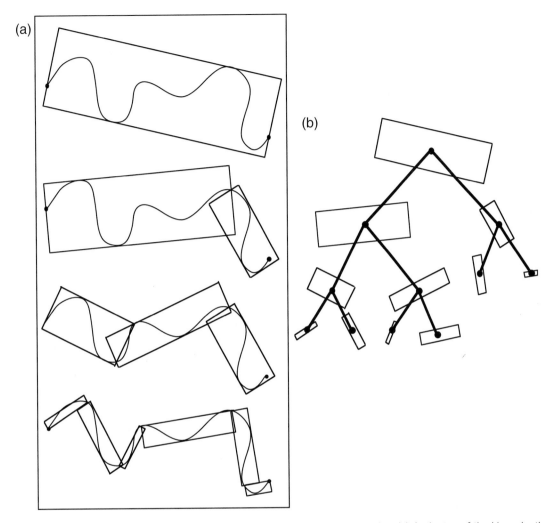

Figure 11.9 Strip trees represent linear features in a hierarchy of rectangular strips. (**a**) At the top of the hierarchy the entire line is represented by a single strip, while lower down, the line is represented in progressively greater detail by narrower strips. (**b**) The strips can be stored in a binary tree data structure, each node of which stores the dimensions of a strip and pointers to more detailed subdivisions. Similar, but more efficient, schemes are used in multi-scale databases.

(Jones *et al.*, 1994) which stores only the vertices of a triangulation, along with triangulation reconstruction algorithms, rather than storing data that explicitly define the structure of the triangulation at any given resolution.

It should be pointed out that the multi-resolution data structures, as described above, are based on the use of relatively crude generalisation algorithms that operate on line, polygon and surface geometry that is approximated by simple subsets of the original set of vertices. They do not encompass generalisation processes that involve major changes in symbolic

form or the displacements that are often involved in map generalisation (see Chapter 16).

Geometric intersection and determination of topological relationships

In this section, we consider the process of geometric intersection between vector-structured geometric objects. As indicated earlier, this is required to determine topological relationships between geometric objects and to find the subsets of geometric data that satisfy particular relationships of equality, containment, overlap, adjacency and separateness. The term intersection is used fairly broadly here to cover the problems of clipping geometry against spatial windows, of polygon overlay and of testing topological relationships. Thus we include consideration of the particular tests for point-in-polygon and line-in-polygon as they may be regarded as determining the result of the intersection between the respective pairs of geometric objects.

We confine ourselves to consideration of the 2D objects of points, lines and polygons. Figure 11.10 illustrates results of the relationships for various combinations of points, lines and areas. Note that in general the result of an intersection is either null (which we do not illustrate) or a geometric object that is of equal or lower dimensionality than each of the objects being intersected.

The procedures that are described in this section are those which, in addition to answering spatial queries, are required for building topological data structures following data acquisition. The preliminary data structuring that may be carried out in conjunction with data acquisition assists greatly in determining the result of data retrievals that we are concerned with in this chapter. However, a spatial query can introduce a spatial window that requires an intersection operation that cannot have been predicted at the time of data structuring. Thus it is not possible to structure data in advance to meet all possible queries except in quite

specific applications that do not require arbitrary clipping against spatial windows.

Geometric intersection: general considerations

The complexity of geometric intersection varies somewhat according to the nature of the objects being compared with the search window. Point objects are perhaps simplest in that they will be found, subject to a tolerance value, to be either inside or not inside the region of search, although a rule must be established to decide whether a point lying on a boundary is to be regarded as included or not. Linear objects may be either inside, outside, or part in and part out. In the last case it is necessary to clip the line at the boundary. For sinuous lines, this could result in fragmenting the line such that more than one part of it was included. The greatest complexity occurs when searching polygonal objects since their intersection with the boundary of the search polygon gives rise to the need, in vector models, to construct one or more included polygons, the edges of which consist of a combination of their original edges and parts of the boundary of the search polygon.

The phrase *polygon overlay* refers to the general case of superimposing one polygonal map on another polygonal map to form a new map in which new polygons may be created from the arcs resulting from the intersection of all boundaries on the two original maps. If certain polygons of each map are regarded as the search window or windows, i.e. the

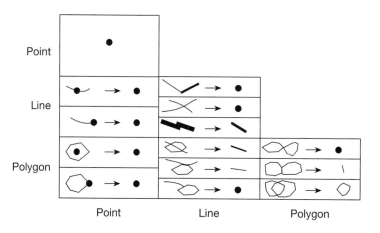

Figure 11.10   Intersection relationships for points, lines and polygons.

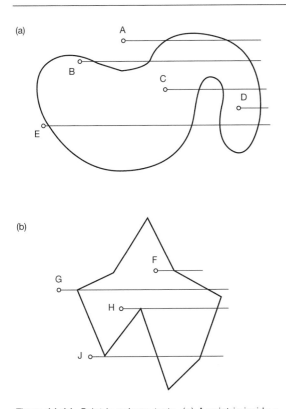

(a)

(b)

areas of interest, then the output required from the overlay operation is that set of polygons which constitute the area of overlap between the respective search polygons.

Point-in-polygon tests

The first stage of a point-in-polygon test is to compare the point with the polygon's minimum bounding rectangle. From the database search point of view, this part of the problem can be treated as a range search. Having found a candidate point, the test for determining inclusion consists of extending a straight line in any direction from the point and counting the number of intersections of this line with the edges of the polygon. An odd number of intersections will be found if the point is inside, and an even number if it is outside (Figure 11.11a). The intersection calculations are simplified somewhat if the extended line is horizontal or vertical. We will assume here that it is horizontal, parallel to the x-axis.

This apparently simple point-in-polygon test is complicated by the fact that if an intersection coincides with a vertex of the polygon it may be counted twice, since the vertex will belong to two successive edges. The problem can be avoided by checking whether each intersection with the horizontal line is with a vertex and counting it if the vertex is at the lower end of the line. Here 'lower' refers to the minimum y value. However, the test will work equally well if we choose the higher, maximum y, end instead. This may be appreciated by reference to the diagram in

Figure 11.11 Point-in-polygon tests. (**a**) A point is inside a polygon if a straight line extended infinitely in one direction from the point intersects the polygon boundary an odd number of times (points B, C and D), otherwise it is outside (points A and E). (**b**) Care must be taken with digitised data to avoid counting the same boundary twice if the search line passes through a vertex (points F and G). If the vertex is a maximum or a minimum however, the problem does not arise (points H and J).

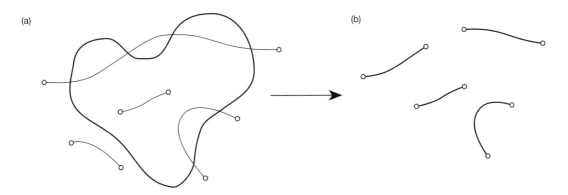

(a)

(b)

Figure 11.12 Clipping a line against a polygon. When the polygon is concave, as in (**a**), then the result of clipping a line feature may be either nothing, part of the line, the whole of the line or multiple parts of the line (**b**).

Figure 11.11(b), where we see that counting vertices either twice, or not at all, at maxima or minima, does not affect the basic rule, whereas for all other vertices, only one of the edges should be counted.

Line-in-polygon tests

Given a linear feature defined by a sequence of vertices, we can say that the output of the line-in-polygon test should be either nothing, if the line is outside, or one or more linear features which consist either of the complete original linear feature, if is entirely inside, or of separate fragments (Figure 11.12a). We must be prepared, therefore, to create new output linear features, the start and/or end vertices of which may be found from the intersection of the original line with the polygonal search window (Figure 11.12b).

As with the point-in-polygon test, the first stage of the test for inclusion of a linear feature involves a comparison with the polygon's minimum bounding rectangle. Once the potential for overlap between the linear feature and the polygon has been established, it is necessary to test each successive straight edge of the line with the polygon. It may be appreciated, with reference to Figure 11.13, that in the general case there do not appear to be any simple rules which allow us to prove that a straight edge inside the minimum bounding rectangle is either entirely inside or outside the polygon. In particular, the fact that both vertices are outside the polygon (cases A, B and C in the figure) does not mean that the whole edge is outside. If both vertices are inside (cases D and E), then it is only possible to say that the line is also entirely inside if the polygon is convex. For most geographical situations, search polygons cannot be assumed to be convex.

Efficient solutions to the line-in-polygon test have been developed for the special cases of convex polygons and for rectangular convex polygons. The test against a rectangular window is used extensively in computer graphics to display graphical data within a screen viewport, and it is frequently required in GIS. The Cohen–Sutherland clipping algorithm was one of the earliest algorithms developed for this purpose. It is characterised by the use of a strategy to categorise end points of a straight-line segment according to their relationship with the window (i.e. above, below, left or right). A set of rules then identifies the simple cases of whether the line is definitely inside or definitely outside the window. Thus if both points are above (such as F in Figure 11.13), below, right of, or left of the rectangle then the line is outside, and if both points are inside then the entire line is inside. If no such condition applies (such as B in Figure 11.13), the line is progressively clipped back to each of the four (infinitely extended) window boundaries, until the remaining line segment does satisfy one of the simple tests. Algorithms for general convex polygons and for concave polygons become progressively more complicated (see Foley *et al.* (1990) for details in a computer graphics context).

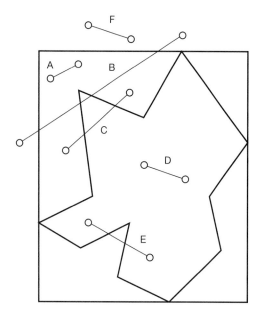

Figure 11.13 Line-in-polygon tests. The minimum bounding rectangle can be used to exclude edges (F) from further consideration. If an edge wholly or partially intersects the rectangle boundary then more detailed tests must be performed to determine whether the edge is inside the polygon. The fact that both vertices are outside the polygon does not prove that the edge does not intersect it (see edges A and C). If both vertices are inside the polygon (edges D and E), the entire edge is only definitely inside if the polygon is convex. If there are concavities then the edge may be only partially inside (E).

Polygon overlay

The previous problem of determining the inclusion of a linear feature within a polygonal search window serves as an introduction to the general problem of polygon overlay. As we saw in Chapter 3, polygon data are typically represented using topological data structures in which individual polygons may be described by a series of links or arcs consisting of

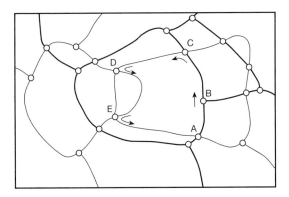

Figure 11.14 Overlaying two polygonal regions requires finding all intersections between the arcs belonging to each layer. Individual polygons can then be constructed by following arcs from each node, such as A, turning left at each successive node encountered (B, C, D, E), until returning to the start node.

chains of coordinates terminated by nodes. A topologically encoded map consists of a network of arcs, each of which may be associated with left and right attribute codes which identify the polygons they bound, either with a polygon identifier or with actual attribute values. Overlaying one polygon map on another requires calculating intersections between the component chains in order to create new links which bound the newly created polygons. The output of the overlay operation then consists of those polygons whose attributes meet the overlay criteria.

It is assumed here that overlay is performed between polygon maps that have been retrieved from within a spatial window on a database. The first stage of the general polygon overlay problem requires that we find all linear fragments which result from the intersection of the two polygon networks (Figure 11.14). We can do this by taking each chain of the first network, drawn in the figure with a thick line symbol, and testing it for intersections with each chain from the second network, drawn with a thin line symbol. If, for an individual chain, no intersections are found, the chain is retained as it is. If intersections are found, however, all chains concerned, from both maps, must be subdivided at the locations of the intersections to create new shorter chains. The new intersection vertices must be denoted as nodes and inserted into a node table which records all nodes, both old and new. The node records store the identifiers and associated attributes and directions of all chains that are connected to them. To assist in the polygon reconstruction process, the references to all chains emanating from each node should be stored in angular order.

To perform the reconstruction operation, let us assume that polygons are to be defined by an anti-clockwise ordering of their component chains. Starting with any node (such as node A in Figure 11.14), we take two angularly adjacent chains, regarding the first as entering the node and the second as leaving it. We then search the other nodes to find one that references the leaving chain, which we regard as entering this second node (which is labelled B in Figure 11.14). The next chain in anti-clockwise order from this second node will be the next exit chain. We continue to step from node to node (C, D and E in Figure 11.14) in this manner until we find the chain which entered the first node. Provided that chains are flagged once they have been processed in this way, it will be possible to continue building polygons until there are no chains left unprocessed. As each new polygon is constructed it may be labelled with the attributes of the left side of the chains of which it is composed.

The result of the above process is a set of polygons created by the overlay. If, as will generally be the case, we are only interested in polygons with particular attributes or combinations of attributes, the process may be speeded up somewhat by only following chains which meet the criteria, in which case only the required polygons will be constructed.

The procedure described assumes that the overlaid polygon networks do not include isolated holes or islands inside polygons. If such islands are present, then the explicit definition of polygons that have internal as well as external boundaries requires additional tests to establish the nesting of islands within their parent polygons (Kirby *et al.*, 1989).

Major problems arise with implementing polygon overlay procedures due to the inherent error in digitised data and the limited precision of computer storage of numbers. It is quite common for maps that are to be overlaid to include features, such as nodes or entire arcs, or parts of arcs, that are equivalent, in the sense that they represent the same real-world phenomena. Unfortunately, due to the error in the representations, these features may not be geometrically equivalent. The result is that the overlay procedure may introduce spurious features. These can include additional nodes where two or more nodes are logically equivalent but different in location. They can also include sliver polygons, where two logically equivalent line segments have been intersected and may cross back and forth along their length.

Some GIS packages attempt to overcome these problems by using distance tolerances which specify that if two features are within that distance of each other they are to be regarded as equivalent. Unfortunately tolerances can result in the amalgamation of features that are further apart than the tolerance, due to the process of creep, referred to in Chapter 5. Thus point A can be merged with point B, to form a point X which is within tolerance of, and hence merged with, another point C, even though C is further than the tolerance distance from A. The solution to problems like this should ideally take account of metadata regarding the quality and meaning of the data, rather than operating on purely geometric principles. As Chrisman (1987a) has pointed out, when equivalent linear features are being overlaid, the best solution may be to select the better quality line and discard the other, rather than produce a merged feature which is degraded relative to the better quality feature.

Plane sweep algorithms

A major part of the computation in polygon overlay is concerned with determining intersections between line segments. The number of intersection tests can be minimised by sorting the data in one or more spatial dimensions. A classic computational geometry approach to the problem of determining the intersections of the bounding lines of polygons is the use of plane sweep techniques. The approach is as follows. The vertices of all line segments of the two maps to be overlain are sorted by their x coordinates. Starting at the leftmost point, a vertical line is moved progressively rightwards while maintaining data records of all line segments that intersect the sweeping line. At any given position of the sweep line, edges of the maps that intersect the sweep line and are adjacent to each, are tested to determine whether they intersect each other. If they do, then the point of intersection is added to the set of vertices in appropriate sorted order. Note that all possible intersections will be found as a result of this procedure, since all pairs of straight-line segments that do intersect each other will at some stage be found to be adjacent to each other on the sweep line. Testing all sweep-line adjacent edges will ensure that all intersections will be found while avoiding testing for impossible intersections.

In the course of a plane sweep algorithm the sweep line moves in steps determined by the successive sorted x coordinates of points (whether original points or found intersection points). In addition to simply determining intersections, the progression of the sweep line can be accompanied by a polygon-building procedure that keeps track of the polygons intersected by the sweep line. At each successive step of the sweep a new edge may be added to a polygon, or it may be completed, or a new polygon may be started.

Grid indexing for overlay

A rather simpler alternative to the plane sweep approach, for the purposes of improving efficiency in intersection calculation, is to superimpose a conventional spatial index on the geometry of the line segments. If this is a simple regular grid, each edge inside a grid cell need only be compared against other edges inside the same cell. Franklin (1989) has found this to be a very effective method and it has the advantage that it lends itself to implementation using parallel processing, since the preliminary search for intersections can be carried out independently for each grid cell.

Summary

In this chapter we have considered spatial data access with regard to techniques for spatial indexing and for geometric processing required to answer spatial queries. Spatial indexing methods are used to provide efficient retrieval of spatial data from a large database. They work by partitioning the data, either by regular decomposition methods which superimpose a grid on the spatial data and record what is present in each cell, or by object-directed decomposition in which the spatial partitions are more precisely adapted to individual spatial objects. In regular decomposition methods, such as those based on quadtrees, addresses must be allocated to each of the partitions or cells. These addresses reflect the spatial ordering that the partitions impose on the data and, because they have a natural sorting order, they can exploit conventional indexing schemes such as B-trees. The object-based decomposition methods have resulted in specialised indexing schemes, notably R-trees, which like conventional, non-spatial indexing schemes, are hierarchical, but use criteria of spatial containment and overlap to determine the branching of the tree data structure.

Some spatial queries can be answered on the basis of the retrieved data, which may include stored topological data obtained by spatial indexing. However, spatial containment searches and phenomenon-based queries referencing several classes of data, depend upon the use of specialised geometric procedures that may clip spatial data against a spatial window, or intersect polygons from multiple layers or spatial objects. These procedures use techniques developed in the context of computer graphics and they also require polygon-overlay techniques adapted specifically to geographical data.

Further reading

The two books by Samet (1990a,b) provide detailed accounts of techniques for spatial indexing, with particular regard to the use of quadtrees. For collections of articles on 3D data manipulation, see Raper (1989), Pflug and Harbaugh (1992) and Turner (1992). Algorithms for clipping and some aspects of polygon overlay can be found in Foley *et al.* (1990) and Rogers (1985). For an account of the development of an early polygon overlay procedure for GIS, see Chrisman and Dougenik (1992). Preparata and Shamos (1988) is one of the standard texts on computational geometry and deals with some aspects of spatial data indexing, as well as describing some of the techniques involved in polygon overlay. Franklin (1984) and Greene and Yao (1986) are notable for bringing attention to the problems that arise due to the limited precision of computational procedures for geometric processing. For an approach that addresses these problems of precision in GIS databases, see Guting and Schneider (1993). For purposes of processing proximal spatial queries, i.e. those concerning neighbourhood, data structures based on Voronoi diagrams and Delaunay triangulations have a lot of potential in GIS, as Gold (1994) and elsewhere has emphasised. Okabe *et al.* (1992) have written a textbook on the subject; see also Okabe *et al.* (1994) for a shorter article. For an example of using constrained Delaunay triangulation for localised search, see Ware *et al.* (1995).

PART 4 Spatial Data Modelling and Analysis

CHAPTER 12

Surface modelling and spatial interpolation

Introduction

Early on in this book we made a distinction between object-based and field-based models, noting that the former were primarily concerned with the form and location of individual phenomena, while the latter were concerned with representing the continuous distribution of phenomena. In this chapter we pursue the subject of the field-based view of phenomena, looking at the ways in which they are represented and analysed in surface models. The term 'surface modelling' is used to here to refer to a range of techniques for manipulating data that are regarded as representing a field. As we have pointed out earlier, in the geographical sciences there is a general distinction between two types of surface model, differing according to the phenomena represented. The first is for the purpose of modelling *physical surfaces* of terrain and geology. The second is for modelling *abstract surfaces* which represent the distribution of less tangible, spatially variable statistics. Many surface modelling techniques are common to these two applications. Terrain models can in fact be regarded as a class of statistical surface model in that they are based on a sample of numerical observations from which the 'real' surface must be inferred.

Important applications of physical surface representation arise in the use of terrain models for landscape visualisation, for hydrological analyses, soil studies, radiocommunications and in civil engineering. Landscape visualisation is required to help in assessing the visual impact of architectural and civil engineering construction works. Hydrological analyses use terrain models to predict the flow of water on the earth's surface and hence determine catchment areas, river volumes and flood zones. Soil scientists use terrain models to predict the stability of soils on the basis of ground surface gradients. Radio-communication networks can be planned using terrain models to evaluate the effects of topography on radio transmission. Civil engineers use terrain models to site engineering works and to determine volumes of earth which may need to be moved. For several decades geologists have been using surface modelling techniques to interpret subsurface data from boreholes and seismic surveys.

Abstract surfaces are used in socio-economic studies to understand the variation in phenomena such as population density, which itself may be used to understand patterns in other factors such as the occurrence of disease or the accessibility of populated areas to given facilities. An early example of abstract surface modelling in geography was to study the variation of land use relative to centres of population (see e.g. Hagget *et al.*, 1977). Environmental studies of air and ground pollution and of the distribution of soil types and minerals also depend heavily upon the use of abstract surface modelling tecniques. Although the surface refers to physical phenomena, the surface model is intended to be an abstract representation of local and broader trends in the variation of the phenomena, rather than a precise description of its distribution in space, which in any event is necessarily three-dimensional. In geophysical studies, surfaces are used to represent potential fields of gravity and magnetics, and are often used in combination with physical surface modelling techniques for the associated geological strata.

In this chapter we start by reviewing various methods for describing surface models. We distinguish between the representations based on discrete samples of points, lines or networks and complete, mathematically defined surfaces, based on triangulations and on gridded sets of points. A section then follows on the construction of triangulated irregular networks (TINs), given an initial set of irregularly distributed point samples. This includes the issue of selecting the most important subset of points for lower resolution triangulations. We then introduce simple techniques of interpolating unknown values from given sampled values, before introducing the concept of spatial autocorrelation, which underlies a theoretical understanding of interpolation techniques. This serves to introduce the widely used and theoretically optimal interpolation technique of kriging. The last section introduces concepts of interpolation between areal units, which is essential to the modelling of socio-economic data.

Surface representation methods

When referring to phenomena distributed on the earth's surface, a surface model is intended to provide a description of the phenomena at all locations within the region of interest. The original measurements are usually discrete samples which are either very localised, at points or along lines, or, if they cover areas, are local aggregations which may homogenise underlying structure or variation. Sample data of this sort are *incomplete representations* of a surface.

Complete representation of a surface may either take the form of a contiguous set of zones, each of which has a single value associated with it, or it may consist of mathematical functions that can be evaluated at all locations in the region. It is notable then that while only *contiguous zones* and *mathematical functions* provide complete descriptions of a surface, it is often convenient to store a surface in a sampled form that can be used, in combination with mathematical functions, to infer a complete surface. Figure 12.1 provides an overview of surface representation methods. We will now look in more detail at the various types of representation, both incomplete and complete.

Sampled (incomplete) surface representations

Surface representations consisting of discrete samples are usually based upon points or lines or a combination of both. Incomplete areal samples that consist of non-contiguous areal zones are not common and will not be discussed further.

Point samples

Point-sample representations may be either irregularly or regularly distributed. The points in an irregular distribution will normally each be defined by x, y and z coordinates. However, the most common point-sampled surface is a regular grid, in which, because of the fixed grid interval, the x and y coordinates are implicit. Thus knowing the origin of the grid and the interval, it is only necessary to store the z value of each point.

If the grid spacing of a regular grid is sufficiently fine to record all significant detail, it is very likely there will also be regions of the surface of low variability that are recorded by an unnecessarily high number of grid points. This can lead to an undesirable overhead in data storage. Increasing the grid spacing, on the other hand, may result in an undesirable error in regions of high variability.

A variation on the regular grid is an irregular grid in which the grid interval changes between different parts of the surface, according to the degree of local variation. This is referred to as progressive sampling (Makarovic, 1973). Grids and irregular distributions may be generated by the original sampling process, such as from a field survey, or from a regular sampling procedure on a photogrammetric device. Rectangular grids are frequently derived by interpolation from originally irregularly sampled points.

Linear samples

Contours, or isolines, are the most common form of linear sample surface representation. They can be generated either manually or automatically by interpolation from regular and irregular point samples. Alternatively they may stem from the original survey method, such as when traced photogrammetrically on a stereo plotter. Another form of linear surface sample is a structure line, which may define ridges, valleys and breaks of slope. Structure lines may also be derived

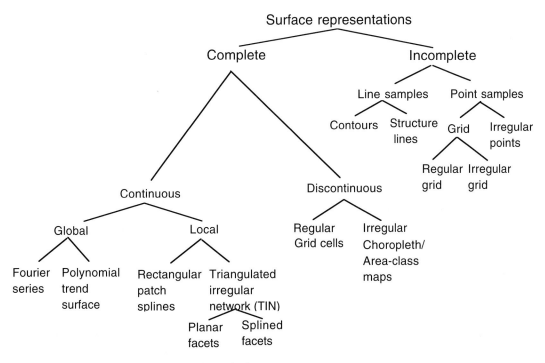

Figure 12.1 Overview of surface representation methods.

from the analysis of regular and irregular point samples, or they may be original surveyed features.

Polygonal networks

Polygon networks may be regarded as interpretations of point-sampled data, in that the points are connected together to form polygons, thereby imposing explicit relationships between upon the points. Categories or values associated with the points may change in some way along the length of the boundary, from one of its bounding points to the next. It is common practice, for purposes of visualisation, to connect the rows and columns of a regular grid with straight lines, to form a pattern of rectangular cells. Such a polygon network can be displayed in the form of a wire net, or mesh, or indeed as individually coloured quadrilateral polygons. However, it should be realised that in three dimensions, the polygons will not in general form planar facets, unless the four corners happen to lie on a plane. Thus the regions inside the polygons are undefined without combining the representation with an interpolation or mathematical surface-fitting procedure.

Complete surface models

Polygonal zones

A polygonal zone surface segments space into polygons, each with a single distinct category. The zones may be irregularly shaped or they may conform to a regular tessellation such as a square grid, i.e. a raster. Due to the possibility of abrupt changes at the polygonal boundaries, it is a *discontinuous representation*. *Choropleth maps* and *categorical maps* are examples of irregular polygonal zone surfaces. The polygonal boundaries may coincide with some pre-existing areal unit, such as an administrative region (county, state, etc.), or they may have been inferred from a study of the pattern of the phenomenon of interest. Examples of inferred zones arise in soil and geological mapping and in land-use studies. In such applications a zone may be allocated on the basis of what appears to be the dominant category, even though it may not be uniform within the zone. The relevant value might be chosen according to several possible criteria, including the mean, the mode, the median or the total of some included values. The most common example of a *regular polygonal zone surface* is provided by the raster data model. The computer storage for a raster may be identical to that of the regular grid point sampled surface. The difference is that each data value is regarded as corresponding to a square cell of uniform value, rather than to a point.

Triangulated irregular networks (TIN)

A very versatile approach to surface representation is to triangulate the sample points to produce a triangulated irregular network (TIN). The resulting triangular facets of a TIN are usually regarded as planar and will therefore constitute a fully defined, continuous model of the surface (Figure 12.2). A major attraction of triangulated networks is the fact that they can incorporate original observations, unlike regular grids and rectangular polygon networks which are typically interpolated from original observations and are therefore more liable to error.

Another benefit of triangulated networks is that the density of sampling is adaptable to the original data source. Thus there may be closely spaced points with small triangles where the original surface is highly variable, while points may be sparse, with large triangles, where the surface is relatively flat or of constant slope. In contrast, regular grids are prone to over- and under-sampling of original surfaces that have variable degrees of roughness and

undulation. Although TINs are usually regarded as defining a surface by planar facets, this need not necessarily be the case and, for purposes of visualisation and interpolation, the TIN can be used as the basis of a set of quintic surface patches that are continuous in gradient from one triangular region to the next (Akima, 1978).

Delaunay triangulation

In triangulating a set of sample points it is usually desirable to create triangles that are as equiangular as possible. In doing so, individual triangles will tend to be locally representative of the value of the surface. Another desirable characteristic of a triangulation procedure is that it should produce a unique triangulation independently of the starting point or orientation of the data set. Thus its results will be predictable and easily repeatable. It has been shown by Sibson (1978) that, in general, the *Delaunay triangulation* meets these objectives, though there are certain configurations of sample data (such as a rectangular grid) that can result in at least locally non-unique solutions.

The Delaunay triangulation is closely related to the *Dirichlet tessellation* of a set of points, which partitions them by a unique set of polygons called Thiessen polygons, or the Voronoi diagram, referred to in Chapter 3. Thiessen polygons enclose each point by regions in which all locations are nearer to that sample point than to any other sample point

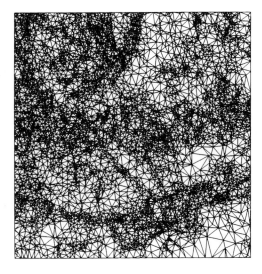

Figure 12.2 (**a**) An example of a triangulated irregular network and (b) the corresponding contour map. Courtesy of D. B. Kidner. ©Crown copyright.

(Figure 12.3a). Each edge of each polygon is a perpendicular bisector separating the enclosed point from a neighbouring point in an adjacent polygon. If all such neighbouring points are connected together by edges, the Delaunay triangulation results (Figure 12.3b). A notable characteristic of each triangle in a Delaunay triangulation is that the circumcircle through the triangle vertices contains no other vertices in the triangulation.

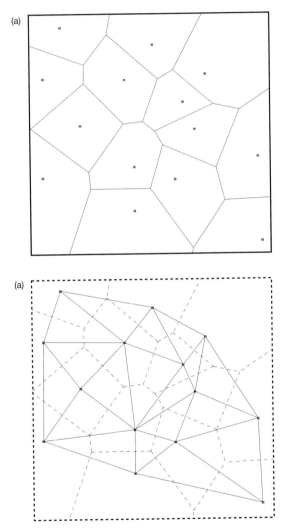

Figure 12.3 (**a**) Thiessen polygons define the regions of space nearer to each of a set of sample points than to any other sample point. The set of polygons is called the Voronoi diagram. (**b**) The set of edges connecting neighbouring sample points define the Delaunay triangulation, which is the dual of the Voronoi diagram.

Selection of critical points and lines from a gridded surface

The construction of triangulated surfaces depends upon the existence of a set of points to be triangulated. The points may be originally surveyed data, corresponding in location, for example, to boreholes or ground-survey stations. Alternatively, if the purpose of triangulation is to create an efficient and convenient data structure, the points may have to be selected from an existing surface representation. This is often likely to be in the form of a regular grid.

Assuming that a storage efficient representation is required, the objective of point selection from gridded data is to find the smallest number of points which, after triangulation, can represent the original gridded surface to within a given tolerance. It is generally assumed that the appropriate points to select are those which are located at maxima and minima and at breaks of slope. Such points correspond topographically to peaks, pits, ridges, valleys and saddle points (passes). In hydrological studies of terrain models, it is of interest to identify such features since they can be used to delineate drainage channels and regions of runoff. If, however, the entire surface must be represented equally accurately, it may be necessary to select less obvious points in addition to these other points. Methods of detecting the critical points can be classified as local and global, depending on how much of a data set is examined in order to classify and select individual points.

Local processing

Most critical point detection methods are local and involve classifying each grid point in terms of its relationship with its immediately adjacent points, typically by means of a 2×2, 3×3 or 5×5 matrix located on the point under consideration. Jenson (1985) has described a simple procedure for identifying the valleys or channels of a drainage network. Using a 3×3 matrix, each point within a gridded data set is examined to determine whether it can be classified as a drainage cell (Figure 12.4). The criterion is that the point be lower in elevation than each of any two neighbouring points which are not themselves adjacent. Jenson's method of creating drainage channels can also be applied to the detection of ridges by looking for points that are higher than two symmetric neighbours.

Peucker and Douglas (1975) experimented with a variety of point-selection techniques and found that the most effective was based on a surface-climbing

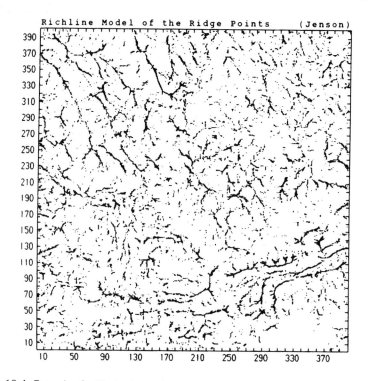

Figure 12.4 Example of critical points selected using Jenson's technique with a 3×3 mask. Courtesy of D. B. Kidner. ©Crown copyright.

approach. The idea is that if the highest points of small sets of cells are labelled throughout the data set, those points which remain unlabelled must lie in channels ('ravine-lines'). Repeating the process by labelling the lowest points will similarly reveal the ridge lines.

All of the above methods tend to produce rather blurred linear features that are often more than one pixel wide. If the intention is to create structure lines then the techniques of thinning, filling, simplifying and smoothing, referred to in Chapter 8, can be applied.

The selection of peak, pit, ridge and valley points may not be sufficient to reconstruct the original surface accurately. A solution to this problem, described in Fowler and Little (1979), is to make an initial selection of points which, after vectorising and simplification, are triangulated. Points in the original grid are then compared with their equivalent values in the triangulation by interpolating within the triangles. If the error exceeds a given tolerance, the point in the grid which identified the largest error within a particular triangle is added to the critical point data

set, which is then retriangulated. The process is repeated until no errors exceed the tolerance. A similar approach is used by Floriani (1989) in creating a pyramidal data structure, each layer of which consists of a triangulation that is nowhere greater than a specified tolerance from the original complete set of data points.

Lee's drop heuristic method

Lee (1991) has provided a useful comparison of TIN construction methods and has presented a quite computationally demanding method of point selection based on the so-called drop heuristic. The approach is the opposite to that of Fowler and Little (1979) and of Floriani (1989), in that rather than starting with a sparse selection of points and adding what appear to be the most significant ones, he starts with a complete triangulation of all points in a grid and progressively eliminates points. The criterion for elimination of a point in this iterative process is that its removal causes the least error when compared with all possible other points. Error here is calculated as the vertical distance between the point after its elimination and the newly triangulated surface without the point. In order to eliminate a single point, each point in a triangulation must be systematically removed, the surface temporarily retriangulated and then the point replaced before choosing that point associated with minimum error. Although the great number of retriangulations required may be performed using a localised triangulation update procedure, this is a computationally demanding procedure. However, it has the definite advantage of constraining the error in the resulting surface.

Mathematical functions

Mathematical functions are very important in the description of surfaces because they provide methods of inferring the presence of phenomena at all locations in space, even though they may only have been measured originally at particular discrete locations (which may be points, lines, areas or volumes). Thus mathematical methods can be used to interpolate between known sample points providing estimates at the unsampled locations, and they can be used to provide approximating surfaces which describe the nature of variation in measured phenomena without necessarily fitting the original samples exactly.

Precise interpolating functions, which honour original sample points, are most useful when the sample measurements are known to be of high accuracy and very valuable in their own right. This usually applies to structural surface measurements such as digital elevation models or highly localised measurements of physical or chemical phenomena. *Approximating functions* are useful when the original observations are generalised somewhat; for example, in the measurement of population in a given region such as a census enumeration district, within which the density of population may in fact be quite variable. Approximating functions may also be applied to model accurate and highly localised data if it is of interest to understand trends or patterns in the variation of the phenomena.

In the context of storing representations of surfaces, mathematical functions may be used implicitly, as indicated in the previous section, in association with discrete representations such as point samples and TINs. Thus a particular mathematical method (such as weighted averages or linear interpolation) is applied to the permanently stored discrete observations. Alternatively, it is possible to store the mathematical representation explicitly, in which case it is the parameters of derived local or global mathematical functions that are stored, rather than (or in addition to) the original sample data.

Global mathematical functions
Polynomial trend surfaces Early interest in the idea of fitting mathematical functions to geographically referenced data was often directed towards the study of trend surfaces (Davis, 1986). By approximating an entire surface by a simple polynomial function, its primary characteristics can be studied and deviations from the trend can be identified by examining a map of residuals which represent the difference between the original surface and the fitted trend surface. The extent to which a trend surface matches the actual surface depends upon the degree of variation in the original data and the degree of the polynomial (Figure 12.5). Clearly, highly variable surfaces need high-order polynomials to model them closely.

The formula for a first-order polynomial surface is

$$z = ax + by + c$$

This is the function describing a plane. The equation for a second-order (quadratic) surface is

$$z = a + bx + cy + dx^2 + exy + fy^2$$

The expression for a third-order surface is given by adding terms for $x^i y^j$, where the sum of i and j is 3, i.e. for x^3, y^3, x^2y and xy^2, hence

$$z = a + bx + cy + dx^2 + exy + fy^2 + g x^3 + h x^2y + i xy^2 + jy^3$$

The coefficients are calculated by the least-squares method whereby the sum of squares of the differences between the z values of the polynomial surface and of the original data points is minimised.

Estimation of the coefficients of the first-order trend surface requires inverting a 3×3 matrix, while for the second order the matrix to invert is 6×6. Higher order trend surfaces become very much more demanding to compute. How well a trend surface fits the original data can be measured by examining the sum of squared differences between the trend surface and the data points. A measure of goodness of fit, R^2, can be found by dividing the sum of squares due to the trend surface (SST) by the total sum of squares of the original data (SSo), i.e.

$$R^2 = SST/SSo$$

where

$$SST = \Sigma \ (z_{trend} - mean(z_{trend}))^2$$
$$SSo = \Sigma \ (z_{obs} - mean(z_{obs}))^2$$

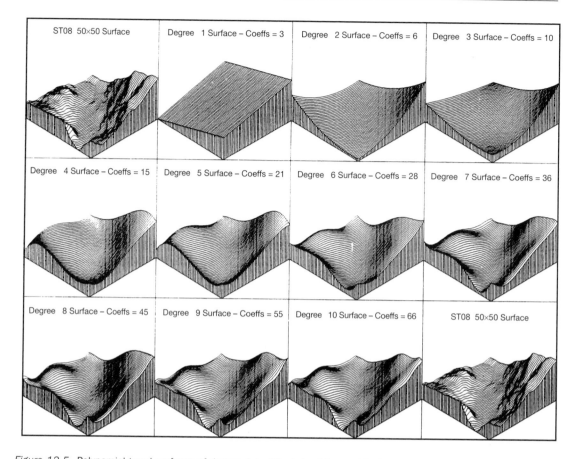

Figure 12.5 Polynomial trend surfaces of degree 1 to 10, derived from a UK Ordnance Survey 50×50 grid of digital elevations. Courtesy of D. B. Kidner. ©Crown copyright.

Another indication of the goodness of fit can be provided by studying the correlation between the residuals of the trend surface, i.e. the values of the original sample points after subtracting the trend surface values at the corresponding locations. If there were no significant correlation between the values then it could be assumed that the trend surface had successfully represented all such correlated variation. Methods of measuring correlation are described below in the context of autocorrelation.

The attraction of global trend surface analysis is chiefly in modelling the broad nature of variations and in studying local deviations from this trend. As a means of representing a surface (other than a very smooth one) to a high degree of accuracy, it is not usually practicable due to difficulties in computation of high-degree polynomials and the fact that polynomials tend to incorporate unacceptable errors at the edges of the surface.

Fourier analysis An alternative to the use of polynomials for representing the global characteristics of a surface is that of Fourier analysis. Fourier series describe variation in space (or time) by a summation of periodic, sinusoidal functions. Terms of the series represent sine and cosine functions of a particular wavelength. The amplitude of a particular wavelength, or harmonic, is indicated by the corresponding coefficients of the sine and cosine functions which, when squared and summed, is related to the variance contributed by the harmonic to the complete data set. A graph of the squared amplitude values for each wavelength, called a *power spectrum*, illustrates the relative importance of different wavelengths. If a data set has some significant periodic element, or characteristic wavelengths, it will be indicated by the presence of peaks in the corresponding part or parts of the power spectrum.

Surfaces are represented by 2D Fourier series in which coefficients are calculated for the x and y directions. Calculation of the Fourier series is usually based on original data sampled on a regular square grid. Examples of applications of Fourier series are to study geological characteristics, such as changes in ore grade (Davis, 1986), and to study scale-variable characteristics of terrain (Clarke, 1988).

Local surface estimation with least-squares polynomials

In general, global polynomials and Fourier series are only of interest for representing trends and analysing periodicities in a surface. However, smooth mathematical functions can be used for close-fitting representations if they are based on a mosaic of surface patches (Jancaitis and Junkins, 1973) that may or may not have gradient continuity constraints at their boundaries (Figure 12.6). Each cell is then approximated by a polynomial patch.

The original application of polynomial patches was to create a smooth surface for constructing con-tours. This involved blending the adjacent patches to ensure continuity of the approximated surface. Kidner *et al.* (1990) found that the approach is effective as a means of data compression, in the context of generating terrain profiles for radio communications planning. Small polynomial patches based on 3×3 or 5×5 grid points are also very useful for interpolating heights within the central cell of the patch for applications such as intervisibility and viewshed analysis (see Chapter 14).

A shortcoming of the polynomial patch method is that it is not possible, for a fixed degree of poly nomial, to ensure that all errors are within a pre-specified tolerance. One solution to this would be to increase progressively the degree of an individual patch's polynomial until the error was acceptable. This complicates data storage somewhat, since the number of coefficients to be stored for each cell is not predictable and may therefore require variable-length records.

An alternative approach is to vary the size of the patch using a fixed degree of polynomial, or at least a single type of some other function. This is the basis

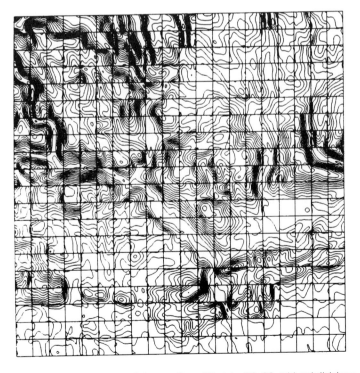

Figure 12.6 Polynomial patches of degree five, fitted to 20x20 grid subdivisions, using Ordnance Survey data. Courtesy of D. B. Kidner. ©Crown copyright.

of the surface patch quadtree of Chen and Tobler (1986). Starting with a large cell size, an attempt is made to fit the data with the given function. If any errors exceed the tolerance, the cell is divided into four quadrants and the process repeated for each quadrant. Individual quadrants continue to be subdivided until all cells meet the error tolerance. The result is a quadtree representation with the usual variable-sized cells, and with a fixed amount of data to be stored for each cell (Figure 12.7).

Spline surfaces

Spline functions are widely used in computer-aided design to provide a smooth, or at least well-controlled, representation of curves and surfaces passing through or near a given set of sample points (Rogers and Adams, 1990). They differ from the polynomial pach methods just described in that they can ensure continuity across the entire surface to be modelled. Typically splines consist of low-order, quadratic or cubic, polynomials which pass locally through a small number of sample points. For surfaces it is common to use the third-order bicubic splines. They may also be regarded as global, however, in that successive polynomials can be constrained to be continuous with their neighbours, thus ensuring a globally smooth surface which still fits relatively tightly to the sample points.

The B-spline and NURBS (non-uniform rational B-spline) methods are notable for providing considerable flexibility in the specification of parameters that control the degree of polynomials and whether they pass exactly or approximately through the sample points. Several different spline methods have been applied to geographical data and they appear to have considerable merits (Hutchinson, 1995), though at present they still do not appear to be widely used.

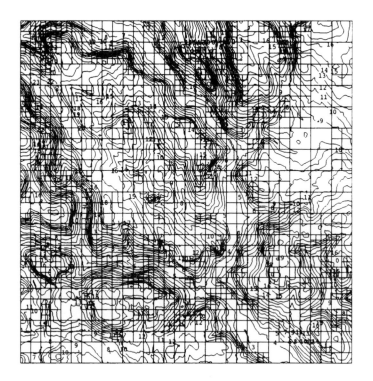

Figure 12.7 Surface patch quadtrees ensure that error is constrained with fixed-degree polynomial patches by varying the size of the individual patches. Note that the region displayed is the north–west area of the regions in Figures 12.2 and 12.6. Courtesy of D. B. Kidner. ©Crown copyright.

Simple weighted interpolation from point samples

Nearest neighbours

One of the simplest methods of estimating the value at a point is to set its value to that of the nearest sample point. It is equivalent to modelling the surface by Thiessen (or Voronoi) polygons, in which each sample point lies inside its own polygon. The approach does not take account of gradual variations across the surface. It may be appropriate for data that represent a phenomenon subject to rapid and abrupt variation, i.e. one with relatively low spatial correlation. As Burrough (1986) points out, it may also be appropriate for qualitative, categorical data for which numerically based interpolation has no meaning.

Averages and moving averages

A common approach to point interpolation is to base the interpolated point value on an average, usually a weighted average, of the neighbouring sample points. The chosen sample points would be the nearest n points, or all points within a given radius. The greater the value of n, the greater will be the smoothing effect of the average. Smoothing can be counterbalanced by using a weighted average, in which the weight attached to a point is inversely proportional to its distance from the interpolated point. The inverse proportionality may be linear or to a power (such as two) to provide an inverse squared relationship. Using a linear inverse distance weighting, the value e of the interpolated point would be given by

$$e = \sum_{i=1}^{n} (s_i/d_i) / \sum_{i=1}^{n} 1/d_i$$

where s_i is a sample point distance d_i from the interpolated point. This formula can be modified to change the distance weighting by raising d_i to an appropriate power. Note that the division by the sum of inverse distances is necessary to ensure that the distances are normalised and do not therefore affect the magnitude of the measured value. Note also that if the interpolated point coincides with a sample point, the formula will result in a divide by zero. If the distance d_i is zero, the interpolated point may be set to the value of the sample point.

A more general expression to describe a weighted average is

$$e = \sum_{i=1}^{n} w_i s_i$$

where w_i is the weight to be applied to each sample point that is used. Typically w_i is inversely proportional to the distance of the sample point s_i from the estimated point, as indicated above. Inappropriate scaling of the estimated value can be avoided by ensuring that the sum of all weights w_i is equal to one.

Distance-weighted averaging can be used for interpolating irregularly distributed sample points to a regular grid. The estimated points then correspond to the locations of the regular grid points and the procedure is referred to as a moving average. Using a moving-average procedure, individual sample points may be used in the estimation of several grid points which lie in their vicinity.

If the sample points are not fairly evenly distributed, it is possible for nearest-neighbour searches to find clusters of points which might all be in one direction from the interpolated point, thereby biasing its value. This can be overcome to some extent by ensuring that a minimum number of sample points are found in each of, say, four or eight sectors around the estimated point. This could also result in a poor estimation if it required using points that were very

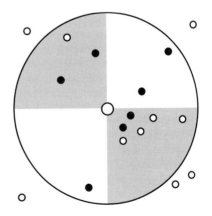

Figure 12.8 When interpolating the value of an unknown point from a set of nearby points, the search for points may be constrained by sampling from all quadrants or octants, within a given maximum distance, to ensure that the interpolated point is not biased to the value of a cluster of points in one direction. A maximum of two points (solid symbol) per quadrant have been selected in the figure.

distant. The problem is alleviated by distance weighting and can also be controlled by imposing a maximum search radius (Figure 12.8). Directionally controlled searches using quadrants or octants may be appropriate when interpolating from geophysical seismic shot points, which are often densely distributed along traverses that are far apart compared to the shot-point spacing along the traverse.

Spatial autocorrelation

The numerous possible ways apparently available for interpolating from known to unknown locations led to attempts to adopt a more mathematically rigorous approach. These methods fall within the broad category of *geostatistics* and are exemplified by the technique of *kriging*. In attempting to understand how to formalise the nature of the relationships between spatially separate samples, we will provide an introduction to the subject of spatial autocorrelation which underlies some of the more advanced techniques of interpolation and surface modelling.

All techniques for modelling a complete surface from a limited set of sample points are based on the assumption that the unknown points and the sample points are, at least to some extent, autocorrelated. In other words, adjacent variables on the surface are expected to be similar to each other; the expectation being that, in general, the closer two points are, the more similar their associated phenomena will be. If this were not so, it would be a waste of time to consider the value of existing sample points unless they coincided precisely with a location that we wished to observe.

Techniques of interpolation from sampled to unsampled points differ in the extent to which they assume correlation, and indeed in the nature of the correlation. The elevation of points on the earth's surface shows strong spatial correlation in general, but adjacent points within one side of a valley will display a different pattern of correlation than the points on the other side, which has a different slope direction. Horizontally adjacent points parallel to the valley direction will probably be more closely correlated than those perpendicular to it. Many sampled observations, such as soil type and human population density may be very similar within areal regions, but quite disparate across boundaries of these regions.

Figure 12.9 illustrates two hypothetical binary raster maps in which there are two major zones, on the left and the right. Within these zones, in the first example (a) there is relatively low autocorrelation, indicated by the fact that adjacent pixels are often different, while in the second example (b), there is relatively high autocorrelation, in which adjacent pixels are more likely to be the same value.

The possibility of changes in correlation across a surface suggests that for the most accurate interpolation of data at unsampled locations, it may be desirable to perform a prior analysis which could involve partitioning into regions. Partitioning is particularly applicable when boundaries or discontinuities have been recorded as part of the original

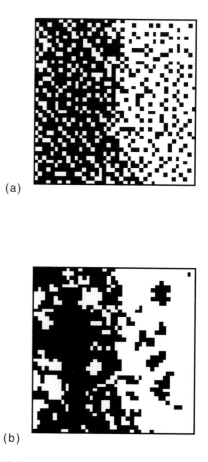

(a)

(b)

Figure 12.9 Autocorrelation. Adjacent cells in the upper map (a) have a high probability of being different from their neighbours, demonstrating low autocorrelation; while in the lower map (b) adjacent cells are more frequently of the same value as their neighbours, indicating a higher degree of autocorrelation.

survey, such as in geological surveys in which boundaries may be clearly visible. In soil surveys, study of the changes in correlation may be used to interpret the presence of boundaries.

As pointed out in Chapter 2, particular difficulties arise in interpolation and in the estimation of correlation when data have been sampled within areal units that have little or no relation to real variability in the phenomenon of interest. This situation often arises in socio-economic studies in which data are derived from administrative units, such as electoral wards or enumeration districts that have been imposed quite independently of the statistic of interest (such as unemployment, property ownership or health). We return to the subject of areal interpolation later in this chapter.

Measurement of autocorrelation

Several methods have been applied to the measurement of spatial autocorrelation. Two related measures that are often referred to in geographical literature are Geary's index and Morans's index (Cliff and Ord, 1981; Goodchild, 1986). They were originally applied to polygonal zones that have numeric attributes recorded on an interval scale. Geary's index measures the attribute similarity c_{ij} between a pair of zones i and j by means of the square of the difference between their respective attribute values z_i and z_j.

$$c_{ij} = (z_i - z_j)^2$$

The spatial similarity between pairs of zones is represented by a matrix $\mathbf{W}$ of values w_{ij}, the elements of which are set to 1 if two zones i and j are adjacent (share a common boundary), and 0 otherwise. These parameters are combined for the data set under consideration into the Geary index c which is defined as

$$c = \sum_{i=1}^{n} \sum_{j=1}^{n} w_{ij} c_{ij} / 2 \sum_{i=1}^{n} \sum_{j=1}^{n} w_{ij} v$$

where v is the variance of the attribute values, i.e.

$$v = \sum_{i=1}^{n} (z_i - z_m)^2 / (n-1)$$

where z_m is the mean of the z_i values, of which there are n. A value of 1 for c indicates that there is no autocorrelation. Values below 1 indicate positive autocorrelation, while values above 1 indicate nega-

tive autocorrelation. This reflects that the fact that c_{ij} will have low values if the corresponding attribute values are similar to each other, and increasing values as the attributes become more different from each other. The effect of the weights w_{ij} is that only the similarity values of adjacent zones with be summed.

The Moran index I measures similar data characteristics to the Geary index c, but positive autocorrelation is indicated by values of I greater than 0, no autocorrelation is indicated by I equal to 0, while negative autocorrelation is indicated by negative values of I. The measure of similarity between a pair of zones is given by the product of the attribute values after they have had the mean subtracted from them. Thus

$$c_{ij} = (z_i - z_m)(z_j - z_m)$$

This value is in fact identical to the *covariance* between the two attribute values. As with the Geary index, w_{ij} is a measure of the spatial proximity of the two zones. The index is calculated as

$$I = \sum_{i=1}^{n} \sum_{j=1}^{n} w_{ij} c_{ij} / v_s \sum_{i=1}^{n} \sum_{j=1}^{n} w_{ij}$$

where v_s is the sample variance, i.e.

$$v_s = \sum_{i=1}^{n} (z_i - z_m)^2 / n$$

In the original use of the Geary and Moran indices, the weights w_{ij} only took on the values 0 and 1, as indicated above. This may be regarded as a somewhat simplistic measure of spatial proximity, and extensions to the binary scheme have been used to take account of factors such the length of the common boundary between zones and the distance between the centroids of the pair of zones (Cliff and Ord, 1981). As pointed out by Ding and Fotheringham (1992), parameters such as these can in principle be derived quite easily from topologically structured polygonal maps.

Provided that it is possible to define a measure w_{ij} of the spatial proximity of pairs of objects, the Geary and Moran indices can be applied to spatial objects other than zones. If they are applied to points the weights may be inversely proportional to the distances between the points. As we saw in the previous section, this is a simple approach to interpolating the attribute of an unknown point from those of a set of known points.

The idea of spatial autocorrelation changing with distance is expressed by means of autocovariograms and autocorrelograms which represent the change in value of indices, such as those of Moran and Geary, as a function of the separation of the pairs of objects for which correlations are calculated. The separation of the objects can be represented in terms of how many intermediate objects separate them, or simply in terms of the measured distance, or lag, between them. If the data are sampled at regular intervals from 1 to $t + h$, where h is the lag, the autocorrelogram r_h is given by

$$r_h = \sum_{t=1+h}^{n} z_t z_{t-h} - \bar{z}_t \bar{z}_{t-h}$$

where z_t and z_{t-h} are the data values at a pair of locations separated by the lag, and $\bar{z}_t$ and $\bar{z}_{t-h}$ are the means of the respective data values calculated over the range of values of t from $1 + h$ to n.

The value of functions such as these is to indicate the manner in which correlation changes with distance (which we will see exploited in the kriging interpolation techniques), and to detect the presence of spatial periodicities, which might for example characterise particular types of landscape.

In simple kriging, in which no trend is assumed, this term is a constant equivalent to the mean of the data. In either case the term $e'(x)$ is a residual from the local value of the drift. A product of the preliminary data analysis is a graph called the *semi-variogram*, which expresses half the mean of squared differences between samples as a function of distance between them, i.e. the *semi-variance*.

For a set of sample points x_i, the semivariance is estimated by

$$\gamma(h) = 1/2n \sum_{i=1}^{n} [z(x_i) - z(x_i + h)]^2$$

where $z(x_i)$ is the data value at point x_i and h is the distance between two points. $\gamma(h)$ can be evaluated in different directions to take account of anisotropy in spatial correlation. Figure 12.10 illustrates an example semi-variogram. Provided sampling is performed over sufficiently large distances of h, it is normal for $\gamma(h)$ to rise from a value near zero before levelling out to form what is called the *sill*. The value of h at which levelling starts is called the *range* and corresponds to the maximum distance for which one data value can be said to influence another. Having plotted the semi-variogram, an analytical function may then be fitted to it.

Kriging

The fact that interpolations are based on the assumption of spatial correlation has already been stressed. The technique of kriging is remarkable for performing a preliminary analysis of the data set to determine the nature of correlation between the given sample points.

It is assumed that the value $z(x)$ of a variable at position x can be expressed as the sum of two components which are the global trend, or 'drift' $m(x)$ of the data and a locally spatially variable term $e'(x)$, which is dependent on the surrounding data values. It is appropriate to consider a non-spatially dependent error term e''. Thus

$$z(x) = m(x) + e'(x) + e''$$

In universal kriging the drift is typically modelled by a low-order polynomial, either linear or quadratic.

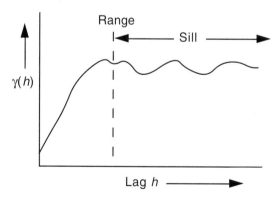

Figure 12.10 Typical form of a semi-variogram used to model variation in correlation between sample points in kriging. The steep slope represents a zone up to the range, in which there is a clear relationship of diminishing similarity between the value associated with points in the data set as their separation (lag) increases. The sill is the region of the semi-variogram beyond the range, in which there is no systematic relationship between lag and semi-variance (see text).

Having done so, the semi-variogram is used to calculate weights which are attached to sample points in order to estimate values at the unknown locations. The weights are calculated in a manner that minimises the variance of the estimated values, taking account of all samples used in each individual estimate. This is done using the method of *Lagrange multipliers* and is computationally expensive (see Davis (1986) for a simple introduction to the calculation procedure). Estimation of an unknown value is then performed by summing the weighted values in a manner comparable to the simple local moving-average techniques described earlier.

The technique of kriging was originally developed for estimating the value of gold-ore deposits between known samples. It is based on the theory of regionalised variables and is described as being optimal. This refers to the fact that the variance of estimates z is a minimum. A valuable aspect of kriging is that the variance of every point is calculated on the basis of the weighted sample points, and is therefore available to assess the confidence of the resulting values.

Despite the mathematical optimality of kriging, which places it above other interpolation methods referred to here, it is still subject to the assumption that, having subtracted the drift or trend value, spatial correlation is fully described throughout the data set by the variogram. For data subject to abrupt changes, such as due to geological or geomorphological structures, this assumption may not hold.

Areal interpolation

It was pointed out earlier in this chapter that there is a common requirement in the social sciences to transform surfaces represented by polygonal zones (i.e. choropleth or categorical maps) to a different set of zones. This may arise when data referenced to a different sets of zones, for the same geographical area, need to be compared with each other. Thus census data are normally accumulated within enumeration districts (EDs). If the need arose to find the population, or characteristics of the population, within a polygonal zone that is not composed of a whole number of enumeration districts, then it is necessary to estimate the contribution of each of the overlapped enumeration districts to the other, target zone.

A simple, though by no means satisfactory, way of apportioning the contents of the enumeration districts is to calculate for each overlapping ED the actual proportion of it that is overlapping with the target zone. This proportion may then be used to scale the contribution of the related statistic. How that apportioning is done depends on the nature of the statistic. If the data consist simply of counts of population, then the count for a source ED could be multiplied by the proportion of it that is overlapped by the target zone, before summing it with the similarly calculated contributions from the other overlapping EDs. If the statistic was a density measurement, being a count per unit area, then the measurement from each source ED would need to be multiplied by the respective overlapped area, to convert it to a count. If, in the latter case, an equivalent density measurement was required for the new zone, it would simply be required to divide the accumulated counts by the area of the zone.

The method just described, though useful, suffers from a fundamental flaw in many situations, in that it assumes that the measured statistic is uniformly distributed within the original zones. In practice this is often far from the truth. As regards human population counts, individuals usually live in buildings which may be very localised within the measured zone. In suburban and rural areas, domestic buildings may be scattered quite irregularly depending on local topography and on the pattern of land use for recreation, agriculture, industry and transport. In Figure 12.11, data from the source zones in (a) are to be interpolated to the target zones in (c), and the values in each zone in the source are derived originally from population counts derived from the buildings (square symbols) marked in the diagram. Assuming a uniform distribution, as indicated above, the target zone L would be given a finite value based on the data for overlapping zones A and D, despite the fact that there are no buildings at all in the zone. Conversely, target zone P corresponds to a high density of buildings, but in obtaining counts contributed from source zones D, C and E it will be given a lower value than the true distribution warrants. Even when, as in a dense urban area, housing may be distributed fairly uniformly across all zones, the people inside may vary enormously in terms of the socio-economic parameters that might be of interest. The study of data for a given area based on differing zonal aggregations can therefore lead to very misleading results. The issue is known as the modifiable areal unit problem, or MAUP (Openshaw, 1984).

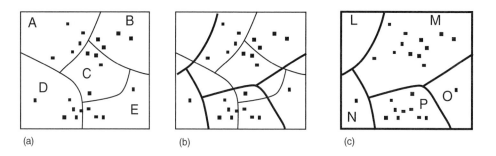

Figure 12.11 Areal interpolation between two sets of areal zones is subject to error if the distribution of the phenomena within the source zones is not uniform. The problem can be reduced by using more detailed knowledge of the distribution of other phenomena known to correlate with the parameter of interest (see text).

There have been some attempts to develop methods that compensate for the non-uniform distribution of data within zones. One approach is provided by Flowerdew and Green (1989), who introduced an additional variable which, though not necessarily of interest to a study, was known to be correlated with a variable that was of interest. As an example, population was interpolated from source zones consisting of local government districts to target zones consisting of parliamentary constituencies. The additional variable of party political representative was used to modify the interpolation between areal units. It was assumed that population density was related to parliamentary representation, based on the known correlation between low-density (rural) population with Conservative voters and high-density (urban) population with Labour voters. Thus if the target zone constituency was represented by Labour, the proportions of the source zone variables were weighted positively to carry over an increased proportion of their source zone population. The method involved ensuring that the total allocation to target zones from a given source zone was equal to the original aggregate value associated with the source zone.

Another example of the use of a correlated variable, when interpolating population values between areal zones, would be the distribution of domestic housing obtained from topographic maps or remote sensing. The assumption is that the distribution of population is closely correlated with the location of the houses (Langford *et al.*, 1990).

Another approach to the problem of human-population estimation for areal units is to produce a refined, regular grid cell (raster) model of the population surface in which known or implied assumptions about the variation in population density are used to interpolate the population values of the grid cells. Martin (1989) and Bracken and Martin (1989) did this by first transforming the polygonal zones to an irregularly distributed point representation in which the points were the centroids of the zones (allocated as part of the census procedures). They then used these points to interpolate to a regular set of grid points using a distance-weighted averaging procedure (see Figure 15.2). The resulting raster data structure had a finer resolution than the original zones and could be used to estimate the population and associated census statistics of arbitrarily specified polygonal zones by rasterising the polygon to determine the overlapped grid cells.

Summary

The surface modelling procedures described in this chapter are founded on a field-based view of geographic reality. Thus the phenomena of interest are regarded as being continuous and uniquely valued at all points in space, though for the purposes of abstract phenomena the space has been regarded as two-dimensional. We have made a distinction between incomplete surface models based on discrete samples usually consisting of points or lines, and complete surface representations in which all locations can be evaluated directly from the model. The complete surface representations include polygonal zone maps (also called area-class, categorical and choropleth maps); mathematical functions such a polynomials and Fourier series; and combinations of discrete spatial data with mathematical functions. The latter include triangulated irregular networks (TIN) in which the spaces within the triangles are represented by planar or quintic interpolating functions; spline functions fitted to a regular grid of points; and approximating procedures, such as distance-weighted averaging and kriging, which work in combination with a set of sample points.

Approximating functions that estimate values of a surface in-between discrete sample data operate on the assumption that the surface is autocorrelated, in the sense that the values associated with nearby sample points are related as an inverse function of distance. We have looked at methods for understanding the nature of autocorrelation, and found that the technique of kriging is based on a preliminary analysis of that correlation, as represented by the variogram.

We have concluded the chapter by considering the problem of interpolating between different-shaped areal units that overlap each other and hence relate to the same regions of space. This is part of the modifiable areal unit problem which arises in the social sciences because most such data are collected in a manner that is independent of the nature of variation of the phenomena of interest, and may vary in time due to administrative changes. When data relating to different zone schemes for the same geographical area need to be compared, one of the schemes must be interpolated to the other to create equivalent zones.

Further reading

Notable early expositions on TINs and triangulations in cartography and GIS are provided by Peucker *et al.* (1978), Gold *et al.* (1977) and McCullagh and Ross (1980). Tsai (1993) has provided a recent review of methods for constructing triangulated surfaces. Goodchild (1986) has published a monograph on the subject of autocorrelation. For a review article on spatial interpolation methods, see Lam (1983). Texts on interpolation methods and geostatistics in the environmental sciences include Watson (1992) and Isaaks and Srivastava (1989). Oliver *et al.* (1989a,b) review the use of geostatistics. Texts on methods of modelling spatial statistics with more emphasis on human geography and the social sciences include Cliff and Ord (1981), and Haining (1990).

CHAPTER 13

Techniques in spatial decision support

Introduction

The capacity of GIS for integrating information from a variety of sources in a spatial context makes them well suited to supporting decision-making procedures that must take account of multiple factors. This chapter is concerned with techniques that employ geographically referenced information to identify locations, paths and spatial interactions that are optimal, or at least in some sense preferable, when measured against the factors of interest. These techniques can contribute to decision-support systems (DSS) which help decision-makers to formulate problems, create appropriate models and evaluate outcomes that the model may predict. The GIS aspect of such systems is sometimes referred to as spatial decision support.

It is very important to realise that in many decision-making situations, the objectives are not always clear-cut and often many of the factors that should be taken into account cannot easily be quantified. Even when a great deal of data are available, they are bound to be subject to some inaccuracy, thus any solution to individual optimisation procedures must also be subject to doubt. Decision-support tools in GIS should therefore be regarded as there to assist in the decision-making process, not to make the decisions. The value of the tools is in helping the decision-makers to evaluate alternatives and to explore certain possibilities, hopefully in more depth and with greater precision than might otherwise have been possible.

The contents of this chapter fall into five parts. In the first part we focus on the problem of evaluating the suitability of land for specific purposes and show

how the simple Boolean overlay techniques described in Chapter 3 can be extended to take account of the relative importance of the different layers of the overlay. In doing so we introduce the methods of multi-criteria evaluation (MCE) and show how they can be integrated with GIS.

The idea of using computers to find optimal solutions to complex problems of planning is well established in management science. Linear programming is one such method that has the potential to help in solving problems with geographically referenced data. Though not currently part of standard commercial GIS, it is described briefly here in the second section of the chapter as an example of a technique that could usefully be incorporated in GIS. It is relevant to problems concerned with allocating limited sets of resources to a given set of locations where those resources are required or can be exploited. Examples include allocating patients to hospitals and selecting efficient strategies of planting crops.

Many areas of decision-making depend upon the accumulated knowledge and experience of experts. Rule-based systems and expert systems have been used to attempt to encode such knowledge in an effort to make it more readily available and to apply it consistently. Some of the most successful expert systems have taken account of the uncertainty in knowledge and in the decision-making process. This has led to the use of fuzzy reasoning and of Bayesian probability methods in such automated decision-making systems. An introduction to these methods is provided here.

A further section of the chapter examines the use of network models for solving problems concerned with determining optimal routes between two or more places. Optimal routeing techniques are required for purposes of navigation, but they may also be integral to solving problems concerned with determining the relative accessibility of different locations. An algorithm to find shortest (or least-cost) paths is described, as are some methods for finding sets of places on a network that are within a specified distance of a given location.

The last section of the chapter introduces spatial interaction modelling as a means to describe and predict flows between locations. These flows might be people migrating between places, in demographic studies; they might be flows of customers, or their expenditure, between their homes and shopping centres; or they might be individuals making use of social facilities such as schools and hospitals. One of the commonest uses of spatial interaction models is in planning the location of new shops with a view to maximising profits.

Land suitability and multi-criteria evaluation

Identification of land sites that meet particular criteria is one of the main spatial analytical applications of GIS. As has been shown by Carver (1991) and others, a systematic approach to site selection is provided by the techniques of multi-criteria evaluation (MCE) and these techniques can be incorporated within GIS. The purpose of MCE is to help decision-makers distin-

guish between several possible options, where the preference for each of the options may depend upon several factors, each of which might be regarded as having different levels of importance by different members of the community. In other words, there may be no right answer, but the approach does at least enable the decision-maker to take account of different levels of perceived importance for different factors.

Boolean overlay

The ability to overlay several thematic layers in a GIS in order to identify regions that combine selected attributes from each of the layers was one of the earliest types of analytical facility provided by GIS packages and was introduced in Chapter 3. When working with a layer-based GIS the method entails creating a set of layers, each of which represents one of the constraints on the solution to the problem. The selected layers might consist of regions representing land (a) above 10 m elevation; (b) of gradient less than 15%; (c) within 200 m of a major road; (d) more than 50 m away from domestic and commercial buildings; and (e) outside conservation zones. The layers are then superimposed to identify regions of space that satisfy all constraints. In a vector-based system, the regions would be polygons resulting from the intersection of the component polygons of the individual layers. In a raster system, the regions would

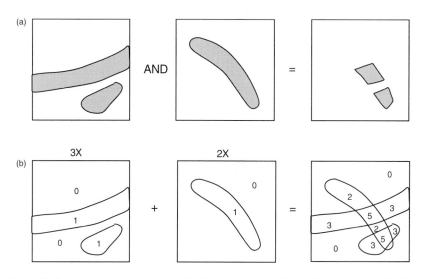

Figure 13.1 Boolean overlay compared with weighted overlay. (**a**) The areas identified as a result of a Boolean overlay, using the AND operator. (**b**) Weighting the individual binary layers results in a more sensitive discrimination between map zones.

consist of those cells that fell inside the regions in all five layers. The combination process is, in its simplest form, a Boolean one in that the result is, in this case, the logical AND of each of the component layers (Figure 13.1a). Boolean logic can be used to find combinations of layers that are defined by conjunctions of AND, OR and NOT operators. A NOT operator, for example, could be applied to layer c to find land that was more than 200 m from a major road. The OR operator could be used to generate a layer in which land belonged to two or more categories, which did not have to be coincident.

Weighted overlay

Boolean operators are appropriate if each factor, or constraint, under consideration is of equal importance. In practice certain factors may be much more important than others and it may be desirable to differentiate between candidate sites according to how well they meet the various criteria. The relative levels of importance of the different types of data can be taken into account by attaching numeric weights to each of the layers in an overlay operation. Regions that meet all the search criteria are then associated with weighted summations of the factors in each layer. Thus for any given location, which is considered here to be either a raster cell or a polygon resulting from intersection of multiple polygon layers, the weighted sum would be given by

$$\sum w_i * p_{ik}$$

where w_i was the weight for layer i and p_{ik} was equal to 0 or 1 depending upon whether the factor was present at the given location k. The weights w_i are assumed here to sum to 1. This procedure is illustrated in Figure 13.1(b), which may be compared with the result of a Boolean overlay for the same layers illustrated in Figure 13.1(a).

The use of weighted constraints only goes part of the way to taking into account the relative importance of different factors. If each layer in the overlay operation itself consists of various types or values within the respective factor represented by the layer, then each type may have a different score according to its perceived importance within the layer. Thus each of a set of land-use types, such as woodland, crops, suburban etc, might be categorised on a scale from 0 to 9 as a measure of its suitability for a particular purpose. Alternatively, continuous-scale scores may be allocated by means of a possibly non-linear function that transforms from one continuous numeric scale (e.g. temperature, rainfall, or distance from the nearest road) to another which was the weight scale, or score, for the particular value within the layer. A weighted overlay operation would then evaluate intersected regions or cells by a weighted sum of the scores, so that each resulting region was characterised by a score measuring its suitability. Thus for a particular location (polygonal region or cell) k, the weighted sum is given by

$$\sum_{i=1}^{I} w_i * x_{kij}$$

where x_{kij} is the score of the value j within the ith factor, or layer, in cell k, and w_i is the weight associated with the ith factor. Figure 13.2 illustrates the application of this method diagrammatically. Plate 13 illustrates the application of the method for coastal zone management in the Netherlands. Distance-weighted buffer zones, for housing, industry, roads and drinking water supply, have been overlaid to create the larger map, in which green areas are the most suitable for the development of a slufter (which is a wet dune environment subject to tidal flooding).

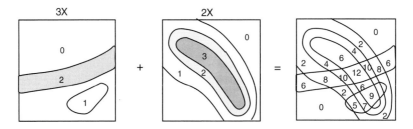

Figure 13.2 Weighted layers and weighted sums of scores for data values within layers. Each data type within a layer is allocated a score which is multiplied by the layer weight before being added to the weighted scores of intersecting zones from the other layers.

Assignment of weights

A problem that arises with weighting of the factors to be considered in a weighted overlay operation, is the selection of appropriate weights. The process may be a subjective one in that the weights are intended to reflect the preferences of the decision-maker (or makers). The selection of weights may be assisted by an interactive user interface designed to solicit weights and to ensure that they become appropriately normalised. A well-established procedure for assigning weights to a set of factors, which may correspond to the layers of an overlay, is Saaty's Analytical Hierarchy Process (AHP), which builds a matrix of pairwise comparisons (ratios) between the factors. Assuming there are m factors, (e.g. elevation, nearness to a road, slope and temperature), the matrix records the perceived relative importance of each factor with each other factor.

The matrix is constructed by eliciting, from the decision-maker, values for relative importance on a scale from 1 to 9, where 1 means that the two factors are equally important, and 9 means that one factor is absolutely more important than the other. If a factor is less important than another then this is indicated by reciprocals of the 1 to 9 values (i.e. 1/1 to 1/9). For consistency, the matrix should ensure that if the importance of an attribute a relative to b is n, then that of b compared to a should be $1/n$. If the user exhibits a small amount of inconsistency in this regard, the values may be modified automatically to force consistency. The process of allocating weights is of course a subjective one and can be done in a participatory style in which a group of decision-makers may be encouraged to reach a concensus of opinion about the relative importance of the factors. An example matrix is illustrated in Table 13.1.

Having created the pairwise comparison matrix **A**, it can be used to derive the individual normalised weights w_i to w_I of the I attributes. Given a matrix **A** of pairwise values, they can be normalised by calculating the principal eigenvector of the matrix. This can be approximated 'manually' by dividing each entry in column i of **A** by the sum of the ratio values in the column. This results in a matrix of values a_{ij} that are in the range 0 to 1 and which sum to 1 in each column. Estimates of weights for each attitubute can then be obtained by averaging the values in each row 1 to I. These weights will now themselves sum to 1 and can be used to derive weighted sums of factor values (i.e. scores) for each region or cell of the map layers.

The process just described relates largely to assignment and normalisation of the weights attached to each layer, or factor. Since values within each layer also have scores, these values must also be normalised to ensure that no one layer exerts an influence beyond that determined by its weight. Unless they are allocated as part of the AHP process itself, then the scores for the values within each layer may either be allocated initially on a standard scale such as 1 to 10, or the raw scores R_i for each may be normalised simply by the formula

$$x_i = \frac{R_i - R_{min}}{R_{max} - R_{min}}$$

It may be noted that if the raw score was a value such as distance from the nearest road, and low distance was regarded as most desirable, then the formula should be modified to ensure that the lower the value, the higher the score; hence

$$x_i = \frac{R_{max} - R_i}{R_{max} - R_{min}}$$

Multiple objectives

Some problems of land suitability require simply that land be identified that meets certain criteria; for example, sites that might be suitable for industrial development. These may be classified as single-objective, multiple-criteria problems, on the assumption that suitability did depend upon more than one criterion.

Table 13.1 Matrix of pairwise comparison of factors.

(a)	Roads	Parks	Schools	Elevation
Roads	1	1/5	1/3	1/2
Parks	5	1	4	2
Schools	3	1/4	1	1
Elevation	2	1/2	1	1

(b)	Weights
Roads	0.087
Parks	0.519
Schools	0.192
Elevation	0.202
Consistency ratio	0.03

Sometimes the problem is to divide up regions of land according to their suitability for different objectives. There may be situations in which it is necessary to find land that is suitable for several uses that can coexist. For example, some types of wildlife conservation areas, such as in deciduous woodland, can coexist with certain recreational land uses. Sometimes, however, multiple objectives are conflicting, in the sense that it is desirable to identify sites that are by definition exclusive, e.g. arable agricultural land and sites for industrial development.

Automated techniques for solving multiple-objective problems have usually involved mathematical programming methods, such as linear programming, which is described below. However, recently some conceptually simpler techniques have been introduced, such as the Multiple Objective Land Allocation (MOLA) module of the **IDRISI GIS** package (Eastman *et al.*, 1993). This method uses an input that consists of the results of the weighted overlay procedure. These suitability values need to have been ranked, whereby the highest-rank values indicate the most suitable locations with respect to the particular objective. Allocation on the basis of the multiple objectives is an iterative procedure which identifies those cells that are ranked highest with respect to all objectives. Some cells may turn out to be ranked highly for one objective but not for any of the others. These cells can be allocated directly to that objective. Other cells may be ranked highly with respect to two or more objectives and may therefore be regarded as in conflict. Such cells are divided between the respective objectives on the basis of whether they are nearer to one objective than another.

The procedure of allocation is illustrated in Figure 13.3, which represents a graph of the rankings of cells with respect to just two objectives that are represented by the two axes. To find cells that are highly ranked with respect to one axis, a line is moved down from the highest values to encompass a pre-specified number of cells, i.e. the top n cells for the axis. The conflict cells are those in the top right corner of the graph which fall into the top-ranked cells for both objectives. Cells are split between objectives by constructing a boundary line from the origin that divides cells on the basis of which objective they meet best. Cells on each side of the line are nearer to the respective 'ideal point' for the respective objection, as indicated in Figure 13.3(b). If the objectives are of equal importance, this line is at 45° (as illustrated), but if one objective was more important than another the line could have a different slope. Allocation is an iterative procedure in that cells are progressively creamed off from the highest ranking zone, going down to progressively lower ranked cells for that objective until the required number of cells has been allocated for each objective.

Linear programming

Linear programming (LP) is a well-established method of finding an optimal solution to problems that require several factors to be balanced against each other, subject to certain constraints. Examples of the application of linear programming are common in manufacturing environments in which the problem could be to determine how many of each of a set of possible products should be manufactured in order to

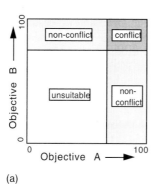

(a)

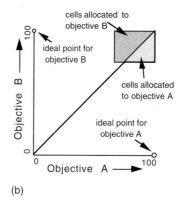

(b)

Figure 13.3 Ranking multiple objectives.

maximise profits. The problem could be constrained by differing sale prices, by limited capacity for production, a limited market for particular products and differing costs of production. Killen (1979) has given numerous examples of geographical problems to which LP is relevant. These include scheduling production to match demand, determining an appropriate mix of economic activities in a region, minimising transportation costs and organising animal grazing. Wright *et al.* (1983) describe the use of LP for allocating land usage. More recently, Chuvieco (1993) has described its use in the context of selecting land usages in a manner that maximises employment in an area suffering from high unemployment.

Although linear programming is more complicated than the technique of multi-objective optimisation described above, it is also more versatile, in that it is possible to take account of certain types of constraint that are not easily dealt with by other methods. An example of such a constraint in land-use allocation is that certain land uses should not be adjacent to others, while it might be desirable that some uses are adjacent. It might also be important to find sets of cells that constituted continuous regions of a minimum area and which were relatively compact in shape.

To illustrate some of the basic principles of linear programming we will examine a simplified land-usage problem (based on an example from Killen, 1979). It may be supposed that a farmer wishes to balance the production of two crops in a way which maximises profit. The crops differ in the quantity of land, labour and water (irrigation) required for their production, while each of these resources differ in their availability. The input information may be summarised as shown in Table 13.2.

A linear program represents the problem by a set of linear functions that are either equations or inequalities. The functions contain decision variables that each represent some particular aspect of the problem. The decision variables in this problem are the quantities (in tonnes) x_1 and x_2 of the two crops that should be grown. The optimisation aspect of a linear program is described by an objective function which is constructed using some or all of the decision variables. The purpose of the linear program is to maximise or minimise the objective function subject to constraints defined by other functions in the program.

Since the objective in the example is to maximise profit, this can be represented by the maximum of z where

$$z = 2x_1 + 3x_2$$

since the profit per tonne for crop 1 is two units of money and for crop 2 it is three units.

The limitations of resources can be expressed by three constraints, for land, labour and water. The land constraint takes the form

$$2x_1 + x_2 \leq 8$$

This represents the fact that each tonne x_1 of crop 1 uses two units of land and each tonne x_2 of crop 2 uses one unit of land and that the total land so utilised will not exceed eight units, which is all that is available. The constraint for labour is that each crop requires one unit of labour per tonne and that the total available is five units, hence

$$x_1 + x_2 \leq 5$$

The constraint for water resources is that each tonne of crop 1 requires one unit of water and each tonne of crop 2 requires two units, and the total water available is eight units, thus

$$x_1 + 2x_2 \leq 8$$

The other constraint is that both quantities of the crops produced must be non-negative:

$$x_1 \geq 0$$
$$x_2 \geq 0$$

The formulation of linear programs is subject to several conditions:

1. the objective function and constraint functions must be linear functions (there can be no terms that are cross products or powers of the decision variables);
2. decision variables must be able to take on real and integer values;
3. the value on the right-hand side of a constraint must be known (i.e. not itself a decision variable);
4. each variable must be defined with respect to its possible sign: it may be either non-negative as in the above example, or unrestricted in sign (urs).

Table 13.2 Input data for crop management problem.

Resource	Units of resource used per tonne of crop grown		Total availability of resource
	Crop 1	Crop 2	
Land	2	1	8
Labour	1	1	5
Water	1	2	8
Profit per tonne grown	2	3	

There are many optimisation problems in which these conditions cannot be met. If the first condition does not apply, which would be the case if variables were dependent upon each other, then it becomes a *non-linear programming* problem. There are many problems in which a whole number of units is required in the solution (e.g optimum number of people to be employed in a company; number of fire stations to be constructed). This gives rise to *integer programs* which, like non-linear programs, are more difficult to solve than linear programs. A particular case of integer programs is a 0–1 program in which the decision variables may only take on values of 0 and 1. This would be relevant to a program that was considering a set of possible facilities, each of which might or might not be required in the solution. 0–1 coefficients could then be used to indicate their absence or presence.

Another example of a particularly common type of linear program is a *transportation problem*, in which the aim is to allocate a limited set of service facilities to a set of demand centres that require them. They are called transportation problems because they are characterised by the fact that a commodity is somehow transported from the supply to the demand centre and that costs are a function of this transportation.

An example of a transportation problem is that of supplying several towns with electricity generated by several power stations (Winston, 1994). The intention is to minimise costs subject to the fact that the power stations each have a limited supply, the towns each have a particular maximum power demand and the cost of supply is proportional to the distance from the power stations to the towns. Decision variables x_{ij} are set up for the power supplies from each of the power supplies i to each town j. Given that there are m power stations and n towns, and that c_{ij} is the cost of supply, then the quantity z that is to be minimised is represented by the objective function

$$z = \sum_{i=1}^{m} \sum_{j=1}^{n} c_{ij} x_{ij}$$

There are two constraints regarding limitation on supply s_i from power stations i and the demands d_j from towns j. These are

$$\sum_{j=1}^{n} x_{ij} \leq s_i \quad (i = 1 \ldots m)$$

$$\sum_{i=1}^{m} x_{ij} \geq d_j \quad (j = 1 \ldots n)$$

A final constraint is that all values of supply x_{ij} must be non-negative:

$$x_{ij} \geq 0$$

Solving linear programs

The constraint functions in linear programs may be regarded as defining a zone of acceptable values, or a *feasible region*, for the decision variables. If there are only two decision variables, each constraint is simply a straight line, one side of which is acceptable, the other is not. The set of constraint functions, including the non-negative constraints, serve to enclose a convex polygonal region inside which the values of the decision variables satisfy the constraints. The objective function is then a line of fixed gradient, but unknown position, that we wish to intersect with the feasible region in a manner which maximises (or minimises) the objective function. The optimal intersection will be at an extreme point of the feasible region, which in the 2D case is one of the vertices of the polygon bounding the region (Figure 13.4). The points on the boundary of the feasible region are referred to as *feasible solutions*. Solving a linear program consists in finding the feasible solution which maximises or minimises z, the value of the objective function.

In all but the most trivial linear programs there may be many thousands of feasible solutions. Thus evaluating each one to find that which is optimal would become highly expensive in computational terms. There is, how-

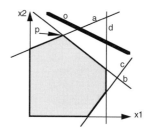

Figure 13.4 Feasible region in linear programming. The shaded region represents the range of all acceptable solutions to the linear programming problem specified by the constraints of the lines a, b, c and d, each of which represents an inequality. A solution which maximises (or minimises) the objective function is found by moving the line o, representing the objective function, in from infinity until it touches a vertex on the boundary of the feasible region, which is point p in the diagram.

ever, an efficient procedure, called the *simplex algorithm*, which finds a solution quickly having examined only a very small proportion of the feasible solutions.

Rule-based systems

There are many situations in which the knowledge required to solve problems is based on a combination of both the ability to differentiate between numerous possible situations, or descriptions of objects, and the need to take action which may depend upon this recognition. Such knowledge, which frequently depends upon the expertise gained in particular application domains, can conveniently be encoded as rules of the form 'IF certain conditions are met THEN perform an action or draw a conclusion'. The condition part of the rule is described as the *antecedent*, while the conclusion part is the *consequent*. Problem-solving systems founded on rules are sometimes described as *production systems*, while the rules are called *production rules*.

Early applications of production rules were in the fields of medical diagnosis, as in MYCIN, and in the evaluation of geological and geophysical data for mineral exploration, as in PROSPECTOR (Hart *et al.*, 1978). All of these systems were used to encode the knowledge of human experts, in recognising particular patterns of data, and knowing how to respond or what conclusion to draw. In recognition of the fact that they employed human expertise, they are called *expert systems*. The principle of encoding knowledge in rule-based systems does not always depend upon the knowledge having been elicited directly from human experts, since quite effective rules can sometimes be formulated on the basis of existing texts and documents and on the basis of automatic or machine learning procedures. Expert systems may generally be regarded as a particular type of rule-based system, though the expression is also sometimes used quite loosely to mean any knowledge-based system that exhibits relatively intelligent behaviour either due to declarative or procedural mechanisms.

An important aspect of rule-based expert systems is that because their knowledge base is encoded explicitly as rules, it is possible to provide the user with an explanation, in terms of the rules, of why particular actions were taken or recommended. This may be essential in a decision-making environment in which the user of the system must ultimately take responsibility for the

decisions and must therefore be able to justify them. The explanations may, in the crudest form, consist of a regurgitation of what could be a long list of rules that led to a particular conclusion. More useful explanations would be expressed in terms of the fundamental principles of the relevant domain of knowledge, since an individual rule cannot be its own justification unless it is itself accorded the importance of being a principle. In practice, this more intelligent level of explanation is likely to be provided, if at all, by commentaries built into the system and associated with particular rules.

While a set of rules may be regarded as a form of knowledge representation in its own right, the implementation of a rule-based system requires other components, in addition to the rule base (Figure 13.5). These include a *knowledge base*, an *inference engine* (or rule interpreter), a *user interface*, a *knowledge acquisition* facility for creating and editing the knowledge base, and may also include a set of programmed *procedures*. Evaluation of the antecedent of a rule necessitates the presence of a database of facts against which the condition of the antecedent can be tested. This database constitutes a description of the current state of the program domain and is sometimes referred to as a *working memory*. It could consist of a simple table, comparable to that in a relational database, but some systems use more sophisticated data structures

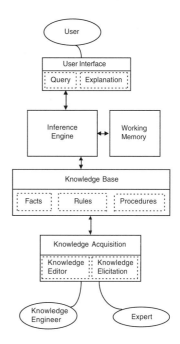

Figure 13.5 Components of an expert system.

based on, for example, the *frame* concept. Frames may be regarded as representing stereotypical objects or situations, and may be implemented in a similar manner to objects in an object-oriented programming system or database. Thus they may belong to a particular class and be related by generalisation and specialisation to other classes. A frame may store both data and procedures, some of which may be executed automatically on attempting to access particular items of data in the frame.

Application of the rules requires the inference engine to cycle through the rule base evaluating the rule antecedents and, when appropriate, performing the operation specified by the consequent. If the antecedent condition of a rule is satisfied, a rule is said to be *triggered*. The act of execution of the consequent is described as *firing* the rule. The process of rule interpretation is rendered quite complicated, however, by the fact that, given a set of rules, several of them may be applicable at any one time. Hence a decision must be taken on which rule to fire first, since the order of firing may affect the result of the inference process. Control of the rule firing is governed by a *conflict resolution* procedure, which may be specified when building or setting up the system for a particular task. The simplest approach to conflict resolution is to determine firing and execution according to the order in which rules are encountered. An expert in some domain may know that certain pieces of evidence or situations may be critical and should take priority. This can be reflected in the ordering of the rules.

In applications of rule-based systems, the evaluation of the antecedent and the execution of the consequent can frequently be expected to entail the use of separately programmed procedures. These procedures might be required to perform measurements or spatial searches upon the contents of the database or, for example, to modify the graphic representation of objects to be displayed on a map. Such specialised procedures would be in addition to those for maintaining the rule base and its associated working memory or database. Rule-based systems can often be seen therefore to combine both *declarative knowledge*, embodied in the rule base, with *procedural knowledge* embedded in these specialised procedures.

The last component of a rule-based system, referred to above, is the user interface. The importance of this facility may depend on the extent to which the system is intended to serve as an automated process or as an adviser to assist a human decision-maker. In the case of expert systems acting in an advisory capacity, such as in policy planning, which could have significant social or political implications, the facility for interrogating the system to determine the logic behind its conclusion is essential. Such applications could require a highly interactive mode of use in which many possible scenarios would need to be explored. Smith *et al.* (1988) have described a graphical user interface that was an essential component of a planning-decision support system.

Dealing with uncertainty in rule-based systems

A major contribution of human expertise is an awareness of the reliability of factual evidence and of the rules that may be used to draw conclusions from pieces of evidence. The need to incorporate this awareness of uncertainty in rule-based systems has long been recognised and many of the earliest examples of expert systems included some capability for representing and reasoning with uncertainty. In a GIS context, Leung and Leung (1993a,b) have described an expert system shell (i.e. programming system) that uses *fuzzy logic*. The general form of a rule in their system is as follows:

> (*rule* <rule_name>
>   *if* <object*1*> <operator*1*> <value*1*> *and/or*
>   <object*2*> <operator*2*> <value*2*> *and/or*
>   :
>   :
>   *then* <object*n*> is <value*n*>
>   *certainty* is <*n*>

Examples of rules in this form are:

> *rule stability 9*
>   *if slope is approximately 30° and erosion is strong*
>   *then stability score is 1*
>   *certainty is 1*
> *rule climatic 5*
>   *if rainfall is substantial and temperature is hot*
>   *then type is SH*
>   *certainty is 1*

The values of objects may be either fixed values, such as a number or a Boolean (true or false), or they may be fuzzy, in which case they are described by a fuzzy set. The value 'hot' for the fuzzy object 'temperature' could be a range of values between 0 and 1, corresponding to a set of temperature values regarded as in the range of hot. This relationship constitutes a fuzzy membership function for the fuzzy set of temperature values called 'hot' (Figure 13.6). Fuzzy sets enable comparison between values that are similar but not

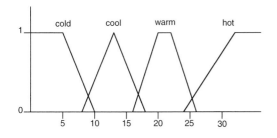

Figure 13.6 Fuzzy variables. The fuzzy object temperature is composed of several fuzzy values, cold, cool, warm and hot, each of which is a fuzzy set. Membership of a fuzzy set is specified by a fuzzy membership function which relates values of the attribute (temperature) to the individual fuzzy set values which range between 0 and 1.

exactly equivalent. Thus in addition to the conventional equality and inequality operators of $=, <, >, \leq, \geq$, a similar set of fuzzy operators are available that may be regarded as qualifying each of the operators with *approximately*. The result of such an operator can be expressed in terms of a certainty value. In addition to objects having uncertainty, the reliability of rules can also be described by certainty values, allocated by experts. Fuzzy reasoning systems can manipulate both conventional and fuzzy objects and generate conclusions that are a function of the certainty of the data and of the rules.

It may be noted that fuzzy logic can be applied to overlay operations in GIS. When combining values for the same location, or cell, from several layers, a fuzzy AND operation takes the minimum of all fuzzy values for that location. Conversely, the fuzzy OR operation takes the maximum of the fuzzy values for the particular location.

Some early expert systems, such as the PROS-PECTOR geological exploration expert system, used *Bayes theorem* to manipulate probabilities. Bayes theorem provides a means of evaluating the probability of some event or situation given prior knowledge of evidence for believing in the event. In the case of geological exploration, for example, certain minerals are known to be commonly associated with other minerals and with particular geological conditions. The presence of these other minerals or geological conditions provides evidence in favour of believing that the mineral of interest may also be present.

Bayes theorem may be expressed by the equation

$$P(H_a|E_b) = \frac{P(E_b|H_a)\,P(H_a)}{P(E_b)}$$

where $P(H_a|E_b)$ is the probability of a hypothesis H_a, such as a particular mineral a being present, given the presence of the item of evidence E_b, namely the presence of b. It is called the posterior probability. $P(E_b|H_a)$ is the probability that, when the hypothesised event H_a is true, the item of evidence E_b is actually present. It is called the prior conditional probability. It is based on existing knowledge of the coincidence of the event (such as the presence of mineral a) and the evidence (such as the presence of mineral b). $P(H_a)$ is called the prior probability and is the overall probability of the event or situation occurring. $P(H_a)$ must also be based on existing knowledge of the frequency with which the event is believed to occur in the field of study. It would be updated as knowledge is improved. $P(E_b)$ is similarly the prior probability of the evidence occurring and could also be subject to update.

Skidmore *et al.* (1991) have described an expert system for mapping forest soils that used Bayes theorem in an expert system. Expert knowledge was used to deduce the type of forest soil on the basis of known occurrences of certain environmental factors. The relevant expertise was encoded in the form of rules that expressed the probability of occurrence of a particular environmental factor at the location of a particular soil type. The rules thus correspond to the Bayesian prior conditional probability $P(E_b|H_a)$ of an item of evidence E_b being present, given the hypothesis H_a that a particular class of soil S_a is present (i.e. both the evidence E_b and the soil type S_a are present). For each environmental variable (item of evidence) such as dry gully, moist gully, wet ridge, very dry ridge, etc., the probabilities of their occurrence with each of a set of soil landscape types (e.g. residual crest, aggraded well-drained slopes) are stored as values between 0 and 1. The inference engine then works on a cell by cell basis in a raster data set, with evidence layers $b = 1$ to k, to find the soil landscape type with the highest probability of occurrence given the item of evidence E_b, using the Bayes formula above.

It is an iterative process in that $P(H_a|E_b)$ is used to update the initial expert-supplied values for $P(H_a)$, the probability that soil class S occurs at the current location. The prior probability, $P(E_b)$, of finding the evidence, such as a dry gully, at the given location can then be updated using

$$P(E_b) = \sum_{}^{n} P(E_b|H_a)\,P(H_a)$$

where a is the soil class of which there are n types.

Network analyses

Major applications of network modelling techniques in GIS relate to navigation, utility planning and management, hydrology and spatial interaction analyses involving communication networks. The purpose of a network model is to represent and analyse flows along interconnecting paths that may correspond in reality to various phenomena such as people or vehicles along a road, gas through a pipeline or electricity in a cable. Despite the apparent difference between the different types of phenomena that may be represented, the analyses required often have much in common. Generic procedures common to several applications include the following:

1. find the least-cost or shortest path between two locations;
2. find the least-cost or smallest set of paths that connect a set of locations;
3. find all locations that are within a given cost of a specified location.

An example of the shortest path between two locations in a road network, corresponding to a solution to case 1, was illustrated in Figure 3.14. Case 2 corresponds to the classic travelling salesman problem in which the salesman is required to visit a set of locations (say customers or pick-up points), while minimising the total distance travelled. The complete path that is a solution to this problem is sometimes referred to as a tour, and an example solution is illustrated in Figure 13.7. Case 3 is an allocation problem. Examples would

be to find all locations that are within a given walking distance of a school, or all locations with 15 min travel time for an ambulance service. Figure 13.8 shows an example of the roads allocated to two centres on the basis of expected travel time.

The expression *cost*, when used in the context of shortest paths, would typically refer to distance or time and would be determined by characteristics of the paths between internal locations on the network. For example, significant factors might include the length, the grade and the speed limit of road segments, and the length and the diameter of pipeline segments. These factors affect the *impedances* associated with each connecting segment of a network. Clearly factors such as the gradient of a road and the direction of flow of a river are different according to which way the relevant part of the network is traversed. Therefore the impedances must, in general, be specified separately for each direction of a network segment. In road networks, there are frequently constraints on which turnings can be taken at which junctions, and how long it takes to turn. These factors may be associated with each junction of the network.

Network data structures

The form of a network can be represented by a network data structure consisting primarily of a set of nodes (or vertices) and a set of edges (or links) connecting pairs of vertices (Figure 13.9). Nodes may correspond to junctions while the edges represent paths between the nodes and might correspond to a

Figure 13.7 Example of a solution to the problem of finding a tour consisting of the shortest path, constrained to visit specified locations. Generated using Arc/Info GIS software.

Figure 13.8 Roads allocated to two specified centres on the basis of travel time as a function of distance.

Node Adjacency Table

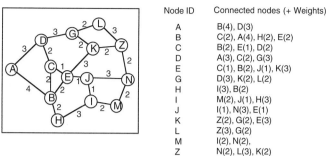

Node ID	Connected nodes (+ Weights)
A	B(4), D(3)
B	C(2), A(4), H(2), E(2)
C	B(2), E(1), D(2)
D	A(3), C(2), G(3)
E	C(1), B(2), J(1), K(3)
G	D(3), K(2), L(2)
H	I(3), B(2)
I	M(2), J(1), H(3)
J	I(1), N(3), E(1)
K	Z(2), G(2), E(3)
L	Z(3), G(2)
M	I(2), N(2),
Z	N(2), L(3), K(2)

Figure 13.9 Network data structure, based on node adjacency table. Each entry in the table stores the nodes connected to a given node in the graph, and the associated weight or imped-ance. Note that in the network illustrated, each link is two-way and symmetric, with the same weight in each direction.

segment of a road, pipeline or electric cable. Edges have a direction of flow and a weight, determined by the impedances, that would be a measure of distance, travel time or some carrying capacity. In the context of spatial interaction modelling, described in more detail below, an edge may also have a measure of supply or demand associated with it, such as the number of people living in houses located on a road segment.

Analysis of paths through a network is assisted by the use of a *node adjacency list*. For each node, the corresponding adjacency list stores the connected nodes, along with the weights of the corresponding edges (given in parentheses in the table illustrated in Figure 13.9). If edges are directed, then a node will only be regarded as connected if there is a possible direction of flow towards it.

Shortest path in a network

One of the major applications of networks, as indi-cated above, is to determine the shortest path between two locations. Viewed more generically as 'least cost', this might also be interpreted as the quickest or the most scenic, for example. Shortest-path algorithms involve building a tree data structure representing specific paths through the network. Branches are equivalent to edges within the network and are only added to the tree or extended if they appear likely to lead towards the intended destination of a path. As the tree grows, certain branches may be abandoned permanently or temporarily in favour of others which are regarded as more promising.

At any one time in the solution procedure to find the least-cost path between two network locations, we have a *tree* consisting of edges already added, a *fringe* consisting of all nodes (and hence edges) that are adjacent to the terminal nodes ('leaves') of the tree, and *unseen nodes* that are not in the tree or immediately adjacent to it (Figure 13.10). The tree grows an edge at a time by selecting from the fringe that node which appears most likely to lead to the destination. Selection of a fringe node is based on a priority which is a measure of the best route. One such priority value, applicable to navigation prob-lems, is the sum of the shortest distance from that fringe node back through the tree to the start node and the straight line (Euclidean) distance from the fringe node to the end node (i.e. based simply on the locational coordinates). Use of this priority measure will tend to ensure that vertices in the direction of the end node are chosen preferentially to others and that they will be adjacent to a branch of the tree that pro-vides a relatively short path from the start node. Each time a fringe node is added to the tree to create a new branch (i.e. edge), those nodes that are adjacent to it become new fringe nodes, while the old fringe node is now part of the tree.

Referring to Figure 13.10, the tree is initialised with the start node A. Of the two fringe nodes B and D, D is chosen as it is nearer to A than B and the Euclidean distance from D to Z is also shorter. Note that having added D to the tree, the vertices G and C are added to the fringe. Node G is chosen rather than C or B because, although the path back to A is shorter in the cases of C and of B, the sum of the

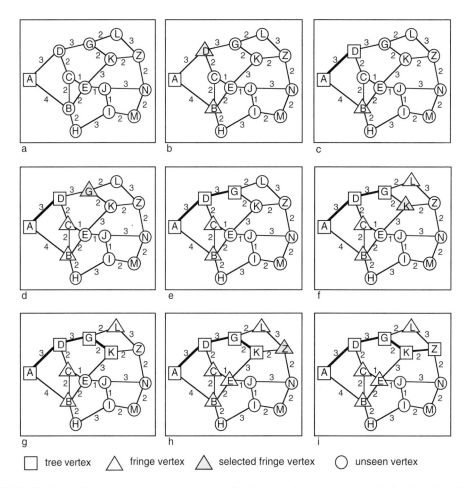

| □ tree vertex | △ fringe vertex | ◮ selected fringe vertex | ○ unseen vertex |

Figure 13.10 Shortest-path procedure in a network graph, for the route from node A to node Z. A tree is built progressively, starting at A, by identifying the fringe vertices and adding to the tree the fringe node that provides the shortest path based on the sum of the network path back to A and the Euclidean (straight-line) distance to Z. After a node is added to the tree, its fringe vertices are found and the process repeated until node Z is found (see text for further explanation).

path back to A and the Euclidean distance to the end node Z gives G the higher priority, it being the smallest distance of the possibilities.

Network allocation

Finding all edges or vertices in a network that are within a specified distance of a given vertex can be done by using the vertex adjacency data to search out from the central vertex accumulating edge distances until the limit is about to be exceeded. One approach to this is to use a *breadth first search*, whereby all vertices immediately adjacent to the start vertex are

examined, then all vertices adjacent to each of those vertices are examined and so on. Once an edge connecting two vertices has been examined it is flagged as 'seen' to prevent going round in circles. The search stops when there are no more vertices that have adjacent vertices within the search distance.

A variation on the above search would take account of demand associated with edges of the network in conjunction with a supply at the central vertex. Thus if there was a limited supply associated with the search centre (e.g. a school), then the search would stop either when a search distance limit was exceeded or when the demand was about to exceed the supply.

Minimum spanning trees

Shortest-path algorithms and network-allocation algorithms may be regarded as special cases of a set of algorithms for identifying subsets of edges in a network or graph data structure that meet various criteria concerned with the connecting nodes. Another example of such a subset is the minimum spanning tree. This consists of the smallest set of edges in a graph that provide a route from all nodes to all other nodes, taking account of the sum of the weights associated with the connecting edges. An example of the application of the MST, in the context of navigation, would be to determine the minimum set of air routes required to enable passengers to travel (by some route, however tiresome) between any pair of a set of destinations. See Mackaness and Beard (1993) for an example of an MST application in the context of map generalisation of road networks.

Spatial interaction modelling

Spatial interaction models are used to assist in making decisions about the optimal location or usage of facilities, such as shops, fire stations, schools and hospitals. The problem is to predict flows of people or goods relative to the proposed facilities, taking account of distances and travel times and the level of demand for the facility under consideration. There is particular interest in such models in retailing and service organisations which may wish to locate new shops or branches wherever they believe will attract the maximum custom or profit. In such situations it is necessary to take account of competing facilities. Thus the models may include several facilities that are fixed in location while one proposed facility may be variable in location, the objective being to determine one of several possible locations that, in the case of shops, maximises profit. In a more complex model it might be required to place more than one facility, each of which would be located at one of several possible places. Another example application would be the determination of an optimal set of locations for health care centres such as to minimise overall the distances that potential patients might need to travel

in order to attend. A further related example would be the choice of a set of locations for fire stations such as to ensure that all properties were accessible within a given travel time for the fire brigades.

Spatial interaction models may be formulated in terms of a set of *service centres* representing facilities and a set of *demand centres* representing the potential users of the facilities. These may also be referred to as *destinations* and *origins* respectively. Individual service centres may be weighted in terms of their attractiveness. In the case of a shop, this might be measured by a combination of factors such as floor space, parking space and some measure of the 'image' of the shop, which might be regarded as affecting customers' interest in using it. For a school or a hospital, the factors might include maximum number of pupils and patients respectively, along with other measures of the range and quality of services provided. Demand centres may also be weighted by productivity factors such as the number of residents, their disposable income and their predisposition to make use of a facility. The latter might be measured by socio-economic factors such as social class, age and type of neighbourhood. For purposes of retailing and marketing, several organisations specialise in classifying residents into numerous classes that can be used in determining these weights.

A further consideration that is essential to the spatial nature of these models is a measure of the distance between the service centres and the demand centres. The earliest versions of spatial interaction models assumed that the flow between demand and service centres was directly proportional to the associated demand and attraction and inversely proportional to the square of the distance between them, giving rise to the concept of a *gravity model*, by analogy with Newton's law of gravity. Clearly in practice this distance-related aspect of a model will vary in effect according to the mode of transport and the ability and expectation to travel. It is better regarded in terms of *accessibility*, which may be expressed as a cost function which takes account of distance. This cost function may be based on prior analysis of typical trip lengths in a particular situation for which data are available.

Because of the importance of travel between origins (i.e. the locations of demand) and destinations (i.e. the locations of the service), spatial interaction models may be combined with network models that can be used to compute optimal paths along various types of route and to allocate the nearest demand locations or links to a single centre.

The complete set of factors relating to attractiveness, productivity and accessibility may be combined to create an interaction matrix T_{ij} representing the individual flows between each origin i and each destination j. Each element t_{ij} of the matrix represents a flow, such as expenditure or number of people between origin i and destination j. It can be modelled by

$$t_{ij} = \alpha_i \beta_j e^{\gamma d}{}_{ij} + \varepsilon_{ij}$$

where α_i is a parameter representing the total production at origin i, and β_j is a parameter representing the total attractiveness of destination j. The expression $e^{\gamma d}{}_{ij}$ is the cost function, where d_{ij} is the distance from i to j, and ε_{ij} is a residual error term relating to the modelling of the flow between i and j. Given a set of real observations of y_{ij} of the flows from i to j, the method of maximum likelihood estimation can be used to calibrate the model and hence find values for the three parameters α_i, β_j and γ that provide a best fit with the observations (Bailey and Gatrell, 1995). This estimation procedure may be carried out in a manner that ensures that the results meet constraints of the observed total flows a_i from the origins and the total flows b_j at the destinations, as well as a constraint c of the total cost. These constraints may be expressed as:

$$\sum_j y_{ij} = a_i$$

$$\sum_i y_{ij} = b_j$$

$$\sum_i \sum_j y_{ij} d_{ij} = c$$

Use of all constraints results in what is termed a doubly constrained model. If the second constraint is omitted, then the model is origin-, or production-, constrained, while if the first constraint is omitted, then it is attraction-constrained. The significance of omitting one of the constraints is that it allows the possibility of representing the productions or the attractions by known factors that are assumed to be closely correlated with the actual production or attraction values.

In practice, in the case of shopping models, initial values used for purposes of calibration of the model may be found for the origin from surveys of family expenditure for given places, or by making an assumption based on socio-economic status. For the destination, spending might be found from the accounts of relevant existing stores or by making esti-

mates based on the type and assumed quantity of the goods sold.

Once the model has been calibrated, it is then possible to modify values for existing origins or destinations, which may be done with a doubly constrained model in order to predict the effect of changes. Alternatively, in an origin-constrained model, it is possible to add in a proposed new destination and hence make an estimate of the flows from each origin to that destination. This can be done by constraining the expenditure from all origins to their current value, while using a surrogate for attraction, such as the floor space, or a combination of this and relevant factors such as variety, price and quality of goods. Thus the existing destinations would, in the model, compete with the proposed destination on the basis of this attraction measure, in combination with the estimated cost function which represents accessibility.

Location-allocation models

Location-allocation models are a relatively complicated form of spatial interaction model. They are concerned with optimising locations of facilities, and sometimes demand centres, and optimising or predicting the allocations of people or resources to the facilities. We have already provided a simple, constrained example of this in the transportation model described in the earlier section on linear programming. In that class of problem there was a fixed number of existing supply and demand centres and the objective was to balance the allocation of resources to demand in an optimal manner. However, the problem becomes more complex when constraints of location are reduced, in that it becomes possible to consider several candidate facility locations from which an optimal subset will be chosen. Location-allocation modelling methods are therefore of particular relevance to spatial interaction problems concerned with locating facilities when there may be many possible sites to be considered.

If the number of possible facility locations is strictly limited then it is possible to envisage finding an optimal solution which satisfies the specified constraints, such as minimum travel costs. When the number of possible locations to choose from becomes high and several of them may be chosen, the problem becomes computationally very demanding and requires the use of heuristic algorithms that attempt to find solutions that are 'good' but may not be optimal. A heuristic adopted by Teitz and Bart (1968) and improved upon by Densham and Rushton (1992) involves evaluating marginal changes in a solution by repeatedly substituting one facility location for another. If the swap results in an improvement, then it is kept; otherwise it is rejected and the substituted facility reinstated.

The use of location-allocation algorithms has been subject to considerable criticism (e.g. Gore, 1991) as the supposedly optimal solutions that they produce may fail to take account of all possible contingencies. In particular they tend to assume a steady state in terms of the distribution of population or of land usage. If such parameters change, possibly as a direct consequence of the planned reallocation of resources, the predictions may be seriously awry. Some proponents of the use of location-allocation techniques advocate that they be incorporated in interactive decision-support systems that enable various scenarios to be considered without assuming that there is any one entirely predictable outcome. It is to be expected that automated decision-support techniques will be controlled through interactive graphics interfaces that make it easy for the user to propose various configurations of resources and to visualise the hypothetical consequences (Armstrong *et al.*, 1992).

Genetic algorithms

A relatively recent approach to solving optimisation problems is provided by genetic algorithms, which can be used to optimise the values of a set of variable parameters, in order to achieve some objective that can be measured in terms of a degree of fitness for a purpose. In a GIS context, an example of their application is to problems of spatial interaction modelling described above. Hobbs (1994) has described their use in determining the optimal distribution of building societies with regard to the actual and potential customer base.

The idea of fitness for purpose is inspired by the Darwinian concepts of evolution. By analogy with generations of an organism that undergoes modification in the content and pattern of genes in its chromosomes, as its function and behaviour adapts to that most suitable for survival, so the elements in lists of data values in a genetic algorithm are modified by combining them in different ways and

modifying their value. This is comparable to the natural processes that accompany mating and mutation. The results of these changes are tested against fitness criteria and the more advantageous combinations are retained and further modified. Ideally, those that survive the successive changes will be nearer the optimum than those that do not, and will gradually evolve towards the optimum state.

The success of the method depends upon allowing diverse forms of the simulated chromosome to continue to evolve to increase the chances of one of them converging on the optimum. The surviving chromosomes may be regarded as corresponding to local maxima in a search space. One of the maxima is assumed to lead to the global maximum that is the solution to the problem. The technique is still relatively new but it does appear to have relevance to problems of optimal site selection and to determining combinations of land suitable for various purposes, as discussed earlier in the context of weighted overlay techniques, and linear programming.

Summary

For many organisations the usefulness of GIS depends on the extent to which the technology can assist in decision-making processes that require access to and evaluation of multiple sources of information. In this chapter we have reviewed several techniques that are found within commercial GIS or are used in association with GIS to help in solving problems concerned with finding sites that meet user-specified criteria, finding routes through networks and with allocating resources. In doing so we have encountered several techniques that are applicable to a wide range of spatial data-handling problems beyond those described here.

In the context of evaluating sites suitable for particular purposes, we have seen that map-overlay techniques can be combined with the methods of multi-criteria evaluation (MCE) by attaching weights to individual layers of an overlay and by representing factors within each layer by scores that can contribute to the weighted overlay operation. This is a simple and quite versatile procedure but it lacks some of the expressive power regarding specification of constraints and objectives that can be obtained using the more complex techniques of linear programming that are well established in operational research and management science.

Many decision-making processes depend upon the expertise of professionals in particular application areas. Expert systems and knowledge-based systems provide some potential for encoding this expertise in the form of rules that can be processed by an automated inference mechanism. The rules can be qualified to take account of uncertainty in their relevance and of uncertainty in the data, using methods such as fuzzy logic and Bayesian probability. They can also be combined with functions or programs that solve subproblems using conventional procedural programming methods.

Multi-criteria evaluation and rule-based systems provide methods that assist particularly in solving ill-structured problems to which there may be no definitive solution. An important class of subproblem in some spatial decision-support systems is that of determining optimal routes in network-structured data such transport networks. We have summarised the principles of network search procedures, such as that to find the least-cost path between two locations and to find the set of locations that are within a given cost of a specified centre, for purposes of solving allocation problems.

Network search procedures can be incorporated within the more complex area of spatial interaction modelling, in which the problem is to study the flows, such as of goods and people between demand centres requiring a service and service centres or facilities that can meet the demand. Retail facilities such as shops or entire shopping centres can be characterised in terms of their attractiveness to different sections of the public and the flow of trade can be modelled §using gravity modelling techniques in which the attractiveness of particular facilities can be quantified as a function of their accessibility or travel cost. When evaluating multiple potential sites for new facilities such as shops or hospitals, the technique of location-allocation analysis can be used to find one or more sites that are optimal relative to other potential sites and when taking account of competing sites.

A technique that has not been described in this chapter, but which is relevant to certain siting problems in which environmental impact is an issue, is that of intervisibility analysis. This is described briefly in this book in the context of visualisation methods in Chapter 15.

Further reading

Voogd (1983) is a widely used textbook for multi-criteria analysis. Some clearly described examples of the application of weighted overlay techniques can be found in the book by Bonham-Carter (1994), which focuses particularly on modelling geoscientific problems. For a clear introduction to optimisation techniques, including linear programming and many other techniques, including those for network analysis, see Winston (1994). Sedgewick (1988) provides a good introduction to the implementation of optimisation algorithms, including shortest path and MST, as well as linear programming. A good overview of techniques in artificial intelligence, including rule-based systems, is Winston (1992), which serves as a useful reference for methods such as neural networks and genetic programming. For an article on uncertainty in expert systems, see Spiegelhalter (1986). Zimmermann (1985) has written a text on the subject of fuzzy set theory, while. Davidson et al. (1994) compare the use of fuzzy set and Boolean overlay methods in the context of land-use evaluation. Books on the subject of spatial interaction modelling include Fotheringham and O'Kelly (1989) and Ghosht and Rushton (1987). Birkin et al. (1996) provide informative accounts of the application of spatial interaction modelling. One of the pioneers of the use of spatial interaction modelling techniques is Wilson, who authored and co-authored several textbooks on geographical data modelling methods with a mathematical flavour, including Wilson et al. (1981) and Wilson and Bennet (1985). Fotheringham and Rogerson (1994) have edited a set of essays on spatial analytical methods, several of which relate to topics in this chapter.

PART 5 Graphics and Cartography

CHAPTER 14

Computer graphics technology for display and interaction

Introduction

Early interest in the field of computer cartography may be attributed to the realisation that computer graphics technology can be used to plot detailed maps much more quickly and more precisely than is possible using manual methods. The graphical quality of the early digital map products often left much to be desired when judged by traditional standards. The technology was nevertheless particularly attractive when faced with labour-intensive tasks such as mapping the distribution of large quantities of statistics, the plotting of mathematically complex geographical projections, and the generation of 3D views. Modern graphics-plotting devices have now advanced to a stage where they are widely used for high-quality cartographic production, while in GIS,

computer graphics is one of the cornerstones of the technology. In emphasising the benefits of computer graphics for cartography, it should be remembered that the advantages are at the relatively mechanical stage of plotting graphic symbols and interacting with graphics displays. The design of maps is a separate issue, and is very far from being fully automated (see Chapter 15).

Before describing the various aspects of computer graphics technology in more detail, we take a brief overview. Considering Figure 14.1, the computer graphics capability in a GIS may be regarded as being maintained by a graphics subsystem that is accessed from the main GIS or cartographic system, or from application programs that form part of the overall system. In general, the graphics software procedures operate upon data that are retrieved from the spatial database, and which may sometimes be

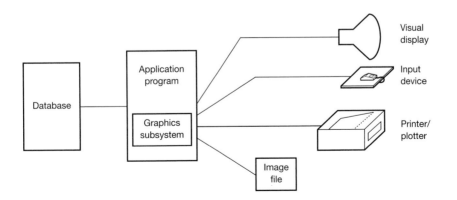

Figure 14.1 Basic components of a graphics system.

transformed by spatial analytical programs before being displayed either on a screen or on a printer. The graphics software is able to direct graphic output to screen displays, to hard copy printers and plotters and to plot files or image files that can store graphical data for subsequent display. An essential part of computer graphics technology is also the capacity to enable the user to input data, and hence to interact with screen displays of graphical information.

In this chapter we will review computer graphics hardware and software, starting with screen display technology and hard copy output technology for printers and plotters. The chapter follows a major division in computer graphics hardware between those devices which produce *soft copy* in the form of a temporary display on a screen, and those which produce *hard copy* in the form of a piece of paper or film on which the graphic is printed or drawn, typically either with some form of ink or by photographic means. A section then follows on the subject of colour specification and methods of transforming between different colour systems that are used in computer graphics. The chapter continues with a review of the interactive input devices that enable the user to control the running of a GIS or cartographic program. The last part of the chapter considers briefly the graphics software systems that programmers use to implement the graphical aspects of GIS and cartographic packages.

Screen display technologies

Cathode ray tube devices

At the time of writing, the majority of screen displays, or *visual display units* (VDUs), employ *cathode ray tube* (CRT) technology, though for the purposes of portable computers they are being replaced by flat panel displays. The CRT consists of a vacuum-sealed glass tube, at one end of which is a source of electrons which are focused into a narrow beam directed towards the relatively flat end of the tube that constitutes the screen (Figure 14.2). The inside of the screen is coated with phosphor, which emits light when it is struck by the electron beam. In addition to focusing, a mechanism must be present to deflect the beam to different parts of the screen. Both focusing and deflection are normally performed by means of electromagnetic coils situated around the neck of the tube. Beam intensity, which governs the brightness of the image, is controlled by a negatively charged control grid adjacent to the metal heating element, or electron gun, which generates the electrons. The passage of electrons in the direction of the screen is ensured by placing a positive (attractive) charge on the inside of the main body of the tube.

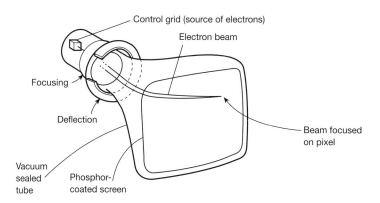

Figure 14.2 Cathode ray tube (CRT). The control grid in the electron gun modifies the intensity of the beam of electrons which are accelerated toward the phosphor-coated screen, via focusing and deflection controls. In a monochrome display (illustrated) there is one electron gun. In a colour display there are three.

Once the electron beam has traced out the pattern of the image, which may consist of many thousands of dots or vectors, it will only remain visible for a very short time. To retain constant visibility of the image it must be redrawn, or refreshed, repeatedly. In practice, *refresh frequencies* vary between different devices, but usually the image is redrawn entirely somewhere between 60 and 80 times a second.

Raster scan displays

Raster scan refresh displays are at present the most widely used type of VDU. They are closely related to televisions in that the path of the electron beam is a sequence of horizontal rows or scan lines and is always the same for any image (Figure 14.3a). The raster scan mode of refresh differs from another earlier, but now quite rare, vector mode, in which the beam can move in any direction, drawing lines as it does so (Figure 14.3b). The reason why raster technology has become dominant is that it provides much greater capacity for high-quality colour, enabling the creation of detailed images, containing either text and graphic symbols, or photorealistic scenes. Vector-mode screens have limited colour capacity, with a tendency for images to flicker if they contain a large amount of detail. Furthermore they would normally have a dark background, which prevents the creation of screen images that resemble equivalent hard copy printed on white or coloured backgrounds.

In a raster scan device, changes in intensity of the beam give rise to different-coloured dots or pixels from which the image is constructed. Typically there are around 72 dots per inch on the screen. The pattern of pixels representing the picture is stored in a *frame buffer* which consists of a 2D array of values corresponding to the intensities of the pixels. A microprocessor, sometimes called the *video controller*, reads through each row of the frame buffer in synchronism with the motion of the electron beam. The value of each pixel element stored in the frame buffer is used to set the intensity of the electron beam at the corresponding position within the scan line.

Single and multiple bit plane devices

In a monochrome raster scan device, the value of each pixel may be either 0 or 1, which correspond to the electron beam being either on or off at that position in the image. In a colour device, a greater range of pixel values can be stored, depending upon the number of bits allocated to each element of the frame buffer. Four bits per pixel gives a range of values from 0 to 15. Many colour displays provide a range of 256 colours, by using eight bits per pixel, while on some devices the range of possible pixel values may extend into millions. If there are n bits per pixel, i.e. n *bit planes*, the range of possible values is 2^n. On displays used for image processing, 24 bits (i.e. three bytes) per pixel is quite commonplace.

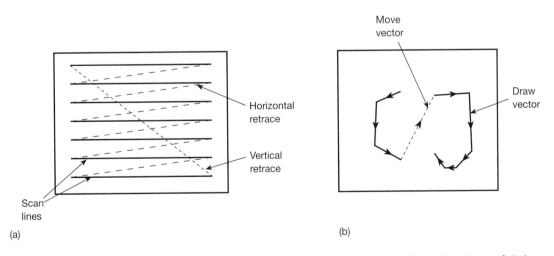

Figure 14.3 (**a**) Raster scan display. The electron beam always traces the same path, creating horizontal rows of pixels on each scan line. (**b**) Vector scan display (rarely to be found!). The electron beam moves in straight lines between pairs of points on the screen. In a draw vector the beam is switched on. In a move vector the beam is switched off.

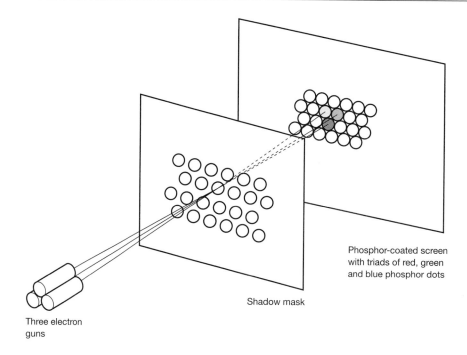

Figure 14.4 Colour raster CRT. Three electron guns, dedicated individually to red, green and blue respectively, are focused on the corresponding phosphor dots of each pixel.

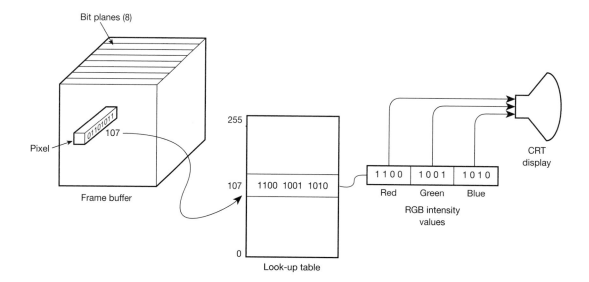

Figure 14.5 Colour look-up table. The pixel value 107 (binary 01101011) serves as an index to the look-up table, which stores the intensity values to be used for red, green and blue guns of the CRT. In the figure, the frame buffer has eight bits, therefore the look-up table must have 256 entries (one for each possible pixel value). The look-up table in the figure has just four bits per primary colour. Often, colour tables have eight bits per primary colour, giving a palette of 2^{24} colours.

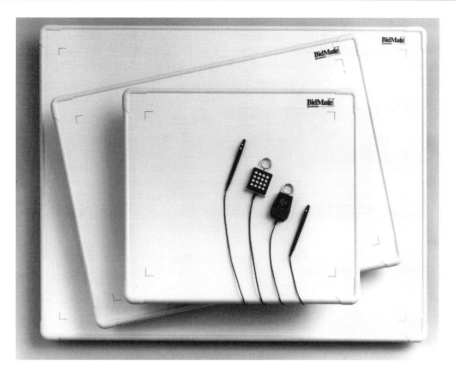

Figure 14.6 (**a**) Digitising tablets with cursors and styluses. Courtesy of Numonics.

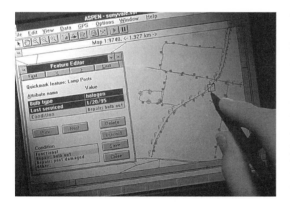

Figure 14.6 (**b**) Flat-panel display with stylus input device. Courtesy of Trimble.

The way in which colour is obtained on a raster scan device is based upon the use of three distinct electron guns dedicated individually to the three primary colours of red, green and blue (Figure 14.4). The phosphor coating of the screen is composed of numerous dots of red, green and blue light-emitting varieties of phosphor. The three types of phosphor dots on the screen are either grouped into triads or in vertically aligned rows of three, and the electron guns are also organised in one or other of these arrangements, referred to as *delta* and *in-line* arrangements respectively. The value of each pixel stored in the frame buffer is used to control the electron intensities to which each different type of dot in the corresponding group is subjected. To ensure that the electron beams only impinge upon one group of phosphor dots at a time, a *shadow mask*, containing carefully positioned holes, is placed on the electron gun side of the phosphor. When focused on a particular triad, the three beams should all pass through the same hole of the shadow mask.

Colour look-up tables

In order to control the intensities of the three electron guns, each pixel value in the frame buffer must be translated into three intensity values. This is usually done by treating the stored pixel value as the identifier, or address, of a colour definition stored in the colour look-up table (Figure 14.5). There will be a colour definition, consisting of red, green and blue intensity components, for each possible pixel value. If the frame buffer had eight bit planes, there would be 256 (2^8) entries in the colour table. If a wide choice of possible colours is required, each colour component in the colour table could be represented by eight bits (i.e. one byte). This would provide a range of possible colours, referred to as the *palette*, of 2^{24} colours. Which colour each pixel value represents is adjustable through software.

The facility to change the colours allocated to pixel values is of particular importance in image-processing applications such as remote sensing. If it was known that all pixel values below a given limit could be regarded as noise, or in other words their variation contributed no useful information, then the corresponding set of pixel values below this limit could all be set to a single background colour in the colour look-up table. This procedure is known as *thresholding*. Another image-processing application of colour look-up tables is that of *contrast stretching*, which is a technique of contrast enhancement. Here it may be known that within the 256 possible values of a digital image, only those within a narrow range are of interest. The pixels within this range would then be set to clearly contrasting colours, while all other pixel values were set to a background colour. The subject of image processing in the context of remote sensing is discussed further in Chapter 6.

Flat CRTs

The dominance of CRT technology amongst display devices is due to its capacity for creating high-quality images, in terms of a combination of high resolution, good colour range and a bright screen which is easy to view from different angles. One of its disadvantages is bulk: they are heavy and occupy a large space, since their depth is usually comparable to the width of the screen. There have been a number of attempts to reduce the bulk of CRTs by making them flatter.

The reason why CRTs are quite deep is that conventionally, if they were otherwise, they would require higher power than they already demand. The power required to deflect the electron beam is proportional to the angle of deflection. The maximum angle of deflection will depend upon the distance of the electron gun from the screen. Thus the further away it is, the smaller the maximum deflection, and hence the less the power required. A flatter CRT can be constructed by exploiting the fact that a low-energy beam can be deflected with less power than a high-energy beam. By placing an *electron multiplier* adjacent to

the screen, an initially low-energy beam can be amplified, after it has been deflected, to the energy level required to generate adequate light at the screen.

Flat-panel displays

Flat-panel displays (Figure 14.6b) work in raster mode. Unlike raster CRTs, not all flat-panel displays need to be refreshed for the image to remain visible. At present, the resolution of flat panels is often similar to CRT displays. However, the image quality tends to be inferior to CRTs with poorer contrast and brightness. This situation can certainly be expected to change with continued improvement in flat-panel technology. There is a variety of types of flat-panel displays, some of the more common of

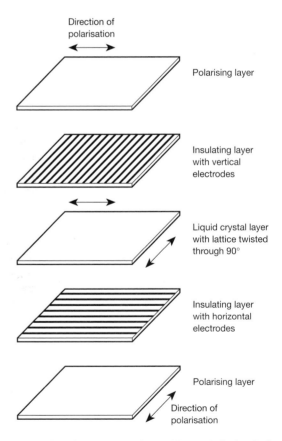

Figure 14.7 Construction of a liquid crystal display (lcd) panel. Applying a potential difference to a pair of wires causes the intervening liquid crystal to be disrupted, preventing light passing through the panel.

which are *liquid crystal displays* (lcd), *plasma panels* and *electroluminescent displays*.

Liquid crystal displays (Figure 14.7) consist of a layer of organic fluid, on either side of which are layers of horizontally and vertically aligned electrodes. These are themselves enclosed by two transparent plates. The plate on one side is polarised vertically while the plate on the other side is polarised horizontally. The liquid crystal layer consists of elongated crystals which are progressively twisted, through 90°, or a multiple of 90°, from one side of the layer (or lattice) to the other, enabling light to be transmitted through the panel. By modifying the voltage on appropriate pairs of wires, the liquid in the zone where the wires intersect becomes disrupted, affecting its capacity to transmit light, with the result that a dark (i.e. opaque) dot appears at the location. Since the lcd panel itself does not generate light, it must either use a light source placed behind the panel or a mirror may be placed behind it to reflect external light.

In an *active matrix lcd*, each pixel location is accompanied by a transistor which can independently maintain the pixel in the on state until a change in the pixel colour is required. In a non-active matrix lcd, each pixel is refreshed repeatedly. Colour versions of lcds have been developed, but their quality, though quite acceptable for many purposes, is not yet competitive with CRTs.

Electroluminescent displays consist of a sandwich of phosphor surrounded by a mesh of wire electrodes enclosed by transparent plates. By passing a current between a pair of horizontal and vertical electrodes, the intervening phosphor is stimulated to emit light. Changing the voltage causes the intensity of the light emitted to vary. It is also possible to create multicolour displays by using different types of phosphor which emit red, blue and green light respectively.

The *plasma panel* is the longest-established flat-panel display technology, but it is relatively expensive and has a high power requirement. Its ruggedness has led to its application in a military context. They also have the capability of being constructed with very large screen sizes. The panel again consists of a transparent sandwich, with transparent plates and vertical and horizontal electrodes (Figure 14.8). The central layer is made up of an array of cells of neon gas. A cell is initially illuminated, or 'fired', by raising the voltage between the adjacent horizontal and vertical wires. The voltage can then be lowered to a sustaining level, as with conventional neon lights. Further lowering of the voltage switches the pixel off. Plasma panels are conventionally monochrome, but colour ones can be expected to be available in future (Foley *et al.*, 1990).

Hard copy devices

Hard copy output devices can be differentiated broadly according to whether they operate in vector or raster modes. We may also distinguish between the majority of printer and plotter output devices, which apply pigments such as inks, powder and wax onto paper, and the less common, though often very high quality, plotters which use light beams to expose photographic film.

Raster printers and plotters

Raster hard copy devices create images from a pattern of single-coloured or multicoloured dots, with the image being built up progressively from rows of dots. The minimum possible distance between adjacent dots is determined by the *addressability* of the device, expressed in dots per unit distance (inches, centimetres, millimetres, etc.). Typical addressability

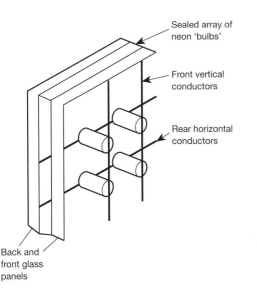

Sealed array of neon 'bulbs'

Front vertical conductors

Rear horizontal conductors

Back and front glass panels

Figure 14.8 Section through part of a plasma panel. By raising the potential difference between a pair of electrode wires, the neon gas between them is made to fluoresce.

is at least 200 dots per inch, with higher quality up to about 600 dots per inch. Considerably higher degrees of addressability are achievable on some laser plotters, which can exceed 2000 dots per inch. The continuity and smoothness of lines in the raster image will depend upon a combination of the addressability and the dot size, since if the dot diameter exceeds the interdot spacing, the dots will merge into each other, creating a smoother line. The *resolution* refers to the maximum number of parallel lines that can be drawn, and be visibly distinguished, per unit distance.

Monochrome printers can be used to achieve grey scale shades, similar to newsprint *half-toning*, by creating patterns of fixed-size dots of different density, rather than the variable-size dots used originally in newsprint. Figure 14.9 illustrates a sequence of ten levels of intensity (0–9) obtained, in this case, with a 3×3 pattern of dots. Note that each successive intensity level is created by adding a dot to the previous pattern. This ensures that small changes in intensity level in the original image do not produce marked contours in the approximated half-tone image. The sequence of intensity levels can be described by an $n×n$ *ordered dither* matrix $\mathbf{D}(i,j)$ in which each element is numbered from 0 to 8, and is 'switched on' if its value is less than the intensity level. The dither matrix for the sequence in Figure 14.9 is

    6 8 4

    1 0 3

    5 2 7

It should also be noted that the patterns for individual intensity levels avoid the introduction of simple lines or blocks of pixels. If they were present, then continuous areas of a particular intensity level would result in a bold and undesirably assertive pattern, e.g. parallel lines. This technique of approximated half-toning can be modified to represent an image with the same number of pixels as in the original rather than with a multiple of n, which results from the use of the dither patterns illustrated in Figure 14.9. The decision whether or not to switch on a pixel in the display is taken on the assumption that the pixel belongs to a larger area of the same required intensity level. Thus for a pixel in the display with coordinates x,y, dither matrix coordinates i and j are calculated with the formulae

$$i = x \text{ modulo } n$$

$$j = y \text{ modulo } n$$

where the modulo function returns the remainder of integer division of x (or y) by n. Provided that there are smooth variations in intensity in the original image, then the fact that two adjacent pixels in the original are not quite the same, will not have a very noticeable effect because successive dither patterns are quite similar.

Colour on raster printers is achieved by combining a limited number of dot colours. Typically, the subtractive primary colours of cyan, yellow and magenta are used in combination with black ink. A separate black ink is used because it usually produces a darker black than the combination of cyan, yellow and magenta, and because, as a frequently required colour, it is more economical. Direct mixing of pairs of the three primary colours can give a total of seven distinct colours. A very much greater range of colour

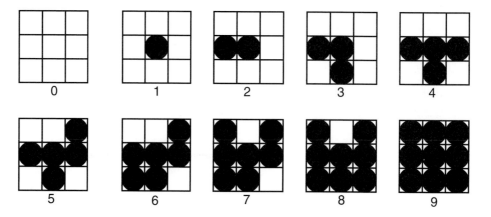

Figure 14.9 A sequence of ten 3×3 dither patterns for approximating half-tones.

can be achieved by using dithering techniques to combine clusters of different-coloured inks. Provided the clusters are small, the colours will appear to merge, to create a wide range of shades.

Dot matrix printers

Dot matrix printers are one of the simplest types of raster printer. A printing head moves across the paper, printing several rows of the raster at a time. The paper is moved perpendicular to the axis of the cylindrical platen. The print head contains columns of metal pins; these strike a ribbon which transfers pigment onto the paper. The use of these printers has diminished somewhat due to improvements and reductions in cost in some of the other technologies described below.

Inkjet printers

Inkjet printers produce raster images in a similar manner to dot matrix printers in that a printing head traverses the paper, generating several rows of a

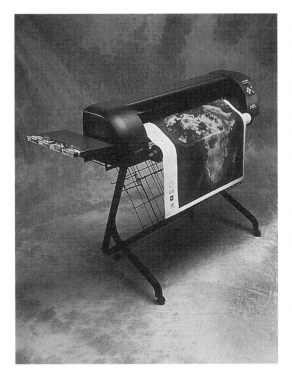

Figure 14.10 Large format Calcomp inkjet printer. Note that the majority of inkjet printers in use are smaller-format (A4).

raster at a time. However, they can produce better quality graphics. Cyan, magenta, yellow and black inks are sprayed simultaneously from the print head directly onto the paper. Resolution for the basic colours is usually about 300 dots per inch though resolutions up to 700 are available in some commercial devices. These are relatively cheap devices, giving good contrast at quite high resolutions. Many of these devices are A4 format, but larger formats are also widely available (Figure 14.10).

Electrostatic plotters

Electrostatic printers are typically used by organisations requiring large-format plotters. The image is obtained by passing the paper over a row, or comb, of electrodes, each of which can place a dot of negative charge. The paper is then passed across positively charged toner which attaches itself to the paper at the negatively charged locations. This produces a monochrome image. Colour electrostatic plotters use cyan, magenta, yellow and black toners, which may require a separate pass for each colour. Resolutions of up to 400 dots per inch may be obtained.

Because the toner may be deposited, to a small degree, even where no charge was placed on the paper, the contrast of the resulting images is not always as good as some other techniques. Large-format devices up to about 2 m in width are available and, for many purposes, have replaced large-format pen plotters because they work much faster to produce complex pictures. They are also able to produce much better colour-shading effects than a vector-made pen plotter, though the smoothness and constrast of linework is not always as good as pen plotters.

Laser printers

The same laser printers that are widely used for text printers, with word processors, can usually also be used for graphics. A laser beam is used to scan a pattern of charge onto a rotating selenium drum. Toner is then flowed over the drum before transferring the resulting pattern onto paper. The normal resolution is either 300 or 600 dots per inch and most devices are currently monochrome. Colour laser printers are available, but at the time of writing are much more expensive than inkjet printers, without producing very much better quality. They use different-coloured toners which are transferred in separate passes. Relatively high contrast and high resolution are thus obtained.

Laser plotters

Very high-resolution graphics can be obtained with laser plotters, which are typically large-format devices in which a laser beam exposes a film mounted on a rotating drum. Colour output is obtained by photographic printing from separately produced films (colour separates). Addressability in excess of 2000 dots per inch can be achieved and hence these devices provide the highest quality output, suitable for purposes of producing published maps, for which the associated expense may be warranted.

Thermal transfer (wax dot) printers

In thermal wax dot printers, coloured pigments embedded in wax paper are transferred to plain printing paper by a row of heating nibs which melt the pigment onto the paper. The image is produced row by row in raster format as the plain paper and wax paper pass under the heating nibs, which are positioned with a density typically of either 300 or 600 per inch. These devices can produce bright, high-contrast images.

Dye sublimation printers

Colour printing which resembles colour photography can be achieved with dye sublimation printers in which the dye is heated and transferred in a gaseous state. The combination of a high addressable resolution and a slight blurring results in a finely defined smooth image in which pixels are not obvious. This is very good for a photographic style of reproduction, but at present they can lack clarity for very fine lines and fonts, when compared with thermal wax and with laser plotters.

Vector mode pen plotters

Pen plotters work directly in vector mode, usually using several (at least three) different-coloured pens, which can be individually lowered and raised to start and stop drawing. There are two main types: *drum plotters* and *flat-bed plotters*.

On drum plotters, a roll of paper is moved over a drum which can rotate back and forth. The pens are mounted on a carriage that moves parallel to the axis of the drum. Thus two perpendicular components of pen motion relative to the paper are provided.

On flat-bed plotters the paper is fixed in position on the plotting surface, while the pens are mounted on a carriage that moves along a gantry which itself moves perpendicular to its length. The paper may be held in place by a vacuum, or by electrostatic means. The dimensions of flat-bed plotters range between A4 format and large devices several metres in extent.

Pen plotters can provide high-resolution, high-contrast output, up to about 1000 addressable units per inch. The pen-drawing operation is controlled by microprocessors which implement line-drawing commands and can generate character fonts, symbols and smooth curves. Note that the microprocessors provide the function of rasterisation in that they must translate vector drawing commands into discrete x and y coordinates which define the path of the pen. A disadvantage of pen plotters, in addition to their limited colour capability, is that they can take a long time (hours) to create complex pictures, plotting time being proportional to the number and length of lines.

The pens in a plotter may be replaced by a light source for exposing photographic film, or by scribing tools for cutting acetate.

Photography

Colour prints with good colour range and contrast can be obtained by simply photographing the screen of a colour monitor using a conventional camera. The resolution of such images is limited by shadow mask technology and, unless a hood of some sort is used, care must be taken to avoid reflections from the screen. Higher quality photographic images can be obtained from a CRT by using *film recorders* which expose the three primary colours of red, green and blue, separately through colour filters. Thus the output device generates pictures in three phases using a single electron gun for each colour component in turn. The principle can be applied to both vector and raster mode CRTs.

Colour specification and transformation

In the previous sections it has become apparent that screen display devices and hard copy devices use two different methods of defining colour. For those familiar with using both visual display units and plotters, this can be somewhat confusing, but more of a problem is the common failure of screen colours to be matched by the hard copy colours for the same image. Thus, in this respect, what you see on the screen rarely turns out to be what you get on the printer. In this section we will provide a short introduction to the theory of colour representation and indicate methods for transforming between the different colour systems in use.

Colour can be described in terms of three dimensions, referred to as *hue*, *lightness* and *saturation*. Hue is the primary means that we use to distinguish colours verbally, as in naming the colours of the rainbow, for example. It is a function of the wavelength of the light that we see. When describing colours, several ranges of wavebands have particular importance, because other colours can be created from combinations of these wavebands. Red, green and blue are described as *additive primaries*, because when viewed simultaneously in different proportions (intensities) they produce different colour combinations. They can be viewed simultaneously by transmitting them onto a single location on a white reflective surface from which we receive them after reflection, or by generating them by stimulating the three different types of phosphor dots on a screen, as described earlier in this chapter.

When coloured pigments are applied to a surface and white light is reflected from them, the pigments absorb different parts of the spectrum and reflect the rest. The three pigments of cyan, magenta and yellow can be mixed to produce a wide range of reflected colours and they are called *subtractive primaries*. The reason for the use of the word subtractive is that, when one of these primaries is combined with another, the effect is an increase in the range of wavebands that is absorbed; hence the greater the mixture, the fewer wavebands that will be seen by the viewer. The combination of all three in equal proportions results, in theory, in black (i.e. absence of colour).

Lightness, as its name implies, is a measure of how light or how dark a colour appears and is a function of the amount of light energy. As the quantity of lightness decreases toward zero, the perceived colour goes towards black. A widely used alternative term for lightness is *value*. Unfortunately the term value tends to be confused with other uses of the word.

Saturation (also called *chroma*) is related to the strength of a colour. A decrease in colour results in paler (rather than blacker) shades of the colour. It is sometimes described in terms of the degree to which a hue is mixed with grey. Taking the example of red, a reduction in intensity results in increasing degrees of pinkness.

The CIE system

A long-established standard for specifying colour is that of the Commission International de l'Eclairage (CIE). It is based on studies of standard sources of light including a tungsten incandescent light and average daylight from an overcast sky, and of the combinations of wavelengths of some additive primary colours required to induce particular colour sensations in human observers. The combined values of these additive primaries required to produce the same colour as a particular wavelength of light are called tristimulus values, X, Y and Z, and they are converted to so-called chromatic coordinates x, y and z by the formulae $x = X/(X + Y + Z)$ and $y = Y/(X + Y + Z)$ and $z = (1 - x - y)$. Within the X, Y, Z coordinate system there is a cone of values that correspond to colours that are visible to humans. Part of this cone is illustrated in Figure 14.11.

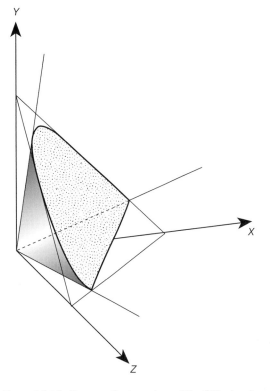

Figure 14.11 The coordinate system of the CIE primaries X, Y, Z. The shaded conical region represents those values of X, Y and Z that produce visible colours. The plane intersecting the cone is that of $X + Y + Z = 1$, and is the plane of the CIE chomaticity diagram (see Figure 14.12).

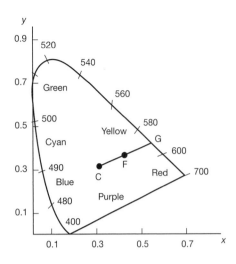

Figure 14.12 CIE chomaticity diagram. The x and y coordinates are chromaticity values. The numbers on the curve are wavelengths of light. The curve represents the spectrally pure hues.

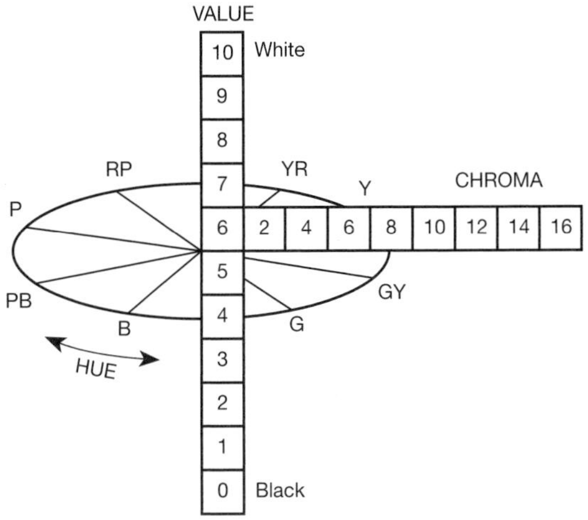

Figure 14.13 Munsell colour system. The letters describing hue refer to Blue, Purple, Red, Yellow, Green and their combinations. Red and BlueGreen are obscured.

The chromatic coordinates are used to construct CIE chromaticity diagrams (Figure 14.12). Note that only x and y are plotted, z being implicit. The chromaticity diagram is drawn within the plane corresponding to $X + Y + Z = 1$ (illustrated in Figure 14.11). The closed curve (spectral curve) in the diagram represents the values equivalent to all hues and hence they correspond to the range of wavelengths of visible light, some of which are shown in Figure 14.12. The point C marks the coordinates of a particular type of white light source. This position is achromatic and values between this point along any radius emanating from it range in intensity from zero at the point to 100% where the radius intersects the curve. A point such as F marks the coordinates of a particular colour. The radius emanating from C, which passes through F, intersects the curve at a point G, which is called the dominant wavelength of the hue corresponding to this particular colour. The ratio of the length CF to CG is called the spectral purity of the colour F. Clearly, the nearer F is to the curve, the nearer the colour is to the pure hue.

The CIE system does provide a consistent means of defining colour, but the diagram does not represent colours in a way that maps easily to familiar human perception of colour. Note that the chromatic coordinates of a colour are obtained using a machine and that the linear scale of these observations does not correspond with human perception of relative strengths of colour. The shortcomings of the CIE system with regard to its failure to use scales of mea-surement that model human perception of colour have been addressed with the CIE LUV system, which is a transformation of the CIE coordinates to produce a perceptually uniform colour space.

The Munsell system

The Munsell system is a colour system that is easily employed by people working with colour design. It describes colour in the dimensions of hue, value and chroma, and uses scales that correspond to human perception, so that equal intervals on the scale correspond to equal perceived intervals regarding changes in value (lightness) and chroma (saturation). Particular equally spaced intervals in each direction are represented by actual samples of the corresponding colour. This standard set of colours provides a common basis for colour specification, provided all concerned have a copy of the standard set! Figure 14.13 illustrates the axes in which it can be seen that hues are defined by points on a circle in the horizontal plane, which is partitioned into ten named colours, between which there are finer divisions. The vertical axis represents value, which is again divided into 10 major divisions, with 0 as black and 10 as white, while chroma is represented by the distance radially from the vertical axis. The central axis represents a range of colours that have zero hue and thus range on a grey scale from black to white. The standard Munsell colours have been translated to the CIE system.

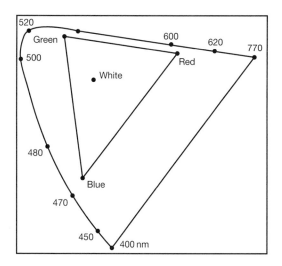

Figure 14.14 The colour gamut, i.e. range of possible colours, on an RGB colour monitor is defined by the region within the triangle, the corners of which correspond to the three primary colours that the monitor produces.

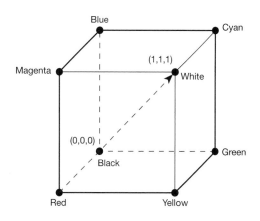

Figure 14.15 RGB colour space.

Colour gamuts

A line between two colour points within the CIE diagram (Figure 14.12) represents all the intermediate colours that can be obtained by mixing (additively) the two colours. The range is known as the *gamut*. Taking three points in a triangle provides a three-colour gamut. Such gamuts can be used to represent the range of colours that can possibly be produced by a display device such as a colour CRT, which creates colours by mixing red, green and blue (Figure 14.14). Note that a triangle based on these three colours does not cover the whole CIE visible colour space, and hence CRTs cannot display all possible visible colours. Furthermore, the colour gamuts of hard copy printing devices are smaller than the CRT colour gamut. Therefore, if it is intended that the screen display should match with a hard copy output, the colours on the screen display should be restricted to those available on the printer.

Colour models

The fact that computer graphics devices create colours either by combining red, green and blue on a VDU, or cyan, magenta and yellow subtractive primaries on a printer, has resulted in the widespread use of the corresponding RGB and CMY colour

models, each of which express colours as combinations of the respective primaries. The RGB colour model can be represented graphically by a cube (Figure 14.15) in which the 3D axes correspond to intensities from 0 to 1 of each primary colour of an RGB device. The origin of the coordinate system at (0,0,0) corresponds to black, while the point (1,1,1) corresponds to white. All points along the diagonal from black to white are shades of grey. The complementary colours of cyan, magenta and yellow occur at respective corners of the cube.

The location of all the colours in an RGB colour space, in terms of the CIE diagram, depends upon the properties of the phosphors used. Thus to translate colours precisely from one CRT to another requires a knowledge of the corresponding chomaticity coordinates (which might be obtained from the manufacturer). For CMY hard copy devices the colour space can be represented in a similar manner to the RGB cube, except that the coordinate system would be the opposite way round, with (0,0,0) giving white and (1,1,1) giving black. Thus increases in intensity of the CMY subtractive primaries result in the removal of colour from white.

Neither RGB nor CMY can be regarded as particularly intuitive, being related as they are to the physical characteristics of the computer graphics devices. This has been addressed by the use of colour models that are based on hue, lightness and saturation (HLS) or, using slightly different terminology,

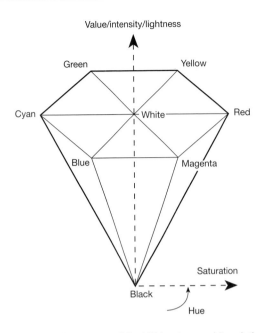

Figure 14.16 Hexcone of the HSV system and its relationship to the RGB colour space.

hue, saturation and value (HSV), that correspond to the more intuitive specifications of colour introduced earlier. HSV is also called HSB, for hue, saturation and brightness.

Referring to the RGB cube in Figure 14.15, it can be seen that it is possible to establish a simple relationship between the hardware-based dimensions and the intuitive dimensions. If we envisage looking toward the RGB origin (0,0,0) from the direction of point (1,1,1) we see a hexagonal cross-section, which in the HSV system is regarded as part of a hexcone, as illustrated in Figure 14.16. The vertical axis is value (lightness), from black to white, giving shades of grey. The radial distance from this axis corresponds to degrees of saturation from zero to full saturation at the external surface of the hexcone. The radial angles around the V-axis correspond to the specification of hue, with zero being red, 60° yellow, 120° green, 180° cyan, 240° blue and 300° magenta. It should be remarked that like the CIE system, the scales used in HSV and HLS do not correspond to human perception, but they do enable a more natural specification of colour.

Transformation between systems such as HSV and HLS and the RGB and CMY systems can be carried out automatically by fairly simple algorithms. This enables the user of an interactive system to specify colours in any of the systems and to have the results displayed on the screen.

Interactive input devices

The history of computer graphics has witnessed the appearance of a somewhat baffling variety of devices which enable the computer user to communicate with the computer. The most important interactive medium of communication from the computer to the user is the graphics screen. Images appear on a screen in response to commands or instructions issued by the software, but the way in which an interactive program behaves is controlled by input received from the input devices operated by the user. Despite the variety of devices available, the functions that they perform can usually be classed as locators, pointers, text (string), choice and valuator devices. Some pieces of hardware do not fall obviously into a single class. An example is speech input, which may be regarded as entering vocal tokens which could represent numbers, choices, displayed objects or text strings.

Locators and pointers

Digitising tables and tablets

Perhaps the most important device for cartographic purposes is the digitising table or tablet (Figures 14.6a and 5.1). Its importance lies in its role in high-accuracy digital data acquisition from existing maps. In that context, it is used principally for recording the coordinates of point, line and polygonal features. Because of its primary importance in digitising data from existing documents, it has been described separately in Chapter 5.

Touch-sensitive screens and tablets

In addition to the input devices concerned with generating precise coordinates, appropriate to data acquisition, we can consider here a set of devices which still generate a location within a device coordinate system, but with lower accuracy. Touch screens and pads are intended to enable the user to indicate position, or objects of interest, by pointing either with their finger or with a probe or stylus. They are commonly used in portable computers (Figure 14.6b) for graphical interaction. They may also be found in publicly located interactive information systems in which the user can either point to menu items or to 'hot' objects on a map or photograph, about which which more information may be available.

Some touch screens consist of rows and columns of infrared light sources that create horizontal and vertical beams which are each detected by corresponding photocells. The array of beams is positioned in front of the display screen so that when a pointer is placed on the screen it interrupts certain of the beams. The average coordinates of the broken beams, detected by the photocells, provide the position of the pointer.

An alternative touch-screen or touch-pad technology consists of a transparent plastic sheet inlaid with electrodes. Pressing the surface results in a change in conductivity of the affected electrodes. Monitoring the location of the change in conductivity provides the location of the pointer. The sheet may be attached to the front of a conventional display screen. A similar technique can be used for touch pads situated beside the screen. Here, positions on the pad can correspond directly to screen position, or the pad could be subdivided into 'button' regions corresponding to particular option selections. Another type of stylus, for use with a screen, emits radio signals which are detected by a transparent, conductive layer on the screen surface. Figure 14.6(b) illustrates the use of a stylus in combination with a portable flat-panel display.

Light pens

One of the oldest graphical interactive devices, which is operated as a pointer, is the *light pen*. Its physical characteristics are such that it identifies displayed objects on the screen, which are detected by the light they emit when the screen is refreshed. The light pen is relatively uncommon now.

Relative position locators

The most common locator device is the *mouse*, which is a hand-held puck on the end of a wire. They are convenient in that the majority of them can be placed on any flat surface, across which they are moved in order to control a displayed cursor. Those capable of working on almost any surface contain a rollerball embedded on their base. Movement of the ball is detected by shaft encoders which translate the motion into digital pulses, corresponding separately to two perpendicular directions.

Some mice only work on a special pad, in which case the pad may contain a fine, regular pattern that is detectable by photocells. Changes in the pattern correspond to units of distance moved by the mouse. Because a mouse can only detect motion, its exact position on the table or pad has no particular significance. Thus it has no local coordinate system and cannot in general be used for digitising documents, though it can be used for free-hand sketching and digitising directly on a screen. Examples of uses for a mouse are pointing to.positions of interest on a map, pointing to items on a displayed menu of options, and drawing, such as the outline of an area enclosing a region of interest for the purposes of a database query.

The *trackerball* is a device closely related to the mouse, except that it has been turned upside down, i.e. it consists of a ball embedded on the upper surface of a stationary box (Figure 14.17). Movement of the ball, by rolling it under the palm of the hand, is detected in the same way as in the type of mouse which contains a ball. Similarly, motion of the ball is accompanied by movement of a cursor on the screen.

There are a number of joystick devices available, all of which perform a similar task to the mouse and trackerball, in that pushing the stick in some direction makes the displayed cursor move in a corresponding direction. The amount pushed in any one direction may also control the speed of the cursor. Some joysticks include a button or a turnable knob on the end of the stick. They may be used respectively for selecting a specific position and for controlling some other value such as the degree of zooming into a display image.

Figure 14.17 Trackerball and button input devices attached to a graphics workstation. Courtesy Laser-Scan.

Three-dimensional input

The majority of locator devices generate coordinates in two dimensions and are therefore of limited use in applications requiring interaction with a 3D world. An early example of a 3D locator used a combination of a sound source attached to a stylus or pointer and a set of three perpendicular strip microphones (Rogers and Adams, 1990). The time delays between transmitting a sound and receiving it at each of the microphones could be used to calculate the coordinates of the stylus. Another approach, exemplified by the Polhemus 3D positioning device (Foley *et al.*, 1990), determines the position of a movable sensor by recording changes in the current in electromagnetic coils resulting from the interaction between three mutually perpendicular transmitter coils and receiver coils, one set of which is fixed in position while the other is located in the sensor.

Devices of the above type have been used for digitising the shape of 3D objects (e.g models of human bodies), the digital representations of which were to be animated. The availability of graphics workstations capable of dynamic 3D graphics, of the sort only previously available in specialised flight or ship simulators, has led to the development of new devices to facilitate 3D interaction. One such device, the *data glove*, embeds sensors into a glove which, when worn on the hand, can detect movements of the individual fingers and, in some versions, may also be able to detect pressure on the fingertips from gripping a real object. The location and orientation of the hand can be detected by a Polhemus sensor attached to the glove.

The potential value of the data glove is greatly enhanced when it is combined with stereoscopic vision, thus enabling the user to manipulate, through the glove, computer-generated 3D objects which are visualised through stereo goggles, in so-called *virtual reality*. The applications are numerous, but in a GIS context the virtual reality scene might consist of a landscape which the user could travel around, using the glove to indicate direction of motion, and to point to objects in the landscape about which information was required or which were to be viewed in more detail. Applications can be envisaged in evaluating planned developments involving new buildings or roads; in tourism, to help in identifying places of interest which might subsequently be visited in the real world, or in geographical education, to introduce students to a wide range of physical or cultural aspects of the landscape which could be studied through other pictorial, textual and aural sources, all of which might also be accessible through a multimedia environment.

Valuators

There are relatively few devices designed solely for entering numeric values. Those that exist typically contain potentiometers operated by dials or slides. Uses include controlling the positions and orientation of displayed objects. They could also be used for adjusting directions from which a 3D landscape was viewed, along with the magnification or scale. The most common method for entering numeric values is via the numeric keys on conventional keyboards, and the use of graphic slides displayed on screen and operated interactively with a mouse.

Choice devices

Physical devices designed for selecting from a choice of program options generally consist of a set of buttons or function keys which may be mounted in isolation in a box or in the form of a pad (Figure 14.17). Alternatively, they may be part of a conventional keyboard. Pressing a button sends a numeric code which the software interprets according to the application. Choice input, however, is usually achieved by means of a pointer or locator device, such as a mouse, to select an option on a displayed menu by moving the cursor onto the option. Typically the currently selected option is highlighted in some way, before the user presses a button to confirm the choice.

Text (string) devices

The keyboard is the standard physical text, or string, device. Depression of a key usually sends a standard (e.g. ASCII) code to the computer. Termination of a string of characters is indicated by pressing a *return* or *enter* key, though this is not always necessary as some computer programs may be able to respond directly to each key depression rather than having to wait until the terminating key is pressed.

Alternative, and less common, approaches to text input are to use a locator device to point to a menu of text characters. Such a menu could be displayed on a screen, or it could be attached to the surface of a digitising table. Any advantage of this technique would depend upon it being less disruptive to continue using the locator, say in the course of digitising, than to move to a keyboard temporarily, before returning to the locator device.

Graphics programming software systems

The facility to control graphics devices from within a computer program is provided by collections, or libraries, of fairly well standardised procedures that can be called from programming languages such as C, Pascal and Fortran. GIS packages make calls to such procedures in order to plot lines and text, and to obtain input from user-operated devices. A major distinction between the various procedural systems is between those that let the programmer specify graphical data in an application-oriented spatial reference system (the world coordinates or modelling coordinates of the original data) and those that request the programmer to specify graphics in terms of the device coordinates (typically pixels or metres) on a piece of hardware. In the latter case the programmer must perform the transformation between world coordinates and the device coordinate system.

The main examples of standard graphics programming systems that use application-oriented coordinates are the *Graphical Kernel System* (GKS), and the *Programmer's Hierarchical Interactive Graphics System* (PHIGS). Device-oriented systems are exemplified by the *X-Window System* and by the *Computer Graphics Interface* (CGI). On PCs, a common means of graphics programming is to use the facilities built in to programming languages such Visual Basic and Visual C++, both of which include tools to assist in the development of user interfaces.

The X Window System and graphical user interfaces (GUIs)

The X Window System (or X Windows) has assumed particular importance in graphics systems because, unlike GKS and PHIGS, it is well adapted to exploiting the capabilities of multi-user networked graphics workstations and to helping the programmer to manipulate the windows-based user interfaces that are the standard on such workstations (Figure 14.17).

X Windows by itself is quite tedious for programmers to use, however, with the result that some higher level systems, or toolkits, have been developed. The toolkits consist of procedures called widgets that are aimed at assisting the creation of graphical user interfaces employing windows with scrolling facilities, pop-up and pull-down menus and the associated buttons and icons. Organised collections of toolkit facilities are described as graphical user interface (GUI) systems and user interface management systems (UIMS). Examples of GUIs are Motif and OpenLook.

A shortcoming of the original versions of X Windows and the user interface toolkits is that they do not provide some of the application-oriented facilities of GKS and PHIGS. This shortcoming has been addressed by the introduction of GKS and PHIGS extensions to X Windows, which help to bridge the gap between the two types of system. The PHIGS extension to X Windows is PEX. It is now common to find GKS and PHIGS systems that have been implemented using X Windows functions and toolkits, and giving the programmer the benefit of the some of the user interface facilities associated with X Windows.

GKS and PHIGS

GKS in its original form was intended for 2D applications, though it has been extended for 3D applications. The standard defines functions relating primarily to the control of workstations, to the creation of graphic output primitives (point, line, area and text symbols); segments, which are logically related collections of output primitives; input functions for interaction with the user; and viewing transformations to control the location and scale of displayed graphics.

PHIGS is closely related to GKS, but with the addition of functions to enable the programmer to build hierarchical models in 2D and 3D. Here standard components called structures (similar to, but more flexible than segments) can be defined and used to create more complex components which may themselves form parts of higher level objects. This hierarchical modelling facility is very useful for many engineering design applications, but it is also relevant to GIS. Thus a map may use standard symbols which, having been described once and stored in a structure, can be used in many parts of the map. In a 3D visualisation, features of the human-made landscape, such as electricity pylons or buildings of a standard design, might then be replicated at various locations in the scene.

Postscript and image formats

Postscript is a graphics programming language which is entirely oriented to the output of hard copy. Its introduction accompanied the widespread use of high-quality laser printers as output devices for word processors and desk top publishing software. The motivation for its development was to provide a standardised method for defining individual pages of output that could combine text and graphics. It is described as a *page description language*.

Graphics systems such as GKS and PHIGS did not, as part of their original definition, include a means for storing a 2D image. They did include the concept of storing pictures, this being the purpose of *metafiles*, but it took some time before a standard definition, in the form of the Computer Graphics Metafile (CGM), was agreed. Postscript is closely related to the metafile concept, but it is extended somewhat in that Postscript is itself a programming language, providing many of the normal facilities of procedural languages.

Many raster printers have the capacity to interpret a Postscript program and hence display the picture or pictures that it defines. The Postscript program will usually have been generated by another program. Graphics systems that use GKS or PHIGS often include a special program to translate a metafile generated by the latter packages into a Postscript program.

The presence of multimedia computing systems has led to a proliferation of digital images stored in a variety of formats. Examples include TIFF (Tagged Image File Format), BMP, PICT, GIF (Graphics Interchange Format) and JPEG (Joint Photographic Engineering Group). These file formats differ from Postscript and from graphics metafiles, in that they represent only an array of pixel values. They do not retain any information on the presence of graphic objects such as lines and text, though some do keep data on colour palettes to be used to interpret the pixel values. They differ particularly in the way in which the data are compressed. The JPEG scheme can achieve good image compression, but it may do so at the expense of some modification to the original pixel values (Lammi and Sarjakoski, 1995). The degree of loss is controllable.

Computer graphics on the Internet

With the now very widespread use of the Internet, and in particular the *World Wide Web*, there is increasing interest in providing network access to geographical information systems and to maps. The contents of pages on a World Wide Web site are usually defined using the hypertext mark-up language HTML. The HTML page definitions are sent via the Internet to the user's computer, which displays them using an interpreter program. The simplest and at present still most common way to present maps on web pages is in the form of static images that are stored in GIF format. Interaction with such maps may be obtained by superimposing graphical buttons or hot spots which, when the user clicks on them, result in access to other web pages that may also consist of images or of text. The locaton of hot spots on a digital image such as a map can be defined in HTML using a so-called *image map*.

Some web sites now provide a more dynamic form of interaction with geographical data by running GIS software in direct response to a user action such as clicking on a hot spot, or selecting from a menu of options. Assuming that this program is run on the web site's own *server*, i.e. a computer local to that web site and supporting the Internet access, then interaction between the user interface and the GIS program can be handled by common gateway interface CGI software (another CGI!). Running the GIS program on the server, via CGI, may result in it generating a new image which represents the product of the user's query. This image may be loaded to a new web page which is then sent for display on the *client* (i.e. the user's) computer.

If interaction with a GIS requires access to a large database, then the database usually needs to be located local to the web site, and access to it provided in the manner just described. Elements of graphical interaction, such as highlighting map features and changing the symbols displayed on a map, may, however, be handled most efficiently (with low network traffic) on the client computer. This can be done using the Java object-oriented programming language, which runs on the client computer, and is similar to the C++ language. Web sites that wish to provide high degrees of interaction can send to the client not just data (retrieved from a server database), but also small computer programs, written in Java, and called *applets*. The applet may then allow the user to interact with and modify the data using, if appropriate, interactive graphics. Thus Java includes functions for displaying graphic primitives such as point and line symbols which the user can select by pointing to them with the cursor. For purposes of presenting specifically three-dimensional data on the World Wide Web, the language called VRML (Virtual Reality Modeling Language) has been developed.

of programming software available on PCs, though on graphics workstations X-Windows and associated higher level toolkits have become a *de facto* standard for device level and user interface programming. Many of the concepts of computer graphics programming are encapsulated within the international standard graphics programming systems of GKS and PHIGS. These systems enable the programmer to specify the picture in world coordinates. In PHIGS it is also possible to define graphical objects in a hierarchical manner whereby an object may be defined in terms of further sub-objects. Several languages and formats have been defined for the purpose of representing pictures and images for transfer between graphics systems or for the operation of printers.

Summary

In this chapter we have reviewed the hardware and software technology for displaying images and for interacting with the displays. At the hardware level we distinguished between screen displays and hard copy devices for providing permanent graphic output on paper or film. Still the most commonly used screen technology is that of the cathode ray tube (CRT), operating in raster mode. Flat-panel raster technology in the form of lcds (liquid crystal displays), electroluminescent and plasma panel displays is however very widely used in portable computers. Hard copy devices are dominated by laser printers for monochrome work, with colour commonly provided by inkjet, thermal wax, electrostatic and dye sublimation printers. Colour laser printers are reducing in price, but at the time of writing are still about ten times as expensive as good-quality inkjet printers.

As regards software, there is a considerable variety

Further reading

An excellent general text on computer graphics technology is Foley *et al.* (1990), which also has an extensive list of references on the subject. For more details on matching colour between monitors and printing devices, see Stone *et al.* (1988). The use of colour in cartographic design is discussed in Chapter 15. Medyckyj-Scott and Hearnshaw (1993) contains a set of articles dealing with many aspects of interacting with GIS technology, under the general heading of 'human factors'. To pursue the subject of programming graphics systems, there are numerous texts; for example, Kasper and Arns (1993) on PHIGS.

Cartographic communication and visualisation

Introduction

The profusion of computer packages for GIS and for digital mapping now mean that maps of some sort can often be created in minutes by anybody capable of operating the program. Assuming that the package contains the relevant digital map data, then the drafting process is highly automated in the sense that the program will plot out the locations of selected map features, at a selected scale, and with whatever symbols the user has chosen. As was pointed out in Part 1 of this book, computer graphics technology enables map data to be plotted faster and with greater precision than is possible using manual methods. Map projections can easily be changed, schemes for symbolisation of data can be modified very rapidly and, once very demanding, visualisations of 3D data can be created with ease.

Automation in cartography has not just speeded up traditional cartographic procedures. It has led to new ways in which to exploit cartography for purposes of communicating information and of visualising – and hence gaining greater understanding of – the spatial relationships inherent in mapped data. Perhaps most importantly, computer graphics technology has provided the opportunity to interact directly with graphic displays, so that the content of the display is changed in response to the user's interests.

Although considerable progress has been made in computer cartography, the automation is only complete in a mechanistic sense that relates to the plotting of graphical symbols. The computer packages provide little or, more usually, no guidance on decisions regarding the design of the map. It is for the user to decide what data should be plotted, and with what projection, symbols, colours, typography and accompanying titles and legends. Cartography is potentially a very powerful means of communication and to be so it depends in general upon the right choices being made regarding the selection of these various parameters.

The purpose of this chapter is to provide an overview of the facilities provided by GIS and mapping packages for displaying spatial data, and to provide some awareness of issues in cartographic design. In the next section, we introduce the basic types of cartographic symbols. We also review issues in map design, which concern the ways in which map symbols can be manipulated to affect the impact of the map. A section then follows on the subject of text placement, which is one of the few areas of cartographic design in which considerable progress has been made in automation. The following section deals with the display of 3D data, with some mention of techniques for helping to achieve realism. This includes issues of hidden surface removal, shading of terrain models and the creation of viewshed maps based on intervisibility analyses. The final section of the chapter reviews briefly some methods available for interactive exploration and retrieval of spatially referenced information.

Graphic symbology

Graphic symbols can be classified according to the type of spatial objects that they represent. In two dimensions this leads to the familiar division in cartographic literature between point, line and area symbols. Thus in each case the assumption is that the symbol is spatially referenced by either a single point (e.g. for a pylon), a line (e.g. for a road), or an area (e.g. for a census enumeration district). Clearly, the type of symbol chosen depends upon the degree of generalisation of the phenomenon being represented. A city could be represented by a point symbol at a small scale, an areal symbol (the city boundary) at a medium scale, and a complex combination of point, line and area symbols at larger scales. Spatial data that are continuously variable, i.e. surfaces or fields, are sometimes represented by symbols that are specifically intended to capture this continuous variation. We will refer to them as surface symbols, in the absence of any other familiar term. They consist in practice of a variable density of points or lines, or of a raster of cells of different colours. Maps usually contain another form of mark, i.e. lettering, to represent names and other textual annotation. In what follows we will concentrate initially on 2D information and its communication through the use of point, line and area symbols and text.

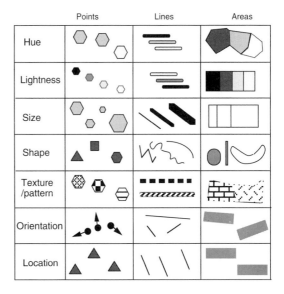

Figure 15.1 Graphic variables. Note that hue refers to colour and hence is not adequately represented in this monochrome diagram.

Graphic variables

The 2D graphic symbols, and the associated lettering, can all be modulated graphically in various ways to help in communicating different types of information. Following Bertin (1983), we can identify several primary graphic variables, each of which can be varied. These are hue, lightness (value), size, shape, texture, orientation and location. Figure 15.1, based on a figure in Bertin (1983), gives some examples of the use of these elements in the context of point, line and area symbols. As we will see, some of these graphic variables are of much more relevance to particular symbol classes, and to particular types of data, than others.

Hue refers to the use of colours as we would distinguish them by names, such as in the colours of the rainbow. As we saw in Chapter 14, hue is one major aspect of colour, of which lightness (or value), and saturation are the others. Hue is commonly regarded as of most use in a map for representing qualitative rather than quantitative information (Brewer, 1994). It is typically used to distinguish between the major classes of point, line and area-referenced phenomena on a map.

As the number of different hues that are employed rises, however, so does the risk of confusion for the map reader. This can be reduced by combining a hue with another graphic variable such as texture, which helps in recognition. Some hues appear closer to each other than others. Thus if there are minor variations in hue, say in a range from red to yellow, it is preferable that they should relate to similar categories of phenomena, while major differences, such as the primary additive colours of red, green and blue, may best be reserved for the important categorical differences (see Plate 14). It should also be borne in mind that certain colours have familiar meanings to certain groups of people. Water bodies are typically represented in one or more shades of blue, while vegetation may be represented in green and so on. Hue is of less use in representing ordinal and numerical (interval and ratio) data, though a common exception to this is in its use to differentiate between elevations on surfaces.

Lightness (or value) refers, as its name suggests, to the lightness or darkness of an apparently uniform area of pigment used to plot a symbol. Because lightness can be varied continuously across the range from very light to very dark, it is well suited to representing ordinal and numerical variation in data associated with map symbols (Figure 15.2). For point, line and area symbols, however, the map reader is likely to have difficulty in memorising more

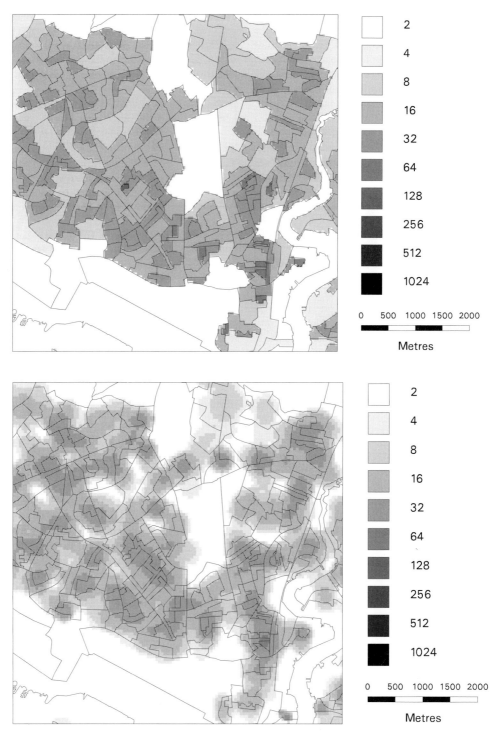

Figure 15.2 Variation in lightness is often used to distinguish between numerical or ordinal values. The figure uses lightness to represent several levels of population density in persons per 50m² cell. (**a**) Chloropeth map in which cell values are constant within enumeration district boundaries. (**b**) Cell values represent a statistical surface of population density interpolated between centroids of enumeration districts. From Martin (1996).

than four or five different levels of lightness, in which case interval and ratio data may need to be transformed to a set of classes. When allocating lightness to an associated attribute variable, care should taken to obtain a natural interpretation if possible. Thus a dark symbol might tend to be associated with a high numerical value. More generally, the higher the contrast in lightness between the symbol and its background, the more important it may appear.

Size is an example of a graphic variable that is applied to some symbol classes much more often than to others. Thus the size of point and line symbols can be varied to distinguish between values of ordinal, numerical and sometimes nominal data. For example, the population of a settlement may be represented by the size of the associated point symbol on a small-scale map, and the width of a road symbol could be varied as a function of traffic flow, or according to the class of road, which is usually related to the actual width of the road itself (Figure 15.3). Point and line symbols that are varied in size according to an associated attribute are described as *graduated symbols*, and it has been suggested (Bertin, 1983; Robinson *et al.*, 1984) that the size variation of such symbols is more effective than variation in value (of colour) as a means of distinguishing graphically between numerical attributes.

Point symbols are most commonly used in this way and the question arises as to how the size of the symbol should be varied as a function of the value of the attribute. A common approach for circular point symbols is to make the area of the symbol proportional to the value, which means that the value is directly related to the square root of the circle radius. Studies of map usage have shown that there is a tendency for map readers to underestimate the value of larger symbols using this method (Robinson *et al.*, 1995). Consequently the radius may be exaggerated slightly, making the attribute value proportional to the radius to the power 0.57 say (rather than 0.5 for a square root). Before deciding on a constant of proportionality, the range of data values to be represented must be determined. If the range is very large, such as for population values between 100 and 1 000 000, then a logarithmic scale may be appropriate. An important point to note here is that if the map reader is expected to make sense of the symbols, a legend must be provided with examples of the meaning of the different symbol sizes. If the data have been range-graded, i.e. classified into a small set of numerical ranges, then it should be possible to represent all different symbol sizes in the legend.

It is not so easy to modify the size of an areal symbol, since to do so results in a modification to the location of itself and other map features. This is indeed what is done in *cartograms*, which introduce intentional location distortion in an effort to assist the map reader to visualise the relative value of some represented attribute (Figure 15.4).

In cartography, the shape of a map symbol can be regarded as a variable, either for the purpose of communicating information about an attribute associated with the location of the symbol, or it can be used in the process of map generalisation, whereby the symbol itself represents the location (the path, the boundary or the internal structure) of a phenomenon, which may need to be simplified as a function of map scale. For the former purpose, i.e. describing an attribute of a location, it is point symbols that provide the most scope for shape modification. There are two main methods of representing the shape of a point symbol. One is to use an abstract symbol such as a circle, a cross, a square or some combination of these and other shapes, which may themselves be modulated by other graphic variables. The other is to use a pictorial symbol, where the shape of the symbol is intended to suggest the phenomenon being represented. Examples would include a tent to represent a campsite, or a car to represent a car park. Some GIS packages, and most graphic design packages, include interactive graphics editors to allow the user to create user-tailored point-referenced symbols, whether abstract or pictorial. In creating such a symbol, it is of course possible to combine many graphic elements. Perhaps the commonest example of using shape as a non-locational variable in a line symbol, is when using extended arrows to represent migrations or flows from one place to another (Figure 15.5).

Modification to the shape of line and area symbols in the course of map generalisation is a complex process, which is discussed in some detail in Chapter 16.

The graphic variable of pattern can be applied to point, line and area symbols. It refers to the internal graphical structure of a symbol, and may also be described in some cases as texture. Thus the inside of a symbol might consist of a distribution of parallel lines, or cross hatching, or a set of dots. As a graphic variable it is probably of most use in the case of area symbols in which there may be a significant amount of space in which to build up a pattern. It is quite common for line symbols to have a pattern, as a sequence of dashes and dots, or as an internal structure, such as a cased road symbol (Figure 15.6).

Orientation may be employed as a graphic variable in several ways. There is a distinction, similar to that of shape in the context of cartography, in that it can either be used to represent an attribute associated with a symbol location (see Plate 15), or it is an

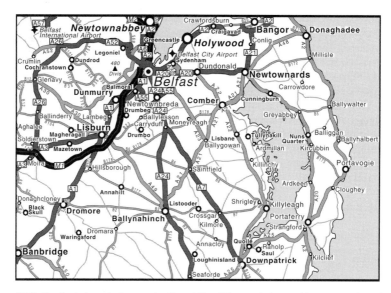

Figure 15.3 Graduated symbol example: the width of the symbols used to represent roads and towns is varied as a function of their class and importance. Produced with Maplex software. Data courtesy Automobile Association.

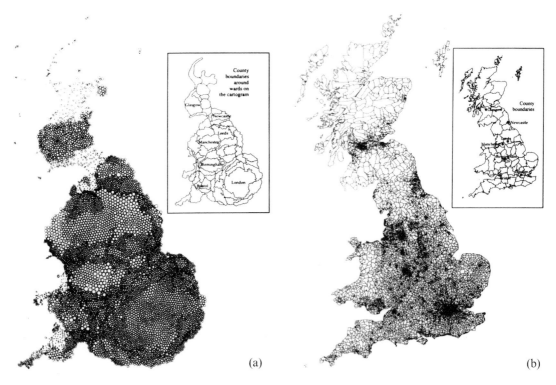

(a)   (b)

Figure 15.4 (**a**) In cartograms, area symbols are modified in size as a function of a mapped variable. In the figure (from Dorling, 1993) circles are used to represent the population of wards in the UK. The circle areas are proportional to population. The untransformed areas, in the National Grid, are represented in (**b**).

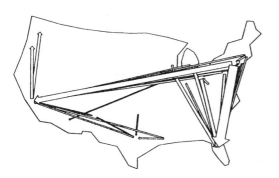

Figure 15.5 The arrow-shaped line symbols represent differing levels of migration between states in North America between 1965 and 1970. From Tobler (1987).

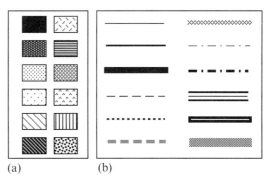

(a)   (b)

Figure 15.6 Patterns used to modulate (**a**) area symbols and (**b**) line symbols.

inherent spatial property of the structure of the spatial object that the symbol represents. In the first case, of representing an attribute associated with the symbol location, it may also be that the attribute is itself one of orientation. Examples would be wind direction, or the direction of flow at a point site. In both of these cases it would be a point symbol that is used, in so far as the symbol is referenced to a single locational point observation. However, the symbol itself would consist of, for example, an arrow, a line or a sector that was oriented as a function of the associated value. Similar point symbols could be used to represent the value of a variable such as percentage concentration of a chemical element, or average income. Figure 15.7, from Carr *et al.* (1992), explains and illustrates the use of ray-glyph symbols for atmospheric sulphate deposition trends. Note that as in Figure 15.7, it may be advisable to use only the orientations from say north, via east, to south, so that up means high (or an increase) and down means low (or a decrease).

The orientation of line and area symbols is in general dictated by the associated locational data, and would only therefore be modified in the course of map generalisation.

The graphic variable of location is the one for which in cartography there appears to be least degrees of freedom, since the coordinates of geometric spatial data provide a location, which should only be modified with great care. Small variations in location may occur as a result of map generalisation, in which case the generalisation operator of displacement may have been invoked in order to avoid conflicts between symbols and hence to maintain the legibility of the map. Significant variations in loca-

tion on a map sheet can arise as a consequence of modifications to the map projection. Such modifications can give rise to variations in shape and size as discussed in Chapter 4. There are certain types of map in which metric location is of less importance than the topological relationships between the mapped phenomena. This is often the case on maps of transport networks, in which the map reader may need to know, primarily, what are the routes available between particular places. It is possible to envisage that the maker of such a map might well move the network nodes, while retaining their connectivity, in an effort to improve legibility.

Issues in map design

Following Robinson *et al.* (1995), we now identify several issues that may be considered when designing a map, in addition to those considered above in the context of individual graphic variables. It should be reiterated that little of the knowledge concerning map design has found its way into computer mapping systems. The reason for raising these issues here is twofold. One is to increase awareness of them as being important in their own right for those concerned with computer cartography. The other is to help to understand the scope of automation (or lack of it) in computer cartography, and hence to provide some background to our subsequent discussion of how some aspects of cartographic design can be automated.

Eastern North America

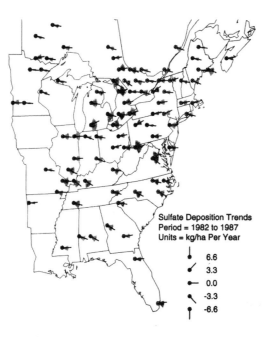

Sulfate Deposition Trends
Period = 1982 to 1987
Units = kg/ha Per Year

6.6
3.3
0.0
-3.3
-6.6

Figure 15.7 Orientation is used here in ray glyph point symbols to represent a variation in a numerical value. Note that upward direction is used to mean high and downwards to mean low. From Carr *et al.* (1992).

Major issues in cartographic design are identified by Robinson *et al.* (1995) as clarity and legibility, hierarchical structure, colour and pattern, visual contrast, figure and ground, balance, and typography.

To achieve clarity and legibility requires a concern for the quality of graphic plotting of map symbols and text, to ensure that they are sharply delineated and that they are of sizes that ensure their visibility. A rough rule for size is that no map symbol should subtend an angle at the eye of less than about 1 minute of arc. This equates to a distance of about 0.3 mm when viewed from 50 cm. This rule can be relaxed somewhat in the case of line symbols, which because of their extensive nature, can be detected when significantly narrower, e.g. 0.1 mm.

Size constraints should apply not just to the symbols of individual map objects, but also to the separation between them. Maintaining adequate separations between map objects is a very important constraint in map generalisation, and may result in symbol displacements to meet a specified tolerance.

Visual contrast is a factor that will affect clarity and legibility, in addition to the issue of size that has just been stressed. It relates to the crispness or sharpness of the distinction between map symbols. It is obtained by appropriate use of the graphic variables. Clearly some colours, such as the primaries, contrast much more with each other than others. Equally, when using graduated symbols, if the different symbols are all very similar in size, they may fail to fulfil their purpose, since lack of contrast results in a lack of differentiation for the map reader. Although contrast is necessary and desirable, there is no doubt that it can become a negative factor if taken to excess, in that a map could become too 'busy' (Tufte, 1983, 1990).

Visual balance concerns the relative weight of the basic graphic shapes that make up a map. Although each of the graphic variables may affect the relative weight of a component of the map, there are some aspects of a map that may be modified more easily than others to affect the balance. Notably some parts of a map, such as the legend, title, key, associated text and the names on the map, may have their location altered in ways that may greatly affect the overall balance. This is another example of an aspect of design for which there is no simple formula for success.

Figure–ground relationships relate to the fact that the modulation of the graphic variables of map symbols can have the effect of bringing certain parts of the map forward, such that some symbols appear to stand out, or be in front, relative to others. By suitable modulation of the graphic variables, the cartographer should be able to manipulate the map reader to ensure that their attention focuses on the important features (assuming that there are some features that are regarded as more important than others). There are various ways in which this can be achieved. These include a clear differentiation between the intended figure and the background, making the former more bold or more bright than the latter; the use of 'closed forms', whereby objects of interest (figures) are represented if possible in their entirety, rather than being cut at map boundaries; the use of internal detail to give interest to the figure; and when possible making the important objects occupy a relatively small proportion of the map (a quarter or a fifth) so that, given appropriate use of other variables, there is not too much other competing information (Robinson *et al.*, 1995).

Hierarchical organisation is concerned with differentiating between broad classes or types of information on a map and helping the map reader to focus on specific themes among several that may be portrayed on a single map. Thus several major categories such as roads, rivers and cities may all have internal subdivisions in terms of levels of importance or of specific subcomponents. There are three types of graphic organisational device – extensional, subdivision and stereogrammic – that help in making these distinctions.

Extensional organisation applies particularly to ordinal distinction in line and point symbols, categories of which can be internally distinguished by means of variations in, for example, value or size. Examples are the thickness of streams in a river network, the width of road symbols in a transport network, and the size of point symbols distinguishing the importance of settlements.

Subdivision organisation applies particularly to areal features that may be internally subdivided, e.g. types of soils, geology and land use. More specifically, a geological map might include three major categories of lithology, consisting of sedimentary, igneous and metamorphic rock types. These might each be allocated a distinct hue, such as blue, red and green respectively. Then within each rock type, subdivisions would be represented by the respective hue, modulated by lightness and pattern. The important design issue is that the major categories should be clearly distinguished at one level of the hierarchy, while each category is then also further subdivided in a manner that helps the map reader to focus on the internal details without causing confusion between the graphic symbols.

Stereogrammic organisation is concerned with altering the apparent visual level of map components. This can be done by means of the techniques referred to in the context of figure–ground relationships, but sometimes with a view to creating several levels, rather than just a binary division between figure and ground. One approach is to use a series of lightness levels to provide depth cues. A related method is the use of variation in colour saturation. For example, a road transport network might all be at a higher level of colour saturation than a river network, on the assumption that for the purposes of this map the former was more important. Thus, as well as the fact that the river network was distinguished by being blue, it would also be a low saturation of blue, rather than a very strong blue, which might otherwise bring it forward to compete with the road network.

Text placement

Text is used on many maps as a primary element in communicating information. It becomes necessary to include text on a map when we need to be able to distinguish between unique members of a class of objects, as in naming individual towns and cities, or when it may be appropriate to remind the map user of a detailed classification, or to remove ambiguity, as in annotating topographic contours or detailed rock and soil classifications.

Placing text on maps is one of the most time-consuming and labour-intensive processes in map production. It has been estimated that it can occupy as much as 50% of the final map preparation time. For this reason alone it is of great interest to find ways of automating name placement. Poor-quality text placement on computer-generated screen displays provides another motivation for automation. Lack of attention to many of the design issues mentioned above in the context of graphic symbols is certainly a problem, but poor text placement can have a particularly debilitating effect on the readability of a map. The reason why text causes such a problem is that, unlike most other map symbols, there is not one location on the map to which it belongs, as it does not itself represent the location of a phenomenon (though it may help to do so). It is an attribute of a location that is usually already represented by a graphic symbol. Thus text is a necessary overhead which demands space on the map. The majority of text refers to point- and line-referenced features, the symbols for which do not in general occupy much space, relative to that needed to display the text. Hence text needs to compete for space with other map symbols and with other items of text. Without proper control over its placement therefore, it may overwrite other text items and important map symbols, rendering the map locally illegible. This arises in particular where information is densely clustered, such as in highly populated areas on a topographic map.

In regions of densely clustered symbols, care must be taken to ensure a clear visual association between the label and the feature it annotates. This may be achieved for point symbols by placing the label closer to its own symbol than to any other symbol, or by maintaining a convention for the position of the label relative to the symbol, e.g. put the beginning of the label adjacent to the symbol. Situations often arise in which, in order to place the name at all, it must overwrite other symbols with which it has no logical association. When this occurs, it is common practice to adopt rules about which classes of map symbol can be overwritten by annotation belonging to other classes. Thus town names may be seen on some maps to be placed across rivers or streams, but rarely across major roads.

Occasionally it may prove impossible to label a particular feature without obscuring another label or overwriting a symbol that has a high priority. In this event the lower-priority label may need to be omitted entirely, or perhaps placed in another part of the map and linked to the associated symbol by means of an arrow.

Automated Name Placement

Since the early 1970s a considerable variety of automated name-placement systems have been described in the research literature (e.g. Yoeli, 1972; Freeman and Ahn, 1984; Doerschler and Freeman, 1989; Cook and Jones, 1990).

It is possible to identify the following components of automated name-placement systems:

1. specification of map features and text characteristics;
2. representation of spatial data for search and display;
3. generation of trial name positions;
4. selection of optimal labels.

Since the visibility of text and map features is a function of the symbols and colour schemes, the name-placement system must store this information.

It is necessary to define the text font, including letter size for each type of name to be used. Having specified the contents of a map in terms of geographical or other thematic features and their symbolisation, it is necessary to specify which of these features are to be annotated.

If it is regarded as desirable to name as many features as possible, while accepting that some features could be unlabelled in locally dense parts of the map, then named features should be ranked in some way to allow the possibility of naming the more important features if conflict arises. This ranking might be based on a combination of class of feature and a numeric attribute, such as population (in the case of settlements), or traffic density (in the case of roads). Another way of controlling placement in dense areas is to specify limits on the minimum size of features that are to be annotated. For example, it could be decided that no road segments (between junctions) shorter than a given length were to be annotated.

Relationship between a label and its feature

The way in which text is graphically associated with the annotated feature varies considerably between different styles of map. In a flexible text-placement system, the user should be able to specify the nature of this relationship. Figure 15.8 illustrates a number of different ways of labelling point, line and areal features.

Point labels are typically horizontal, though they may also be angled or curved. Their position relative to their point symbol may be restricted to a few positions, such as upper right, lower right, upper left and lower left, or they may also be allowed to

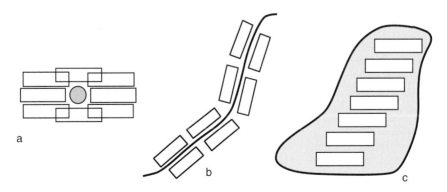

Figure 15.8 Various possible relationships between point, line and area-referenced names and the feature they annotate.

occupy a range of intermediate positions around the feature.

Line-referenced names may be horizontal, straight and parallel to the line, or curved and parallel. In all cases the name may be allowed to occupy a range of positions along the length of the line. For each of the curved, straight, horizontal and parallel options, there may also be a parameter to indicate the offset of the label from the line feature.

When labelling a network of linear features, the presence of junctions in the network, between lines symbolised in the same way, could introduce ambiguity. In the Namex2 system (Jones *et al.*, 1991), this potential problem is addressed by generating at least one label for each link in the network between nodes connecting three or more line segments that are symbolised in the same way. The labelling could, however, be overridden if a minimum link length for labelling was specified.

Areal feature names may be required to follow the trend of the area's shape. Alternatively, they may be constrained to be horizontal. If the areal feature was small relative to the name, rules should be specified to determine whether, for example, the name was to be part in and part out, or if it should be located outside but closely adjacent to the feature.

Since most allowable label configurations will provide the possibility of several candidate or trial label positions, it is appropriate for these to be ordered in terms of preference. The best final position for a label will depend not just on these preferences, but also on the degree of overlap with other features and possibly other labels.

The relationship between names and other features and names

Label positions that satisfy the specifications referred to in previous sections are highly likely to overlap other map features which are in the vicinity of the named feature. In general, overlap with other names is avoided as far as possible, but it may be tolerable in certain circumstances (such as small dark lettering overlapping large lighter-shaded letters). Assuming that it is usually impossible to avoid overlap with other features, specification should be provided for the relative priority of map features. Thus a weight can be attached to each class of feature to indicate the extent to which it is to be avoided by names other than its own.

Spatial data structures for trial position generation and overlap detection

Detection of overlap between names and map symbols and between pairs of names can be achieved efficiently using a raster, or grid cell, representation of the map symbol distribution, as has been done in several automated systems (Yoeli, 1972; Basoglu, 1984). To find what, if anything, a label overlaps in a rasterised map, it is only necessary to calculate the range of pixel locations that the rasterised label occupies and then read the values of those pixels in the raster data structure.

Both Freeman and Ahn's (1984) and Cook and Jones' (1990) systems use vector structures for storing linear features which are to be labelled. The reason for this is that the vector structure is convenient for detecting certain aspects of shape, such as the presence of segments of low curvature which may be appropriate for positioning labels on linear features. Vector structures can also be of value in recording the exact positions of labels, either in terms of their overall bounding rectangles or of the bounding rectangles of the individual letters.

Several vector-based strategies of varying degrees of sophistication have been developed for generating area name positions. In general, a suitable place for an area name is along the path of the main section of a skeleton fitted to the area. A simple method of creating a form of skeleton involves triangulating the polygon representing the area and then joining together successive mid-points of the triangle edges that cross from one side of the area to the other. The letters may then be placed along the path of the skeleton or the path of a trend line fitted through it (Figure 15.9). Freeman and Ahn (1984) used more complex methods to generate a skeleton from an areal feature, before placing the letters of the name initially at equal intervals along a circular arc fitted to the main branch (or 'primary path') of the skeleton. The skeleton may also be used to generate alternative candidate label locations that are parallel to it. See Chapter 8 for raster methods of creating a skeleton.

Conflict resolution and optimisation

A considerable variety of strategies have been developed to attempt to find, for each labelled feature, the particular label position that is satisfactory when all other features and names on the map are considered.

It is possible to envisage a strategy that aims at an optimal solution whereby all labels are placed at their best position when a global view of the problem is taken. Realisation of a genuinely optimal solution (assuming optimality could be defined in an acceptable way) would probably require unacceptable computing times for maps containing high densities of features. Certainly most documented name-placement systems only attempt to find a solution that avoids the worst types of graphic conflict and, in the better systems, succeeds in placing the majority of names in fairly good positions. For certain types of maps (e.g. road atlases), the results of these systems are comparable in quality to published maps produced by manual name-placement techniques (Figures 15.10 and 15.13).

It has already been indicated that a common factor in name-placement systems is the generation of several trial label positions for each name. These trial positions may be determined in an initial pre-processing stage (e.g. Freeman and Ahn, 1984; Cook and Jones, 1990) or, less commonly, in the course of resolving conflicts between names (Hirsch, 1982; Jones, 1989b). Typically the trial positions are ranked in order of an initial set of priorities based on the relationship of the labels to their feature (e.g. Yoeli, 1972; Langran and Poiker, 1986) or, additionally, taking account of overlap with adjacent features (e.g. Freeman and Ahn, 1984; Cook and Jones, 1990).

The methods of selecting the trial positions that are best, when taking account of other labels, differ somewhat between various systems. The iterative schemes, such as those of Hirsch (1982), Doerschler and Freeman (1989) and Langran and Poiker (1986), place all names initially in some way, which may represent a best first try, before repeatedly checking all names and making adjustments where necessary in an attempt to improve the quality of placement.

Visualisation of 3D data

Most of the display facilities in GIS and mapping packages are concerned with data represented in two dimensions. To some extent this reflects the fact that most digital geographically referenced data are indeed 2D. There is, however, increasing interest in 3D data display (Turner, 1992). As usage of GIS has technology has become more widespread, so has an awareness of the possibilities of exploiting the technology. Thus it is now possible to create simulations of 3D scenes for purposes of evaluating potential new developments that may impact upon the landscape and for helping to understand essentially 3D structures and processes on the earth's surface and in the subsurface. It can be expected that as environmental modelling techniques become more sophisticated, so GIS technology will improve in the provision of facilities to display 3D spatial data both statically and dynamically to take account of the dimension of time.

Computer graphics technology for viewing 3D data has been available for a long time relative to the history of GIS, but the main application areas have been in computer-aided design for mechanical engineering, in medical imaging and in molecular modelling. In the earth sciences it has also been used extensively, particularly in the oil industry, for visualising geoscientific data based on geophysical surveys and on data

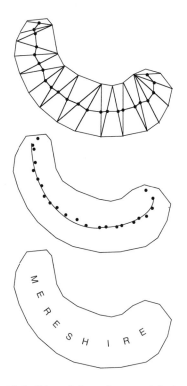

Figure 15.9 Triangulation of an areal feature may be exploited to generate a skeleton which may be used to locate an area label.

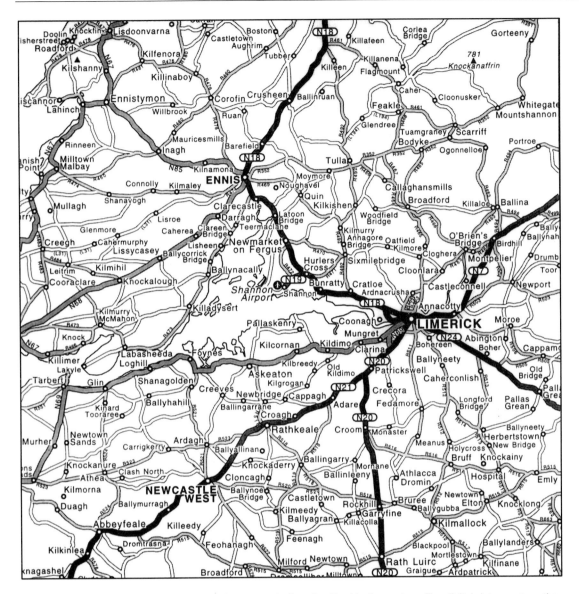

Figure 15.10 Example of text placed entirely automatically using the Maplex system. The digital data courtesy the Automobile Association.

derived from boreholes (Hamilton and Thomas, 1992). The creation of realistic 3D scenes from digital data is computationally very expensive, when compared with 2D graphics. This is particularly so when the application demands real-time updates of the scene, as in flight and ship simulators. Since the late 1980s the term 'virtual reality' has been applied to such dynamic 3D graphics. Here the user can interact with the computer graphics to control the animation. This burst of interest in virtual reality coincided with

a drop in the cost of the computer hardware required to produce it. The cost is still coming down, with the result that good-quality 3D computer graphics technology is no longer largely confined to industrial and scientific organisations with significant funds available for investment in the technology.

The main type of geographically related data that represent phenomena in three dimensions is that of digital elevation models. Three-dimensional visualisation of landscape features can be achieved in some

GIS by superimposing, or draping, a 2D representation of topographic features, such as roads, rivers and forestry, onto a digital elevation model. The various topographic features are then displayed by plotting them on the elevation surface with distinctive colours or symbolisation. Draping involves finding an elevation (z) value for each x,y coordinate in the 2D representation. Values can be obtained fairly easily by extracting directly, or interpolating, the z values from the corresponding x,y locations in the digital elevation model. The same approach can be used to attach z values to the pixels in remotely sensed imagery prior to creating a 3D view (Figure 15.11). Height information on shaded and draped surfaces can be enhanced sometimes by plotting annotated contour lines on the shaded surface, enabling the viewer to keep track of absolute heights.

Draping is often performed solely for purposes of visualisation. There are some applications however, such as hydrology and geology, in which the integration of originally 2D data with a digital elevation model is an important step towards creating a permanent model both for visualisation, and for measurement and analysis. In the context of landscape visualisation, terrain model data can also be integrated with 3D models of buildings and other structures such as pylons and wind turbines.

Techniques for 3D data display

Coordinate systems and transformations for 3D graphics

Most general-purpose computer graphics packages assume the existence of application models defined in a 3D rectangular world coordinate in which the same linear units of measurement are used for each axis. Given that it may be desirable to view the application model from any direction and, for perspective views, from any point in space, it is necessary to transform from the world coordinate system to a viewing coordinate system. The purpose of the viewing coordinate system is to locate the application model relative to the viewing device. If we assume that the scene is to be projected onto a 2D screen, or *viewing plane*, then the x and y coordinates of the viewing coordinate system lie within this view plane, while the z axis is perpendicular to it and is referred to as the *view plane normal*.

To represent the 3D view on a 2D viewing surface, such as the screen of a VDU, a reduction in dimensionality, or *projection*, from 3D to 2D must occur. It is also necessary to scale and translate the 2D viewing coordinates to the 2D device coordinates.

Projections

The projection of a 3D object into 2D is defined by the intersection with the view plane of rays, or projectors, emanating from the object. For an object composed of straight edges of lines and polygons, it is only necessary to project rays from the vertices defining their edges. The projected image is then constructed by joining up the points on the view plane corresponding to the projections of the vertices of their respective edges. Thus for a view plane that is planar, straight lines in 3D will project to straight lines in 2D (Figure 15.12).

Clipping in three dimensions

In computer graphics systems, a region of interest in the world coordinate system is defined by a rectangular window which is transformed to a viewport in the device coordinates. Any parts of the world coordinate data, or application model, extending beyond the window are clipped to the window boundary and only this part of the model is displayed in the viewport. In 3D viewing, the region of interest is defined by up to six bounding planes, and the region confined by these clipping planes is called the *view volume*.

Hidden surface removal

When we view real, non-transparent solid objects, we can only see those parts of the object which face towards us and are not obscured by other parts of the object, or other objects in the scene. In the context of landscape views, we cannot normally see beyond the horizon and, furthermore, there may be many local horizons due to landscape features such as trees, buildings and hills on the viewer's side of the uppermost horizon.

In general, plotting parts of objects which would, in the natural world, be invisible to the viewer causes confusion and may render an image unintelligible. If we are to retain any sense of realism, the obscured parts of a scene should not be plotted. This is the function of the hidden surface removal process.

One of the most important factors in hidden surface removal algorithms is the use of sorting. Whatever technique is adopted, it must succeed in determining an ordering in depth, or distance from the viewer, of the components of the scene, since, if two or more objects overlap, relative to the viewer, the nearest one

Figure 15.11 Remote-sensed imagery can be draped over a digital elevation model. The image (top left) has been combined with the digital elevation model (top right) to produce the projected 3D image (bottom left). Courtesy ERDAS.

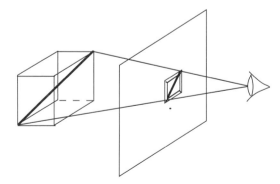

Figure 15.12 Projection of a 3D model is achieved by finding the intersection between projection rays emanating from the object and a view plane on which the 2D image is to be constructed.

must be found and displayed, and others somehow discarded. The fact that depth comparisons need only be performed if two objects overlap relative to the viewer gives rise to the use, in some algorithms, of lateral sorting to assist in determining which objects should be compared with respect to depth.

Hidden surface removal algorithms differ in their use of *object space* and *image space* operations. Object space operations take place at the resolution of the original object definition in the 3D viewing coordinate system. Image space operations are performed at the resolution of the pixels on the display device. Both types of operation may be employed in a single algorithm. For example, object space operations may be used to sort all the facets of a scene in depth order relative to the viewer. Then an image space procedure may be used to scan convert the facets into the frame buffer in reverse order of depth, thus causing the nearer facets to paint out those more distant ones that are invisible to the viewer.

Viewshed analysis

The 3D graphics techniques are designed to create images of a given spatial model as seen from a particular viewpoint. An important application of terrain models in GIS, however, is to ascertain the set of locations from which particular features are visible. The purpose of this is to assist in planning the location of potentially obstrusive phenomena such as new roads, buildings, electricity pylons and wind farms. The result is a viewshed map which displays those regions on the ground from which a particular feature is visible. Viewshed maps may well be used in combination with simulated views (as in Figure 1.7) using the computer graphics techniques described above and in the next sections.

Viewshed maps may be created from grid-based and TIN models of the terrain (Lee, 1991; Floriani and Magillo, 1994). Grid models lend themselves to particularly simple methods of deriving the viewshed, but they are limited by the resolution of the grid itself. The basis of the methods is that from each cell in the grid, a line of site is constructed in the direction of the feature or features, the visibility of which is under consideration (Figure 15.13). If the line of sight intersects a part of the terrain model between the viewpoint and the feature, then the feature is not visible. If the line of sight does not intersect the terrain model except at the location of the feature, then the feature must be located on the horizon relative to the viewpoint and will be highly visible, provided it is within a visible range. If the line of sight intersects the terrain beyond the feature, then the feature is within the field of view but not on the global horizon. The efficiency of execution of viewshed analyses can be improved by progressing outward from the viewpoint in a concentric circle pattern, such that different lines of sight that pass through the same cells need not result in repeated calculations of the visibility of that cell.

There are major problems in the reliability of the results of viewshed analyses. Part of the problem is in the error in the measurement of elevation of the terrain surface. Fisher (1993) has produced maps showing the considerable levels of uncertainty that can be introduced in viewshed analyses when the height error is taken into account. Equally important is the fact that terrain models tend to be based on ground level elevations, while the actual visibility of phenomena is very sensitive to the presence of vegetation and buildings on the terrain surface. High-quality viewshed analyses need to incorporate data on the height of these surface features and to be aware of the fact that the opaqueness of vegetation is seasonally variable.

Surface shading

Our ability to visualise 3D shapes from 2D projections is greatly enhanced by surface shading. Certainly the outermost boundary or silhouette of an object is also of use in identification, but in many circumstances it only provides limited information about the details of form. In the natural world, the intensity of light reflected from a surface to the eye of the viewer depends upon the orientation of the surface relative to the light source. Other factors which affect shading are the texture of the surface (shiny, matt, etc.) and, if it is shiny, its orientation relative to the viewer as well as to the light source. The natural characteristics of surface shading can be simulated in computer graphics. Mathematical descriptions of the effect of shading are referred to as *shading models*.

One of the most generally applicable models is the diffuse shading model based on *Lambertian shading*, which assumes a point light source of given intensity I_p. For matt (i.e. non-shiny) surfaces the intensity of reflected light is proportional to the cosine of the angle α between the normal vector N of the planar surface and the vector L giving the direction to the source of light (Figure 15.14). The constant of proportionality is given by the coefficient of reflected

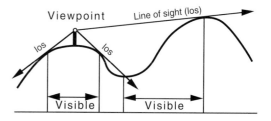

Figure 15.13 Lines of sight used in intervisibility algorithms.

diffuse light k_d. This is a measure of the diffuse reflectivity of the surface. The resulting intensity of diffusely reflected light I_d is given by

$$I_d = I_p.k_d \cos(\alpha)$$

Using this model, intensity I_d will be a maximum when $\alpha = 0$ ($\cos(\alpha) = 1$), which corresponds to the point light source being directed straight at the surface. If the light source is directed obliquely, intensity drops off until it is zero when $\alpha = 90$ and the light source is directed parallel to the surface. In natural circumstances involving a point light source, a facet of an object will often appear to be reflecting light even though it is facing away from the light source. This is due to the presence of ambient light reflected from other objects or facets in the scene. The effect of ambient light of intensity I_a can be included in the shading model by assuming a coefficient of reflectivity for ambient light k_a. Thus

$$I = I_a k_a + I_p\, k_d \cos(\alpha)$$

This quite simple shading model can be further modified to take account of a number of other factors. One of these is the distance that the object to be viewed is from the light source. This is relevant when considering artificial light sources, e.g. street lights. When visualising larger-scale scenes, in particular those in which human-made objects are represented, the sense of realism can often be improved by introducing a component of specular light reflection in addition to diffuse reflection. Thus, objects which are relatively smooth, constructed from metal or plastic, for example, reflect much of the incident light at an angle equal to the angle of incidence. The viewer will only see specularly reflected light if they happen to be looking in the same direction as this angle of reflection, giving rise to local highlights.

Fractals

The use of shading models, such as those described above, to display landscapes represented by digital terrain models often results in an impression of surface smoothness which, though sometimes appropriate, is liable to create a very artificial impression. The sense of realism of natural landscapes can sometimes be enhanced artificially by using fractal methods of roughening the digital terrain surface to simulate the appearance of rocks. If

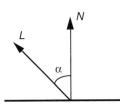

Figure 15.14 Relationship between the normal vector of a surface N and the direction of light L.

the terrain surface is represented by a rectangular or triangulated polygonal mesh, this can be achieved by subdividing, in a somewhat randomised manner (Figure 16.12), the quadrilateral or triangular facets into smaller pieces (Fournier et al., 1982).

Fractal methods can also be used to 'grow' plants and trees, many species of which appear to conform to fairly simple generating rules of growth and subdivision. They can also be used to create clouds. Such artificial vegetation and atmospheric effects can be combined with a terrain model to improve the sense of reality. Other techniques for improving realism include the use of *particle systems* which can represent phenomena such as fire and smoke, as well as vegetation, by particles rather than surfaces (Foley et al., 1990).

Interactive cartography

The facility to interact with graphical elements of a displayed map, provided by computer graphics technology, has introduced challenging opportunities for the future of cartography. The way in which this interaction is exploited may depend upon the roles of the map creators and the map users. If we regard communication as a process of transmission of messages, we can start by making a distinction about the information content of the message. Raw data on the locations of some spatially variable statistic, such as plant species or people working in retailing industries, may start with little information content and the purpose of producing the map could be to help in attaching meaning, or structuring the data. In this case the map maker and the map reader could be one and the same person.

Having made the map and hence communicated visually the locational data, the map reader can use it as a visualisation device, in which the map helps them identify spatial structures, patterns and correlations, thus raising the information content of the data. Using interactive techniques the map reader may be able to interrogate particular aspects of the data by manipulating displayed symbols and modifying the content of the map. This may be regarded as a process of exploratory data analysis, a concept promoted by Tukey (1977) in the context of statistical data analysis. Figure 15.15 illustrates the way in which elements of the mapped data may be linked to other statistical graphing devices such as histograms of the data that are shown in the map.

Cartography and information retrieval

The majority of uses of geographical information systems are focused on professional and industrial application areas. A new direction for GIS, in which interactive computer cartography might play an essential part, is that of providing access, at publicly or privately located terminals, to a wide variety of multimedia information sources. The rapidly improving status of the Internet is providing a hardware and software infrastructure that can support such access to computer-based information bases.

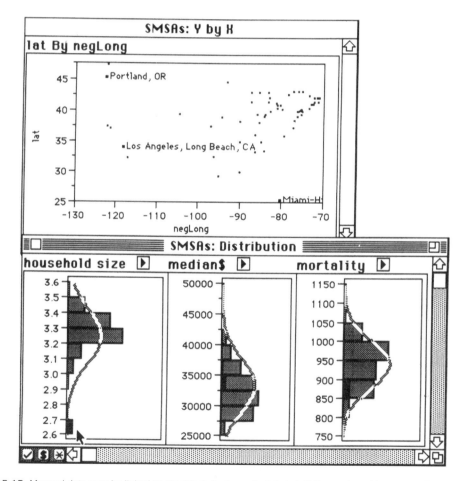

Figure 15.15 Mapped data may be linked to graphical displays of related statistics such as histograms. Data in the graph may be highlighted in response to interaction with the map. From MacDougall (1992).

Public information access

There are many types of information retrieval in which it can be envisaged that map-based interfaces could help the user in identifying information of interest and placing it in a geographical context. In the last few years many regional and municipal authorities have attempted to improve the awareness of local citizens to services provided by the local councils, as well as providing a general information service of relevance to the community and to potential visitors. These are sometimes referred to as community information systems (CIS). The types of information that can be made available include that relating to education, social services, health centres, entertainment, sports facilities, local societies, shops, banks, cinemas, parks, historical buildings, transport facilities, hotel accommodation and public meetings. Because so much of this information has a geographical context, maps at various levels of abstraction may be one of a number of particularly effective means of highlighting the existence of certain facilities and of enabling the user to enquire about them and to find out about where they are located.

Typically the user interfaces of public information systems make use of hierarchical menus and of A to Zs which the user can browse. Alternatively the user may type in key words relating to their interest. It may be envisaged that these user interfaces could be extended to make more use of maps that change their content in response to user actions. An initial screen might include relatively stylised representations of important classes of phenomena, such as hospitals, the town hall, schools and cinemas. If the user pointed to one such symbol, then the map might modify its contents to include closely related phenomena, while associated menus or other non-graphic interactive devices would be focused on related subject matter.

Creating user interfaces that anticipate user needs is a major challenge and it does appear that maps can play an important role in improving the ease of access to multiple and potentially complex sources of information. For the map to be effective in this context it is essential that it be linked with software that can traverse the information space in a manner that exploits knowledge of the conceptual relationships between different types of data. Put another way, we need a geographical information system in which there is as much emphasis on semantics as there is on locations and the structure of space.

Appropriate semantic modelling of information should enable the intelligent user interface to guess that, if the user points to a bus station on a map, or typed in (or spoke) the word 'bus', the map might be expected to respond by highlighting bus routes, bus stations and linked transport systems such as railway stations and airports. Associated with the map data could be menus of related terms. The user might then point to newly displayed map features and to textual or symbolic menu items in order to refine the search. To achieve this, the information system must maintain knowledge of classification systems recording, for example, that a bus station is a part of a transport infrastructure, which includes other components relating to travel by road, rail, air and sea. It must also maintain aggregational structure recording the physical and administrative components of typical classes of phenomena. Such semantic models may also usefully be combined with thesauri which record terms that are similar in meaning.

Summary

Technological developments in cartography have gone a long way towards automating the map production process both for screen displays and for publication purposes. However, little progress has been made in automating the map design process. Here we have reviewed graphic variables of hue, lightness, size, shape, texture, orientation and location, each of which can be used to communicate information in a map. We then discussed the way in which the use of these variables can affect the quality and effectiveness of map design. Text placement is one of the few aspects of design that has been automated to a considerable extent and we have seen how automated systems generate a set of trial positions, which a search procedure uses to find a suitable set of actual positions that are non-conflicting.

Computer graphics has been of particular use in helping to automate the production of 3D displays of map data, using digital terrain models on which topographic features may be superimposed. Realistic viewing of a 3D spatial model requires a process of hidden surface removal, based on identification of the visible parts of the model. It also requires techniques for shading the surface in a natural way that takes account of the light sources and the character of the terrain surface. Terrain models are notable for their use in viewshed analysis in which maps are produced to show which locations are visible from a particular viewpoint.

In the last part of the chapter we emphasised the importance of interactive map use in which the information displayed may change in response to user actions. This can be used simply to help appreciate the form of a landscape, or it may help in exploring a set of spatially referenced statistics, in which case it may be referred to as exploratory data analysis. An area for potential growth in the use of interactive maps is in networked public or community information systems in which the map serves as part of the user interface, to help in exploring information which is spatially referenced. A notable application is in providing information about the facilities available in a municipality.

Further reading

The early parts of this chapter have made frequent reference to Robinson *et al.* (1995), to which the reader is referred to for more detail on issues of map design. The books by Tufte (1983, 1990) provide interesting examples of map design, both good and bad. Brewer (1994) has written very informatively on the subject of colour selection in cartography. For an article on the automation of colour selection, see Wang and Brown (1991). The collected articles in MacEachren and Taylor (1994) and Hearnshaw and Unwin (1994) provide a useful source on visualisation techniques. Imhof (1975) is a classic text on guidelines for name placement, providing many examples of rules that are candidates for automation, while Freeman and Ahn (1984) is a landmark paper on automation of text placement. For reviews of techniques in name placement conflict resolution see Jones (1990) and Christensen *et al.* (1995). Animation in cartography has attracted considerable attention in the early 1990s: see, for example, DiBase *et al.* (1992), Dorling (1992) and Monmonier (1992).

CHAPTER 16

Map generalisation

Introduction

Of all the tasks faced in computer cartography and GIS, few are so fundamental to the process of map making and so difficult to automate as that of generalisation. Because a map is an assemblage of graphic symbols that present a view of some aspect of our understanding of the world, it is necessarily an abstraction of knowledge. It is not simply a collection of facts, rather it is a caricature of these facts which, in their transformation to map symbols, assist the map reader to understand spatial form and structure and to distinguish important characteristics of the phenomena that are represented. The process of generalisation requires the selection of those features that are essential to the map's purpose and the representation of them in a way which is clear and informative. Both selection and representation can be expected to involve a degree of information reduction relative to what is known. A major constraint to the information content of a map is the scale, in so far as it dictates the space available for map symbols. Thus map generalisation may be regarded as a scale-dependent process of information abstraction.

Selection of relevant information from a geographical database implies powers of abstraction that depend upon an understanding of geographical concepts. We can describe this aspect of generalisation as *semantic generalisation*. It is concerned with the meaning and function of a map and it depends upon being able to identify hierarchical structure in the geographical information. Graphic representation requires *symbolisation* of the selected information, which involves both the scale-dependent transformation of geometric data

and the choice of graphic and textual elements to communicate the real-world meaning of the data. The aspect of generalisation concerned with geometric transformation is termed *geometric generalisation* and is dictated by the interplay between semantic generalisation, symbolisation and the constraints of map scale.

Geometric generalisation may be regarded as a process of increasing the level of graphic abstraction relative to the original surveyed form of spatial phenomena. The construction of a large-scale map based directly on surveyed points and lines is a relatively mechanical process in which points may be plotted on a map grid coordinate system and, where appropriate, joined together to form lines and polygons (though the resulting map is still a gross caricature of the real world it represents). As map scale decreases, the relationship to original surveyed points and lines becomes very much more tenuous. There is no longer space on the map for true scale plotting, as the symbols for points and lines must, in order to remain visible, be exaggerated compared to the extent on the ground of the objects they represent. It also becomes necessary to select only the items of greatest importance to the map theme, to simplify their graphic representation and, very often, to move the graphic symbols from their true scale location in order to avoid overlapping or obscuring the symbols of adjacent map features. Figure 16.1 gives examples of the generalisation of UK Ordnance Survey maps that has taken place between the large-scale map (or plan) at a scale of 1:1250 and a derived map at a scale of 1:10 000.

When a cartographer simplifies a graphic representation, it is done in the context of a considerable body of knowledge and preconceptions about the

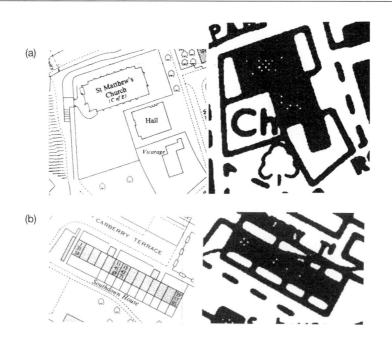

Figure 16.1 Comparison of a generalised Ordnance Survey topographic map at 1:10 000 with a large-scale plan at 1:1250. (**a**) Separate buildings have been merged. (**b**) A block of terraced houses has been merged and the boundary fences reduced in number. ©Crown copyright.

properties and characteristics of the phenomena which the map symbols represent. Thus areal symbols for buildings may have a rectangular or blocked appearance which distinguishes them from, for example, lakes or islands, and this character should be retained even when the outline has been greatly simplified. Linear symbols for major roads are usually smoothly curved, reflecting the fact that the actual roads have restrictions on the degree of curvature, whereas the linear symbol for a coastline may be distinctively crenulated in a manner that varies from one geomorphological region to another.

Successful automation of generalisation concerned with caricature requires a computer program to simulate knowledge of geographical and cartographic structure in order to recognise the cultural or geomorphological nature of the map features. Application of this *structural knowledge* (Armstrong, 1991) is dependent upon automated pattern recognition and is certainly not very advanced in present GIS technology. In addition to retaining a knowledge of the entities that are being generalised, an automated system must possess the equivalent of a human's visual capacity to organise the map symbols in a manner which ensures their legibility and aesthetic acceptability. When it becomes necessary

to displace objects to avoid graphical conflict, the human cartographer is able to see many things simultaneously and to hypothesise about the effects of possible shifts in position. Making decisions about appropriate actions to take in generalisation is part of the high-level management of generalisation which we may regard as process control (Brassel and Weibel, 1988). At the time of writing, process control is not well automated, being heavily dependent upon human skills of planning, structure recognition and design. Most effort has been put into automating procedures for simplifying individual geometric objects, principally lines and polygons. However, progress in knowledge representation, automatic reasoning and computer vision bodes well for future improved capabilities for automated generalisation that are required for many applications of GIS.

In the remainder of this chapter we elaborate on semantic generalisation. In discussing geometric generalisation, we describe procedures for the tasks of elimination, reduction, enhancement, typification, collapse, amalgamation, exaggeration and displacement. The final section provides an overview of the stages involved in process control in automated generalisation.

Semantic generalisation

Determination of the information content of a map, which the semantic phase of generalisation involves, is one that is greatly facilitated by the existence of hierarchical structures in the source data. As Nyerges (1991) has pointed out, geographical meaning may be encoded by conceptual hierarchies that, when combined, form a heterarchy of concepts. This heterarchy may be regarded as a multidimensional framework for knowledge in that it may be traversed from top to bottom, bottom to top and side to side, thereby enabling changes in the level of information content and translation between geographical concepts. Appropriate abstraction may involve selecting a variety of different levels of individual hierarchies in order to create the right balance between types of information. For example, a tourist map might highlight the presence of small towns and villages that happened to include historic buildings, while placing less emphasis on other settlements which, though more significant in a hierarchy based on population and local government administration, were not considered to be of significant interest in terms of historic monuments. The same map might show all major highways in the area, but might also give prominence to particular minor roads that happened to lead to places of tourist interest.

Two types of hierarchical structure are particularly relevant to generalisation. These are *classification* and *aggregation*, both of which have been introduced in the context of semantic modelling for databases (see Chapter 10). Classification is the most familiar hierarchical structure in that there are numerous established classification schemes applying to a wide range of areas of human interest. In the topographic domain, one finds major class divisions such as hydrography, transport systems and settlements, each of which may have several levels of subclassification. Hydrography could be subdivided into oceans, rivers and lakes, which might be further subdivided according to criteria based on, for example, geomorphology, dimensions or chemistry. Classification schemes can themselves be classified according to whether they are qualitative, like those just referred to, or quantitative. Quantitative schemes may be based on interval or ratio scales of measurement. Thus settlements could be classified on an integer numeric scale according to their population size. Such a ratio scale could be transformed hierarchically to an ordinal scale by introducing class intervals for the population (e.g. $a = 1$ to 500, $b = 501$ to 2500, $c = 2501$ to $10\,000$, etc.).

Once data have been classified hierarchically using concepts relevant to the application area, it is possible to devise rules for selecting information on the basis of position within one or more hierarchies. Thus a road map at $1:2\,000\,000$ covering the whole of the UK might only be expected to show roads classed as motorways and primary routes, along with settlements classed as large towns and cities, while larger-scale maps would progressively introduce classes of roads and towns that were lower down their respective hierarchies.

Aggregation hierarchies are concerned with the composition of particular phenomena in that they can be used to represent objects in terms of their constituent parts, which may themselves be further subdivided. For example, a city might be composed of administrative districts, each of which was in turn composed of a network of streets separated by blocks of domestic, commercial, industrial and leisure-based development, each of which might be further composed individually of buildings, passageways and open land areas.

When selecting information, aggregation can, in principle, be used in a similar manner to classification hierarchies to provide initial access to the relevant data. For example, if a town map was to be created at a scale at which it was appropriate to represent entire street blocks, rather than the individual properties that made up the blocks, pre-existing aggregation data could be used to find the geometric data belonging to each street block. If the data were represented originally at the level of individual properties, the geometric representation would be too detailed for graphic display at the street block level. It would therefore be necessary to transform the detailed geometric representation to a less-detailed, block level form, for the purposes of subsequent cartographic display. Problems may be increased if the database contained detailed geometric representations that had not previously been categorised in terms of the aggregations of interest. If this was the case it might be necessary to use a structure recognition procedure to 'find' the street blocks, given the individual constituent buildings. Automatic recognition of hierarchical structure in geometric data is a problem in computer vision and it is not a trivial task, though some commercial GIS systems do now include some facilities to assist with it.

Geometric generalisation

The geometric representation of geographical information may be subject to a wide range of modifications in the course of generalisation. Some of the modifications result directly from the consequences of semantic generalisation. Thus particular pieces of geometry, such as the boundaries between properties in a residential block, may be omitted entirely if they represent phenomena too low down a classification hierarchy or an aggregation hierarchy. Most types of modification arise from attempts to meet objectives of good cartographic symbolisation, particularly relating to clarity and ease of visual communication. Thus, as indicated earlier, reduction in scale is usually accompanied by reduction in detail of the representation of individual objects, as well as the exaggeration or enhancement of individual objects to make them more clearly distinguishable. Changes in form, such as those resulting from exaggeration, may lead to overlap between the geometry of adjacent objects. The objective of visual clarity may then lead to displacement of the objects in order to separate them on a map.

Various categories of geometric generalisation have been developed (e.g. Shea and McMaster, 1989). Here the following set are employed as a basis for further discussion:

- *elimination* of point, line and area geometry;
- *reduction* in the detail of lines, areas and surfaces;
- *enhancement* of the appearance of lines, areas and surfaces;
- *amalgamation* of lines and of areas;
- *collapse* of areas to lines and points;
- *enlargement* or *exaggeration* of area and line objects;
- *typification* of line, area and surface objects;
- *displacement* of points, lines and areas.

Figure 16.2 provides a graphical overview of these operators, which we will now describe in more detail.

Elimination

One of the simplest operators is that of elimination. It has the effect of removing point, line and area geometry representing point-referenced, linear and areal features. In a small-scale representation of settlements, each settlement might be referenced to a single point. If

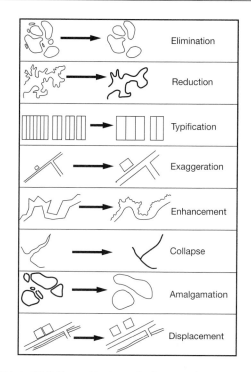

Figure 16.2 Geometric generalisation operators.

the scale of the required map was expected to result in excessive clustering or overlap of symbols and text representing the settlements, then individual places might be eliminated if they were likely to obscure an adjacent, more important settlement. In a similar way, lines representing, for example, minor streams and rivers in a drainage network, and areas representing small lakes, might be eliminated for purposes of cartographic representation on the basis of reducing visual clutter.

Note that semantic generalisation might be expected to have eliminated some features on the basis of classification or of a type-related attribute such as population. Geometric elimination may be performed simplistically on the basis of the size of line and area features respectively. For example, river tributaries shorter than a given length and lakes smaller than a given area might be eliminated. For areal features, elimination might depend upon a combination of falling below a given area and a given maximum dimension.

The importance of features such as streams, lakes and small settlements is a function not just of their classification and their size, but also of their degree of isolation. Thus geometric elimination of features should be tempered by a measure of proximity to other features of the same group. Measures of proximity must be available in combination with semantic measures such as settlement population to ensure that isolated phenomena that are regionally important are not eliminated inadvertently.

The latter problem is complicated by the fact that a few small features may be in close proximity to each other but far from any other feature of the same class. Appropriate handling of such situations may require a form of cluster analysis to ensure that at least one of the features is retained in a representative capacity (see Mackaness and Fisher (1987) for a discussion of the role of clustering in map design).

Control of elimination procedures should depend upon knowledge of an appropriate density of information on the map. Empirical studies of the change in the number of map objects as the scale of the map is changed, on standard map sheets, have resulted in some rules of thumb, of which *Topfer's Radical Law* is a prominent example (Topfer and Pillewizer, 1966). It presents relationships between the ratio of the number of features on the source and derived maps and the change in scale between the two maps. The basic rule for determining the number of objects that should appear on the derived map is given by

$$nd = ns \sqrt{(Ms/Md)}$$

where *nd* is the number of derived objects, *ns* is the number of source objects, and *Ms* and *Md* are scale denominators for the source and derived maps, respectively. This form of the rule applies to large-scale maps in which the symbol dimensions are directly proportional to the corresponding feature dimensions. There are several variations on the formula, for small-scale maps and for point, line and area objects respectively. A problem with Topfer's Law, however, is that it does not indicate which objects should be selected and it does not take account of local variation in the density of phenomena. For purposes of automation it may be better to let the number of objects be determined by applying rules regarding minimum separation of objects in combination with rules for selection of the most important semantic categories appropriate for the required map. It is then perhaps possible to envisage using Topfer's Law, or variants of it, to indicate whether the resulting map differed greatly from what some might regard as a normal density for the scale of map.

Reduction

Rather than removing spatial objects, as in elimination, the process of reduction is concerned with decreasing the constituent detail of lines, polygons and surface representations. In the case of point objects, reduction is equivalent to elimination. Reduction of lines, areas and surfaces generally consists of removing individual points that are used in the original geometric description of the objects. Thus the points defining a line or the boundary of an area would be selectively removed, producing a simplified version of the line or boundary with fewer wiggles than its original representation. We will now consider relevant techniques for lines, areas and surfaces in more detail.

Line reduction

Line reduction is commonly referred to as line simplification and line generalisation, and we shall use the terms interchangeably here. Many of the methods for line reduction were motivated as much by the need for reducing the volume of digital data as for performing cartographically sound generalisation. A number of the algorithms are governed by approaches which provide control over the error in deviation of the simplified line from the original line, though there are still very few techniques which actively seek to retain distinguishing characteristics of the mapped feature.

Techniques for line reduction may be classified as follows:

1. arbitrary point selection;
2. local direction and distance processing;
3. local tolerance band processing;
4. global tolerance band processing;
5. curvature processing; and
6. curve function fitting.

This classification is a modification and extension of that of McMaster (1987).

Arbitrary point selection
Perhaps the crudest method of line simplification is to select every *n*th point along the line, where the degree of simplification increases as *n* decreases. The points will then be selected quite independently of their significance in representing the shape of the line. The approach could have some use in reducing or thinning a data set with a surfeit of points at some particular scale, but it has no cartographic merit, since for the purposes of shape preservation, it is random.

Local direction and distance processing
A number of simplification techniques have been developed which proceed sequentially along the digi-

tised line, selecting those points which meet various criteria constraining one or both of the distance between successive selected points and the angle between successive selected edges. Tobler (1966) describes a method (attributed to A. V. Hershey) based on distance only, in which a point is selected whenever its distance from the last selected point exceeds some minimum value. This value may be related to the resolution of the plotting device.

A simple method (Figure 16.3), based only on angle, selects the second of three consecutive points (P1, P2 and P3) if the angle between the directions of edges P1–P2 and P2–P3 exceeds the given tolerance (Jenks, 1981). Distance and angle can be combined, such that P2 will be retained if the angular tolerance is exceeded and the distances P1–P2 and P1–P3 both exceed respective pre-set tolerance values (McMaster, 1987).

Local tolerance band processing
The above techniques are characterised by the fact that processing is typically based on either two or three successive points. A further class of techniques examine more extended sections of the line. Typically these techniques ensure that the selected points lie within bands or strips which are locally superimposed on the source line. In general, a point is selected and a new band formed whenever the line moves out of the confines of the current band. This approach is exemplified by the Reumann and Witkam (1973) algorithm in which two successive points on the line define the direction of the band, which is centred on the two points (Figure 16.4). The first of the two points is selected while the next selected point is the last one, in sequence, to remain within the strip. This point and its successor define the direction and centre of the next strip.

A modification of the Reumann–Witkam algorithm is provided by the Opheim (1981) algorithm, in which the direction of the band is determined by the line joining the initial point to the last point that lies within a specified radius (of the first point) or, if there is no point within the radius, the next point on the line. Opheim's technique also limits the length of

the band, so that if no edge leaves the band within that distance, the last point within the band will be the one selected.

Deveau (1985) has produced a band algorithm which provides options for a centred band and a floating band approach, as well as providing explicit control over the retention of promontories and small islands. This differs a little from those just described in that the direction of a band may be adjusted as each new point of a current set is considered. In general, the start of a band is centred on the initial point of a local set. As new points are added, an attempt is made to find an orientation for the band such that all points in the set are accommodated within the tolerance. When this condition cannot be met, the last point which could be included is selected and then becomes the first point of the next set.

It may be noted that a relatively early algorithm described by Lang (1969) corresponds fairly closely in practice to Deveau's floating band technique, except that Lang imposes an initial maximum number of points to be considered and works backward towards the initial point, until all intervening points lie within a specified perpendicular tolerance distance. The distance is measured relative to a line joining the first and last of the set under consideration.

Global band processing
Perhaps the most widely used line-simplification algorithm is that of Douglas and Peucker (1973), also known as the Douglas algorithm. It is particularly attractive to cartographers because, unlike many other algorithms, it is designed to retain extreme points that are often also the most important ones for retaining the characteristics of shape (Marino, 1979). It has also been shown that there is a strong correlation between the points selected by the algorithm and those selected by cartographers (White, 1985).

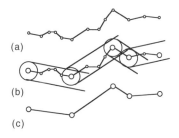

Figure 16.3 Line reduction using local distance and angle processing. Point P2 is selected if angle α exceeds a specified tolerance.

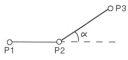

Figure 16.4 Line reduction using local tolerance band processing. The large dots represent points selected using local band processing in the manner of the Reumann–Witkam algorithm.

The Douglas algorithm operates in a global manner, in that it starts by considering the entirety of the line to be simplified (see Figure 16.5). Initially the start and end points of the line, termed the anchor and floater points, are treated as defining a straight line, from which the perpendicular distances of all other points are measured. If the most distant point lies within a pre-specified tolerance, the anchor and floater are taken to represent the entire feature. Otherwise the floater point is stored and the most distant point becomes the new floater. This most distant point therefore splits the line into two parts, each of which will be treated in the same way as the initial line.

The approach is essentially recursive, as each sub-segment of the line is treated in the same way as the parent from which it was derived. Recursion terminates, at least locally, when the anchor and floater can represent a line segment, after which an adjacent segment of the line is examined. When the latter situation arises, the current floater becomes the anchor point and the new floater is set to be the last floater to have been saved. The implementation of the algorithm may differ somewhat depending on whether the implementation language incorporates recursion.

A example of the results of using the algorithm is illustrated in Figure 16.6.

Despite its widespread use, the Douglas algorithm has been criticised for the fact that it may result in self-intersection of the simplified line and it can produce spikey artifactual representations (Visvalingam and Whyatt, 1990). The algorithm can be modified to avoid cross-overs (Muller, 1990), and the spikey shape can be reduced by smoothing the line after simplification. However, the quality of the simplification based on the algorithm will depend greatly upon the tolerance used, and comparisons of other algorithms with the Douglas algorithm are often difficult to assess due to inappropriate tolerance selection. As Buttenfield (1987) has remarked, appropriate selection of tolerance parameters may depend upon the geometric character of the line to be generalised. Jones and Abraham (1987) have described a method of automatic parameter selection which involves a prior analysis of the relationship between tolerance and the number of points selected by the algorithm for a particular class of linear feature, combined with a heuristic based on Topfer's Law to determine the percentage change in points for a particular scale change.

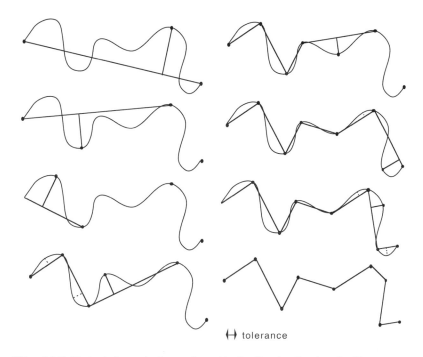

⊢⊣ tolerance

Figure 16.5 Stages in line reduction performed by the Douglas–Peucker algorithm.

Figure 16.6 Results of applying the Douglas–Peucker algorithm to data in (a) for the coast and administrative boundaries of Wales using 3 tolerance values in b, c, d and e, f, g respectively.

Curvature processing

Thapa (1988) has approached the problem of line point selection, and hence line reduction, by adapting techniques developed in the field of image processing. His method involves selecting points which correspond to *zero crossings* in a transformation of the line to be generalised. Zero crossings were first used in image processing by Marr and Hildreth (1980) as an automatic means of finding edges which delineated important features in 2D images. The expression 'zero crossing' refers to the graph of the second derivative of a function representing an abrupt change, i.e. an edge (Figure 16.7). Thus the second derivative exhibits a sharp oscillation as it crosses between positive and negative values (the corresponding first derivative (16.17b) is a peak). This phenomenon can be observed in an image or a digital signal by studying the result of convolving the signal with a filter which is the Laplacian of the Gaussian distribution. The Gaussian part of the filter serves to smooth the signal such that irrelevant or uninteresting variation (noise) at the required scale is removed, while the Laplacian component finds the second derivative.

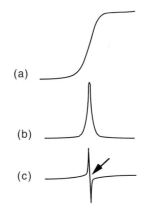

Figure 16.7 The zero crossing is the point at which the second derivative (**c**) of a signal representing an edge (**a**) changes sign between positive and negative.

The signal representing the curve is based on a modification of the Freeman code of the line which transforms it to a sequence of curvature elements. The Freeman codes $e(i)$ are values from 0 to 7 giving directions in each cell through which the line passes. The curvature elements $s(i)$ are given by

$$s(i) = (e(i) - e(i-1) + 11) \bmod 8 - 3$$

The expression 'mod 8' means the remainder after division by 8. The reason for adding 11 is to avoid generating negative values. Note that successive values of $e(i)$ that are very similar will result in low or zero values of $s(i)$, while abrupt changes in e will result in high values of $s(i)$. Therefore a change in the direction of the line will produce a step in the graph of $s(i)$.

An important aspect of using the Gaussian function is that it provides the parameter σ (standard deviation), which is a measure of the width of the Gaussian peak. By using a low value for σ, fine detail will be retained, while higher values will smooth out the detail. The lower the value, the more zero crossings will be found. It is therefore a useful parameter for specifying the degree of generalisation. Thapa (1988) points out that the zero crossings found, and hence the selected points, do not necessarily correspond to the critical points referred to by Marino (1979) and White (1985). Rather they are points which may result in a relatively smooth version of the original line though they do correspond to points which lie on the original line. Thapa contrasts this with the critical points selected by the Douglas algorithm which, as indicated above, can result in a spikey simplification for larger-scale changes (e.g. greater than a factor of four). Thapa's method has not yet been fully evaluated relative to other methods.

Raster methods for line and area boundary reduction
Rasterised representations of linear features can be reduced in detail by using the mathematical morphology image processing techniques of dilation and erosion (see Chapter 8). Dilation expands objects and in the process obliterates local detail, which will tend to be filled in by the accretion of additional pixels. Having generated a less sinuous representation it can be thinned back to its original width by the erosion process. A related method is described below in the context of object amalgamation.

Area boundary reduction
On a topologically structured map, polygons are typically defined by chains of coordinates that terminate at nodes. Individual chains can be simplified using any of the techniques referred to for line simplification. Application of such techniques is relatively straightforward for simple polygon networks that do not include holes or islands. The nodes would normally always be selected and would remain unchanged by the algorithms. The Douglas–Peucker algorithm meets this condition and has been applied to polygon boundaries, though it can be expected to result in a change of the area of the enclosed polygons. Williams (1987) proposed a technique for modifying the simplified chains such that the original areas are reinstated. The method involves shifting all edges by a fixed distance, which is calculated separately for each chain.

Simplification of the chains of a polygon boundary can result in the undesirable side-effect of modification to the topology of the polygon due to the introduction of overlap between simplified chains that are near to each other. This can be rectified by checking for overlap and reintroduction of selected vertices that were previously eliminated (Jones *et al.*, 1994).

Boundary reduction for rectangular objects
The techniques described above have been applied largely to natural features. However, they are not appropriate for highly regular features such as buildings, since they will tend to degrade straight edges and right-angle bends.

Lichtner (1979) described a number of techniques (attributed to Hake and Hoffmeister, 1978; Staufenbiel, 1973) specifically intended for generalising buildings. Several tolerances are used for controlling generalisation. These include a minimum side length, a minimum area and a minimum gap between objects. An example of the result of such automated building generalisation techniques is given in Figure 16.8, produced using the CHANGE automated generalisation system. Note the removal of relatively small extensions and concavities and the forcing of alignments.

Surface reduction

Techniques for surface reduction depend upon the method used for representing the surface. As we saw in Chapter 12, the principal methods are regular grids, local functions and splines, contours, form lines, and triangulations.

Regular grids can be very easily simplified by selecting every nth grid point in the x and y directions. The value of n would typically be two or three. This is an arbitrary method, comparable to nth point

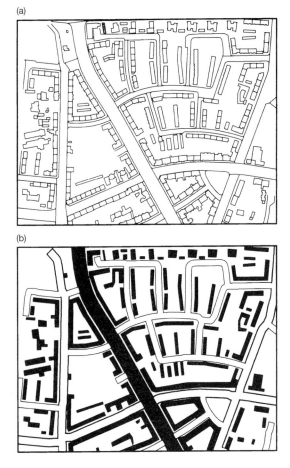

Figure 16.8 An example of automated building simplification using the CHANGE software. (**a**) Part of the town map 1:1000 of the City of Hannover; (**b**) the result of an automated generalisation using CHANGE. Scale values refer to the scale of the original maps. Courtesy, Institut für Kartographie, Universitat Hannover.

line simplification, but for $n = 2$ it has the advantage of reducing the data volume by 75%.

Function fitting using local polynomials lends itself to simplification, while maintaining a smooth form for the surface (assuming that the latter is desirable). The error associated with polynomial approximation is inversely proportional to the degree of the polynomial. Hence, reducing the degree reduces detail. A shortcoming of the method is that, in general, errors cannot be guaranteed to lie within specific bands, unless the degree of the polynomial is variable. An approach which enables the polynomial degree to remain constant, while retaining control over error, is the use of quadtree patches (Leifer and Mark, 1987). Thus cell or

patch size is variable, with quadtree cells being subdivided into further patches until the error criterion is met (see Figure 12.7).

A global mathematical function method which appears superficially to lend itself to surface generalisation is the use of Fourier series, in which the surface is represented by a summation of sinusoidal surfaces of varying wavelength. Reduction in scale can be achieved by omitting (filtering) high-frequency terms from the summation (Clarke, 1988). Use of Fourier series is complicated by the fact that the importance of particular frequencies varies between different surfaces and that, as with polynomials, error levels cannot be guaranteed for a preselected number of terms. Another problem is the poor approximation obtained at the boundaries of a data set. By examining the power spectrum resulting from Fourier analysis, it is possible to identify the most significant frequencies, and hence to select a limited number of frequencies to approximate the surface. Only by iterative testing of combinations of frequencies would it be possible to ensure that error lay within a particular limit.

Contour representations of surfaces can in principle be generalised by applying line reduction techniques to the individual lines. The basic method is fundamentally flawed in that the vertical error introduced by particular 2D line tolerance parameters would be variable along the line, depending on the slope, and could not therefore be predicted without reference to the adjacent contours. It can also lead to cartographic anomalies in that generalised versions of adjacent contours might overlap each other. In general, if a contour representation is required, it is probably better to generalise a digital terrain model (grid or TIN) before regenerating the contours (Pannekoek, 1962).

Triangulations provide perhaps the most attractive model for the controlled reduction of surfaces in that it is possible to apply vertical error tolerances to the selection of important points used in the triangulation (Floriani, 1989; Heller, 1990) and to apply line generalisation procedures to linear features and form lines that are integrated with and constrain the triangulation (Ware and Jones, 1992; Weibel, 1992). Techniques for selecting points from surfaces are discussed in Chapter 12. Figure 16.9 illustrates two generalisations of triangulated terrain models constrained by geological boundaries. The triangulations were derived directly from the MTSD multi-resolution database (Ware and Jones, 1992), which categorises terrain and line vertices on the basis of a prior analysis using surface and line reduction algorithms.

(a) 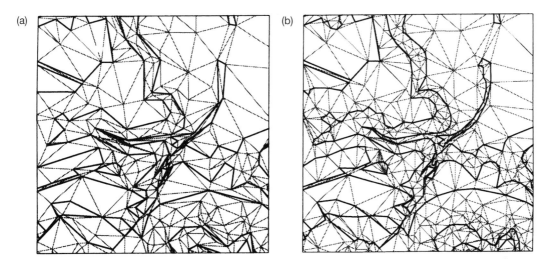 (b)

Figure 16.9 Constrained triangulated surfaces at different levels of generalisation. The linear features embedded in the terrain model are geological outcrop boundaries.

Enhancement

Computer-plotted cartographic lines normally depict a sequence of straight edges. Ideally the vertices should be sufficiently close that the line either appears to be smoothly curved, or at least 'natural' in its representation of the mapped features. It is not unusual, however, for the straight lines to be visually discernable, either due to inadequate or error-prone sampling of the original digitised features or as a side-effect of the geometric generalisation processes described here.

Smoothing

Angularity in plotted lines can be avoided by using various smoothing procedures that either adjust the positions of existing points or which fit digitised curves passing exactly through the existing points or along a path which approximates the path of these points (Figure 16.10). In the context of generalisation, smoothing is normally applied, if at all, after the application of a reduction process. Shea and McMaster (1989) have suggested that in addition to post-reduction smoothing, it may be appropriate to smooth data before reduction.

A simple approach to smoothing is to apply a moving-average operator, which results in shifting the existing points. The coordinates of each point are then adjusted by substituting them with the actual or the weighted average of the coordinates of a point and its immediate neighbours. This would typically consist of either a three- or five-point average, centred on the point to be adjusted. Excessive displacement is avoided by weighting the central point to give it more importance than its neighbours. Such operators can be applied to lines, with a 1D moving-average window, and to gridded surfaces using square windows (or kernels). Weibel (1992) describes and illustrates application of the technique to gridded terrain models for purposes of 'global filtering' (Figure 16.11).

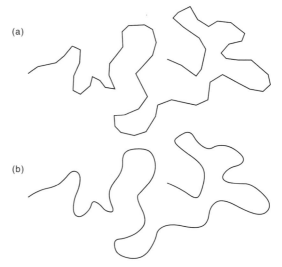

(a)

(b)

Figure 16.10 Smoothing a digitised line.

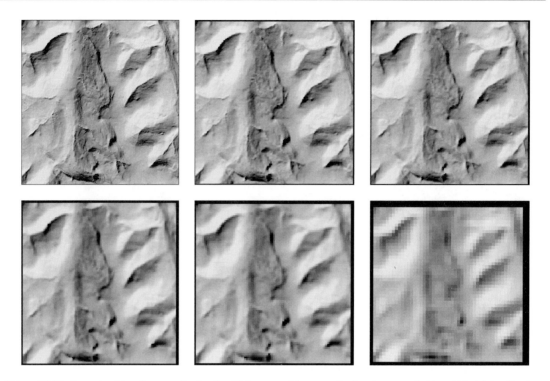

Figure 16.11 Example of terrain generalisation in which a smoothing operator has been applied to gridded data. Courtesy of D. B. Kidner. ©Crown copyright.

The above methods perform smoothing by increasing the convex angles between successive edges. A greater impression of smoothness is given by adding further points (as in Figure 16.10). This can be done mathematically by means of spline curves which are fitted either exactly or approximately through the existing points. The B-spline is particularly versatile in that explicit control is provided over the degree of the fitted curves (quadratic, cubic, etc.), the level of continuity between them and the extent to which they pass through the existing points (Foley *et al.*, 1990; Rogers and Adams, 1990). Having defined spline curves, it is also a matter of user control as to how many digitised points are used to plot them. Thus it is possible to ensure the appearance of a very smooth line.

Fractalisation

Smoothing is undoubtedly useful in improving the appearance of lines that would otherwise include unnatural-looking straight sections. In certain cases, such as roads and meandering rivers, this added smoothness may well reflect the real character of the feature represented. However, there are many natural features that are expected to appear rough. Richardson's (1961) work on measuring the length of coastlines at different scales highlighted the almost infinite wiggliness of such lines. This led to the suggestion by Mandelbrot (1982) and others that they are self-similar in form and can be modelled by fractal processes. The important property of fractal lines is that patterns apparent at low resolution are replicated approximately when the line is represented at successively finer resolution.

Dutton (1981) applied the concept of self-similarity to the enhancement of digitised curves by adding high-resolution angularity. The degree of wiggliness of a line can be measured by its fractal dimension, which varies between one and two. Dutton's technique attempts to preserve the fractal dimension of the original line. The process is a recursive one in which vertices, starting with the original ones, are moved such that the angle that they subtend with the mid-points of their adjacent edges is standardised. Moving a vertex involves creating new vertices at the adjacent edge mid-points. Thus for one 'pass', the total number of straight edges doubles. It is notable that although fractalisation can tend to increase the sense of realism, Dutton found that it was still desirable, for aesthetic reasons, to apply a final smoothing phase.

A similar principle to that described for lines can be applied to surfaces. Assuming that the surface representation consists of planar facets, each such facet may be subdivided by applying displacements to the midpoints of the facet edges (Figure 16.12). Fractalisation procedures of this sort have been used to generate artificial landscapes, having started with quite simple primitive forms, the character of which is reproduced in the self-similar fractalisation process (Fournier *et al.*, 1982; Clarke, 1988). As pointed out in Chapter 15, the approach can be applied to surfaces represented by rectangular or triangulated polygonal meshes.

Amalgamation

The process of *amalgamation*, or *combination*, entails merging originally distinct or entirely separate objects into single representative objects. It arises directly from the competition for space that occurs on symbolised small-scale maps. Groups of adjacent buildings may be amalgamated to a single block of buildings. Adjacent lakes or islands may be amalgamated into single lakes or islands respectively. It is normally applied to a set of adjacent areally referenced objects that are of the same class.

In the case of linear objects, amalgamation would only take place if the linear objects to be amalgamated belonged conceptually to a higher-level parent object. This would apply to the separate carriageways of a major highway or to the adjacent, approximately parallel, but connected, channels of a single river.

Choice of the appropriate objects to amalgamate may become complex and depend on the presence of associated principles for semantic aggregation, as described in the previous section. In other words, the amalgamation should make sense with regard to the classification and the implied structure of the resulting amalgamated object. For example, it might be appropriate to amalgamate buildings that are part of a single block bounded by streets. It might make less sense to amalgamate buildings on opposite sides of an important street. As indicated earlier, major streets might partition aggregations of multiple object classes (buildings, paths, small gardens, etc.).

If two areal vector-represented objects are adjacent and share a common boundary, then amalgamation can be achieved simply by removing (dissolving) the

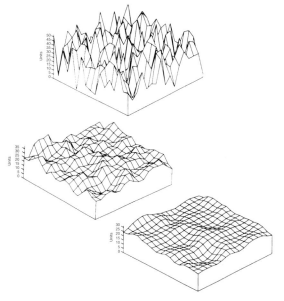

Figure 16.12 Surface enhancement by a fractalising process in which areas are subdivided to create greater detail. From Clarke (1991).

boundary. This is an example of a geometric procedure that may be used in conjunction with semantic aggregation and classification. Thus two similar objects, such as houses in a terraced block, may be amalgamated as part of a semantic aggregation operation, while classification could entail amalgamating adjacent subtypes (e.g. gardens and houses) of a higher-level areal class (e.g. built-up residential areas).

A raster procedure for area amalgamation of separate objects applied by Lay and Weber (1983) involves a succession of pixel accretion and thinning operations (equivalent to *dilation and erosion*, and also referred to as *expansion and contraction*) applied to the rasterised area boundaries. Initial accretion expands the objects, such that nearby ones will merge into each other. Subsequent thinning reduces their size while retaining contiguity between the amalgamated objects, which may be further accreted to reinstate the original areal extent. The procedure is applied separately to the boundaries of primary objects and to the clearings or holes within these objects. Thus the clearings will expand in size initially and merge together where they are nearby, rather than tending to be flooded by the initial boundary expansion phase. Schylberg (1993) has continued with research using related techniques (Figure 16.13).

A similar approach has been applied to vector representations by Muller and Wang (1992), whereby the vector boundaries are expanded and contracted. They combined the method with one for the elimination of areal objects that fell below a given threshold.

A vector method for amalgamating rectangular objects was described by Lichtner (1979). Objects closer than a given gap limit were closed together by moving the smaller towards the larger in a direction perpendicular to the length of the gap. The effect of this is that the boundary may become more complex and require reduction in the manner described earlier for rectangular objects.

Another approach to amalgamating separate objects is to identify the space between them and to reattribute the space with the classification of the amalgamating objects. Ware *et al.* (1995) and Jones *et al.* (1995) have described how triangulated data structures can be used to perform this process, since neighbouring objects are often directly connected in the triangulation, and if not can be found by a simple search procedure through the triangulation. Triangulation can also assist in finding appropriate vectors to be used in moving nearby objects towards each other in order to close the gap between them (see Figure 16.14).

Collapse

Competition for space on a map may dictate that objects cannot be represented with their true-scale areal extent. The collapse operation solves this problem by reducing the dimensionality of the geometric representation of an areal object to a line or to a point. A city that was represented at one level of generalisation by a polygon would be collapsed to a single point at a smaller scale of representation. On the map the city would then be represented by a symbol centred on the given point. Typical examples of collapse from areas to lines arise with conceptually linear (or ribbon-form) features such as roads and rivers that in a detailed representation could be described geometrically by their boundaries, but at a smaller scale would be collapsed to a line that was centred on the path of the road or river (Figure 16.15). Again, the line geometry would then be used to locate an appropriate linear symbolisation on the map. The simplest method of collapsing an area to a point is to generate a centroid for the area and use the resulting point to locate a point-referenced symbol.

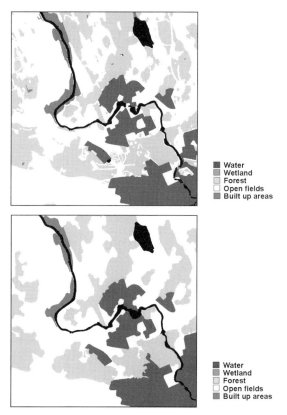

Figure 16.13 Area generalisation in raster mode (from Schylberg, 1993). Objects in map **(a)** are merged and, if they fall below a threshold size, deleted, to produce map **(b)**. A tolerance distance is used to control the amalgamation process.

Medial axis transformation and skeletonisation

The process of reducing the dimensionality of an area to a line is referred to as a medial axis transformation and the resulting line is sometimes described as a skeleton (Montanari, 1969). The skeleton may be regarded as being obtained by transmitting waves inward from the boundary and noting where opposing waves meet. The skeleton has the property of being equidistant, at all points along it, from at least two points on the boundary of the original polygon. Raster-based methods for thinning and hence skeletonisation were described in Chapter 8.

The geometric procedure for generating a correct skeleton from vector data can be fairly complex. However, we may note that there is a relatively simple procedure for generating a form of skeleton by joining up the circumcentres of the triangles of a Delaunay

(a)

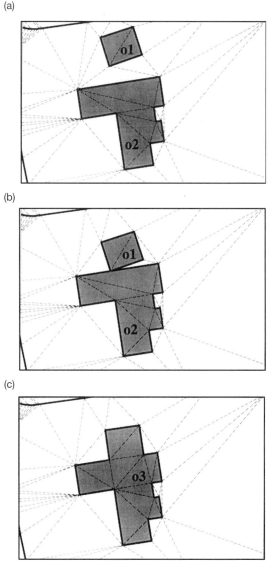

(b)

(c)

Figure 16.14 Areal object merge using a triangulated data structure to snap together nearby objects. The larger of the two triangles between objects o1 and o2 in (**a**) is collapsed (**b**), before rotating o1 to align it with o2 (**c**) and retriangulating. Data from Ordnance Survey, ©Crown copyright.

triangulation of the polygon (Chithambaram and Beard, 1991). Alternatively, the centres of triangulation edges joining opposite sides may be connected, as was illustrated in Chapter 15 in the context of finding paths for areal text (Figure 16.15). Having generated a skeleton, it may be simplified or enhanced if necessary and used for a linear symbolisation.

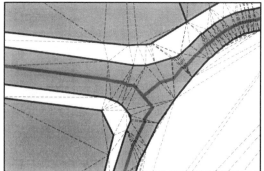

Figure 16.15 A ribbon-shaped feature can be collapsed by finding its skeleton, carried out here using a triangulated data structure. Data from Ordnance Survey, ©Crown copyright.

Enlargement (exaggeration)

Reduction in scale can result in some map features becoming too small to be clearly discernible. If these features are regarded as too important to be eliminated from the map, perhaps because they are isolated or otherwise significant buildings, then they must be enlarged. Another reason for enlargement is to ensure clear distinction of symbolised lines that may not be true scale. Thus if both sides of a road are represented in a cased symbology (Figure 15.3), these boundary lines may provide an exaggerated width relative to real-world dimensions.

Typification

The motive behind typification is that of communicating the form of an object when map space does not permit a geometrically accurate representation. As an example, a row of fourteen physically connected 'terraced' houses might be represented on a large-scale map by an elongated rectangular outline of the entire row, subdivided internally into rectangular sections indicating the individual houses (Figure 16.1b). At a smaller scale, clear symbolisation of the internal subdivisions might not be possible without considerably enlarging the block at the expense of equally significant nearby features. The cartographic solution may then be to draw a block with only four subdivisions, thereby communicating the character of the building, but not the precise detail.

Another example of typification is the small-scale representation of a winding mountain road that has many tortuous bends. Geometric reduction of a line representing the road might appear to straighten the road and lose its real character. A more useful cartographic solution would be to represent the road clearly by a few very obvious wiggles that imparted the character but not the precise path of the road.

It might reasonably be argued that the enhancement processes of smoothing and fractalisation described previously are examples of typification in that they intentionally alter the geometry in an attempt to improve the impression of realism. However, typification processes may be regarded as a particular form of reduction in which the geometry is intentionally altered to provide a specific spatial form. Enhancement procedures may then have the effect of increasing the geometric detail. It should be remarked that apart from the enhancement methods described above, little progress has been made in automated typification with the notable exception of Plazanet *et al.* (1995) in the context of line generalization.

Displacement

The need for displacement arises as a consequence of the fact that the graphic symbology of adjacent generalised geometric representations may overlap each other or become too close to be clearly discernible. When map symbols come into conflict with each other in this way, the solution is either to eliminate or amalgamate one or more conflicting objects, or to move them apart. Resolution of conflict by displacement may be a joint process between two or more objects, in which they all move, or it may be solved by unilateral displacement of one of the objects. Displacement is one of the least well automated aspects of generalisation. One reason for this is that irregularly shaped objects may need to be displaced in a plastic mode whereby parts of the objects are displaced more than others. This is computationally complicated. Another reason is that resolution of conflict between the members of one set of objects may result in propagation of conflict to other objects that were not initially involved.

Nickerson (1988) was one of the first researchers to address problems of map feature displacement, in the context of linear features. The displacement of linear features is subject to several constraints, although their relative importance may differ according to the map's purpose. First, the topology of the map should be unchanged. Thus objects, whether linear or otherwise, should not move from one side of a line to another, and the original connectivity of linear features should be retained. Secondly, the shape of objects should be retained as far as possible. This may be achieved by propagating displacements smoothly rather than imposing significant local shifts that may distort an object from its original character. Thirdly, the amount of displacement and the resulting locational error should be minimised to that required to meet the previous constraints along with the requirement for effective graphic communication.

In his MAPEX system, Nickerson recognised that certain features may be regarded as being more movable than others. He allocated priorities to complete line segments, connected at nodes, and to the vertices that compose each segment, or *linel* as he called them. The linel priorities allocated were, in decreasing order of resistance to movement, hydrography, railroads, roads, power lines and contours. Each node of a linel was given a priority equal to the highest priority of the linels connected to it. Each vertex of a linel was given a priority, depending on the priority of and distance from the two terminal nodes.

The process of applying displacements in MAPEX was a gradual one in that an intermediate map was constructed by adding linels one at a time and resolving any conflicts that arose before adding the next one. Displacements were calculated for each node and then for each vertex as a function of the displacements associated with the nodes. The amount that each vertex was displaced reduced with its distance from the nodes. Conflicts between overlapping or excessively close linels were detected by testing for intersections between parallel-sided bands centred on the linels. Displacement vectors, used to push lines apart, were calculated as a function of the relationship between conflicting linels. Nickerson recognised the fact that a complete solution to conflict resolution does require displacement processing to be combined with amalgamation and with elimination. His system went some way to indicating how this could be achieved, but did not combine all processes in an automated environment and it did not work with a combination of all spatial data types.

In the MAGE system (Bundy *et al.*, 1995; Jones *et al.*, 1995) the triangulated data structure is used to detect conflicts and to generate displacement vectors to resolve them. Conflicts introduce singularities in the triangulation, whereby triangles become folded over each other. The heights of these inverted triangles are used to generate the displacement vectors (Figure 16.16).

Process control

One of the reasons for the relatively slow progress in automating map generalisation is that, though there is a finite set of semantic and geometric generalisation procedures that can be identified, few of them can be treated entirely independently of the others. Successful generalisation requires a holistic approach in which the interaction between cartographic objects can be monitored. The consequences of particular operators need to be predicted and conflicts resolved in the manner that is most likely to meet the semantic and cartographic objectives of generalisation. At present, no successful automatic solution to map generalisation exists. One of the few commercially available systems for map generalisation (Intergraph's Map Generalizer) is notable for adopting an interactive approach which provides a toolkit of generalisation operators, the effects of which must be monitored and controlled by the user.

Several researchers have proposed storing cartographic knowledge of how to perform and control generalisation within knowledge-based systems (Buttenfield and McMaster, 1991), but there is little evidence of working examples of the application of knowledge-based approaches. One of the reasons for this is that cartographers' knowledge of how to perform generalisation depends to a significant extent upon recognising the presence of combinations of circumstances, characterised by particular feature types and particular spatial relationships. Standard feature codes such as those of the Ordnance Survey are not sufficient to detect automatically the presence of, for example, a row of semi-detached houses or the classic case of parallel roads, railway and river at the base of a valley. Thus before any of the generalisation operations can be executed, structure recognition procedures must be performed. At present this is usually achieved by the human eye.

Despite only limited success in automating generalisation, it is possible to identify a hypothetical agenda for automated map generalisation, that would be the subject of process control:

1. specification of objectives;
2. selection of relevant information;
3. specification of required symbolisation and annotation;
4. retrieval of relevant information;
5. preliminary generalisation of retrieved information;
6. assembly of symbolised objects and their annotation;

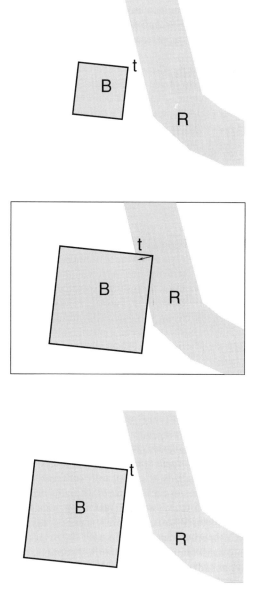

Figure 16.16 An example of a local displacement operation using triangulations. (**a**) Object B is expanded, resulting in conflict with object R (**b**). The form of the inverted triangle *t* is used to resolve the conflict (**c**).

7. identification and resolution of conflicts subject to quality evaluation criteria.

The specification of objectives may arise out of a query to a GIS. This specification might be very short, as in a query in car navigation to plot the shortest route from Cardiff to Canterbury. Other queries might be much more specific, with constraints on exactly what data are to be displayed at a given scale and perhaps including the types of graphic symbols to be used, as in the case of producing a standard map series.

The selection of relevant information to service the query is likely to be dependent upon the application of rules that distinguish between different types of query and encode knowledge of appropriate types of data to be retrieved. As was indicated in the earlier section on semantic generalisation, the rules could refer to the use of particular levels in classification and aggregation hierarchies. Thus for a road map of a particular scale, it would be possible to determine from such rules the classes of road and, for a specific route map, the ones that should be retrieved, e.g. all classes of roads connected to the route, and a limited range of road classes on the remainder of the map. Similarly, settlements for retrieval might vary in range of importance according to whether they are in the immediate zone of the route.

The specification of required symbolisation and style of annotation could be controlled by rules that encode standards or familiar symbology in a particular application area. In the case of the route map, the standard symbols might be modified in order to highlight particular roads that constitute the route generated by the system.

The methodology for processing a database query that requires map generalisation will depend upon the type of data stored. A GIS capable of performing generalisation might store multiple representations of geographical reality at significantly different levels of detail, possibly in combination with multi-resolution data access methods as discussed in Chapter 11. The query processor for such a database would attempt to match the query specification with the most appropriate representation of individual objects, taking account of the target scale and, by implication, the allowable locational error.

Preliminary geometric generalisation of retrieved information would be necessitated if the level of generalisation of retrieved data did not correspond to that required or implied by the user objectives. At this stage of processing, the intention would be to simplify individual geometric representations to a level consistent with the objectives. The correct level of generalisation might vary between different classes of feature according to their relative importance. Automation of this stage is particularly challenging, since it implies some knowledge of how to select the

appropriate generalisation operators and how to apply them to produce a suitable degree of generalisation. Accumulation of this knowledge is a current research issue. Sources of the relevant knowledge include manuals produced by mapping agencies; analysis of the contents of existing generalised maps; and by interviewing and observing cartographers.

Preliminary geometric generalisation is followed in our agenda by the assemblage of all retrieved data. This step enables an assessment to be made of the degree of competition for map space and the location of particular conflicts. It requires a knowledge of symbolisation parameters such as line thickness and point-referenced symbol sizes, along with an assessment of the space required to place text associated with the annotated features. The identification of conflicts requires a data structure that enables detection of overlaps and measurement of separations, since objects closer than particular cartographic tolerances are regarded as being in proximal conflict.

Fully automated conflict resolution is one of the least researched aspects of map generalisation. Its successful implementation requires that a balance be struck between the need to communicate particular types of information and the constraints of map space and effective cartography. Conflict resolution may be achieved through a variety of strategies including elimination of mapped objects, amalgamation and displacement. Because conflict-resolution actions may have knock-on effects of further conflict, the process may be regarded as an iterative one in which it may be possible to consider numerous possible scenarios for placing symbolised map features and their annotation. Just as with the preliminary stage of generalisation, execution of this stage in a manner that accords with conventional cartographic practice will require encoding and applying cartographic knowledge. As proposed by Weibel (1991) and others, monitoring the activities of cartographers may help in identifying appropriate strategies that could be automated. The monitoring can be carried out by logging a cartographer's actions when using interactive computer-assisted generalisation tools.

Considerable success has been achieved in solving the conflict resolution subproblem of text placement. One of the chief characteristics of successful name-placement procedures is the preliminary generation of multiple candidate solutions for each piece of annotation, which can be evaluated in a search procedure against criteria such as overlap between names and between names and map features. It is possible that a similar approach could be adopted in generalisation, though the process of generating multiple possible states of each mapped feature is computationally demanding and, as Mackaness and Fisher (1987) have remarked, there must be adequate criteria for evaluation of the candidate solutions.

An approach to conflict resolution that may have some potential is that based on the use of multiple autonomous software agents. In such multiple agent-based systems, each map object would be capable independently of assessing local conflict and effecting resolution actions such as displacement, elimination (self-destruction!) and amalgamation. Software agent architectures also offer the possibility of negotiation between potentially conflicting agents, the results of which may be based on the relative priority of individual featues and their degrees of freedom to move. The result would be a dynamic process in which the desired 'high-level' results of generalisation might be achieved largely through suitably constrained 'low-level' operations of the individual agents.

Summary

Map generalisation is a fundamental process in graphic communication of spatial information, and as we have seen in this chapter it has so far defied successful automation. It is concerned with information abstraction and may be considered in terms of semantic generalisation, which focuses on the meanings to be communicated, and in terms of geometric generalisation, which addresses the graphic and symbolic representation of the spatial data. Semantic generalisation employs processes of aggregation and classification to structure the meaning of information in an hierarchical manner, often involving overlapping hierarchies (hence heterarchies). Geometric generalisation and the associated issue of symbolisation can be viewed in terms of several main types of operation including the elimination of objects; a reduction in the spatial data required to represent individual line, area and surface objects; the amalgamation of nearby objects, either of the same class or of different classes; a collapse in the dimensionality of areal objects to a point or to a linear form; the exaggeration of objects that are important but which would otherwise be too small to be clearly visible; the enhancement of the form of features to make them appear more realistic, such as by smoothing or by fractalisation; typification whereby characteristic shape is represented often at the expense of precise locational and quantitative veracity; and displacement whereby nearby symbols that would otherwise overlap are moved apart to maintain graphic clarity.

Automation of map generalisation requires the capacity not just to implement each of these operations, but to be able to decide which operators should be used under particular circumstances, and how the results of the operators can be evaluated. This is what is sometimes called the process control aspect of map generalisation and its automation, to emulate conventional cartographic practice, will require encoding knowledge of acceptable cartographic design criteria. It will also require the design of computational methods to emulate the human capacity to identify contexts in which specific generalisations operations should be applied and to resolve conflicts that arise from the individual operators.

Further reading

A useful review article on generalisation is that of Brassel and Weibel (1988). McMaster (1987) gives a good summary of the work on line generalisation up until the late 1980s. For some of the limited amount of literature on displacement procedures, see Nickerson (1988), Monmonier (1987) and Mackaness (1994). A notable article on terrain generalisation is that by Weibel (1992). Recent work on raster-based generalised is exemplified by Schylberg (1993). The potential for using triangulated data structures in map generalisation is illustrated by Jones et al. (1995). The collections of articles edited by Buttenfield and McMaster (1991) and Muller et al. (1995) give a fairly extensive overview of the state of automated map generalisation in the early 1990s.

References

Abdelmoty, A.I., M.H. Williams and N.W. Paton (1993), Deduction and deductive databases for geographic data handling. In D. Abel and B. C. Ooi (ed.) *Advances in Spatial Databases*. Lecture Notes in Computer Science, 692. Berlin: Springer-Verlag, pp. 441–464.

Abel, D. and B.C. Ooi (ed.) (1993) *Advances in Spatial Databases*. Lecture Notes in Computer Science 692, Berlin: Springer-Verlag.

Akima, H. (1978) A method of bivariate interpolation and smooth surface fitting for irregularly distributed data points. *ACM Transactions on Mathematical Software*, **4**(2), 148–159.

Allen, J.F. (1983) Maintaining knowledge about temporal intervals. *Communications of the ACM*, **26** 832–843.

Allen, K.M.S., S.W. Green and E.B.W. Zubrow (ed.) (1990) *Interpreting Space: GIS and Archaeology*. London: Taylor and Francis.

Anon. (1990) Newsfront article. *GPS World*, November/December 1990, 16–22.

Armstrong, M.P. (1991) Knowledge classification and organization. In B.P. Buttenfield and R.B. McMaster (eds), *Map Generalization: Making Decisions for Knowledge Representation*. London: Longman, pp. 86–102.

Armstrong, M.P., P.J. Densham, P. Lolonis and G. Rushton (1992) Cartographic displays to support locational decision making. *Cartography and Geographic Information Systems*, **19**(3), 154–164.

Ashkenazi, V. and G. Ffoulkes-Jones (1990) Millimetres over hundreds of kilometres by GPS. *GPS World*, November, 44–47

Bailey, T.C. and A.C. Gatrell (1995) *Interactive Spatial Data Analysis*. Harlow: Longman.

Ballard, D. (1981) Strip trees: a hierarchial representation for curves. *Communications of the ACM*, **24**, 310–321.

Ballard, D. and C.M. Brown (1982) *Computer Vision*. Englewood Cliffs, New Jersey: Prentice Hall.

Band, L.E. (1986) Topographic partitioning of watersheds with digital terrain models. *Water Resources Research*, **22**(1), 15–24.

Basoglu, U. (1984) An in-depth study of automated name placement systems. PhD thesis, University of Wisconsin.

Beard, M.K. and N.R. Chrisman (1988) Zipper: a localised approach to edgematching. *The American Cartographer*, **15**(2), 163–172.

Becker, B., H.-W. Six and P. Widmayer (1991) Spatial priority search: an access technique for scaleless maps. *ACM SIGMOD Record* **20**(2), 128–137.

Bentley, J.L. (1975) Multidimensional binary search trees used for associative searching. *Communications of the ACM*, **18**(9), 509–517.

Bertin, J. (1983) *Semiology of Graphics* (translated by W.J. Berg.) Madison: University of Wisconsin Press.

Birkin, M. (1995) Customer targeting, geodemographics and lifestyle approaches. *GIS for Business and Service Planning*, GeoInformation International, pp. 104–149.

Birkin, M., G. Clarke, M. Clarke and A. Wilson (1996) *Intelligent GIS Location Decisions and Strategic Planning*. Geoinformation International.

Blakemore, M. (1984) Generalization and error in spatial databases. *Cartographica*, **21**, 131–139.

Bocca, J.B. (1991) MegaLog – a platform for developing knowledge base management systems. *International Symposium on Database Systems for Advanced Applications*, Tokyo, pp. 374–380.

Bolstad, P.V., P. Gessler and T.M. Lillesand (1990) Positional uncertainty in manually digitized map data. *International Journal of Geographical Information Systems*, **4**(4), 399–412.

Bonham-Carter, G.F. (1991) Integration of geoscientific information using GIS. In D.J. Maguire, M.F. Goodchild and D.W. Rhind (ed.) *Geographical Information Systems*, Vol. 2. Harlow: Longman, pp. 171–184.

Bonham-Carter, G.F. (1994) *Geographic Information Systems for Geoscientists: Modelling with GIS*. Oxford: Pergamon.

Bracken, I. and D. Martin (1989) The generation of spatial population distributions from census centroid data. *Environment and Planning A*, **21**, 537–543.

Brassel, K.E. and R. Weibel (1988) A review and conceptual framework of automated map generalization. *International Journal of Geographical Information Systems*, **2**(3), 229–244.

Bregt, A.K., J. Denneboom, H.J. Gesink and Y.van. Randen (1991) Determination of rasterising error: a case study with the Soil Map of the Netherlands. *International Journal of Geographical Information Systems*, **5**, 361–368.

Brewer, C.A. (1994) Color use guidelines for mapping and visualisation. In A.M. MacEachren and D.R.F. Taylor (eds), *Visualisation in Modern Cartography*. Oxford: Pergamon, pp. 123–147.

British Standards Institution (1992) *BS 7567: Parts 1, 2 and 3. Electronic transfer of geographic information (NTF)*.

Brown, P.J.B. (1991) Exploring geodemographics. *Handling Geographical Information*. London: Longman.

Buchmann, A., O. Gunther, T. Smith and Y. Wang (ed.) (1989) Symposium on the Design and Implementation of Large Spatial Databases. Lecture Notes in Computer Science 409. Springer-Verlag.

Bundy, G.L., C.B. Jones and E. Furse (1995) A topological structure for the generalization of large scale cartographic data. In P.F. Fisher (ed.), *Innovations in GIS*, Vol. 2. London: Taylor and Francis, pp. 19–31.

Burrough, P.A. (1986) *Principles of Geographical Information Systems for Land Resources Assessment*. Oxford: Clarendon Press.

Buttenfield, B.P. (1987) Automating the identification of cartographic lines. *The American Cartographer*, **14**(1), 7–20.

Buttenfield, B.P. and R.B. McMaster (ed.) (1991) *Map Generalization: Making Rules for Knowledge Representation*. London: Longman.

Calkins, H.W. and D.F. Marble (1987) The transition to automated production cartography. *The American Cartographer*, **14**(2), 105–119.

Carr, D.B., A.R. Olsen and D. White (1992) Hexagon mosaic maps for display of univariate and bivariate geographical data. *Cartography and Geographic Information Systems*, **19**(4), 228–236.

Carver, S.J. (1991) Integrating multi-criteria evaluation with geographical information systems. *International Journal of Geographical Information Systems*, **5**(3), 321–339.

Chance, A., R.G. Newell and D.G. Theriault (1990) An object-oriented GIS – issues and solutions. *EGIS'90 First European Conference on Geographical Information Systems*, Amsterdam, EGIS Foundation, pp. 179–188.

Charlwood, G., G. Moon and J. Tulip (1987) Developing a DBMS for geographic information: a review. *Auto-Carto* 8, Baltimore, Maryland: ASPRS/ACSM, pp. 302–315.

Chen, Z.-T. and W.R. Tobler (1986) Quadtree Representation of Digital Terrain. London: *Auto-Carto London*, Vol 1, pp. 475–484.

Chithambaram, R. and K. Beard (1991) Skeletonizing polygons for map generalization. *Technical Papers ACSM-ASPRS Convention*, Baltimore, pp. 44–55.

Chrisman, N.R. (1984) The role of quality information in the long term functioning of a geographic information system. *Cartographica*, **21**(2), 79–87.

Chrisman, N.R. (1987a) The accuracy of map overlays: a reassessment. *Landscape and Urban Planning*, **14**, 427–439.

Chrisman, N.R. (1987b) Efficient digitising through the combination of appropriate hardware and software for error detection and editing. *International Journal of Geographical Information Systems*, **1**(3), 265–277.

Chrisman, N.R. and J.A. Dougenik (1992) Lessons for the design of polygon overlay processing from the Odyssey Whirlpool algorithm. *5th International Symposium on Spatial Data Handling*, Charleston, International Geographical Union, pp. 401–410.

Christensen, J., J. Marks and S. Shieber (1995) An empirical study of algorithms for point-feature label placement. *ACM Transactions on Graphics*, **14**(3), 203–232.

Chuvieco, E. (1993) Integration of linear programming and GIS for land-use modelling. *International Journal of Geographical Information Systems*, **7**(1), 71–83.

Clarke, K.C. (1988) Scale-based simulation of topographic relief. *The American Cartographer*, **15**(2), 173–181.

Clarke, K.C. and D.M. Schweizer (1991) Measuring the fractal dimension of natural surfaces using a robust fractal estimator. *Cartography and Geographical Information Systems* **18**(1), 37–47.

Cliff, A.D. and J.K. Ord (1981) *Spatial Processes: Models and Applications*. London: Pion.

Colwell, S. (1991) GIS and GPS: a marriage of convenience. *GIS World*, **4**(6), 36–38.

Cook, A.C. and C.B. Jones (1990) A PROLOG rule-based system for cartographic name placement. *Computer Graphics Forum*, **9**, 109–126.

Corbett, J.P. (1979) *Topological Principles in Cartography*. US Bureau of the Census.

Couclelis, H. (1992) People manipulate objects (but cultivate fields): beyond the raster-vector debate in GIS. In A.U. Frank, I. Camperi and U. Formentini (ed.) *Theory and Methods of Spatio-Temporal Reasoning in Geographic Space*. Lecture Notes in Computer Science 639, Heidelberg: Springer-Verlag, pp. 65–77.

Cui, Z., A.G. Cohn and D.A. Randall (1993) Qualitative and topological relationships in spatial databases. *Advances in Spatial Databases*. Lecture Notes in Computer Science, 692, Springer-Verlag, pp. 296–315.

Curran, P.J. (1985) *Principles of Remote Sensing*. London: Longman.

Dahlberg, R.E. (1993) The design of photo and image maps. *The Cartographic Journal*, **30** 112–118

Dale, P.F. (1990) Land information systems. In D.J. Maguire, M.F. Goodchild and D.W. Rhind (ed.) *Geographical Information Systems*, Vol. 2, Harlow: Longman, pp. 85–99.

Dangermond, J. (1990) A review of digital data commonly available and some of the practical problems of entering them into a GIS. In D.J. Peuquet and D.F. Marble (eds), *Introductory Readings in Geographic Information Systems*. London: Taylor and Francis, pp. 222–232.

Date, C.J. (1995) *An Introduction to Database Systems*, 6th edn. Reading, Massachusetts: Addison-Wesley.

David, B., L.Raynal, G. Schorter and V. Mansart (1993) GeO2: Why objects in a geographical DBMS. *Advances in Spatial Databases (SSD'93)*. Berlin: Springer-Verlag, pp. 264–276.

Davidson, D.A., S.P. Theocharopoulos and R.J. Bloksma (1994) A land evaluation project in Greece using GIS and based on Boolean and fuzzy set methodologies. *International Journal of Geographical Information Systems*, **8**(4) 369–384.

Davis, J.C. (1986) *Statistics and Data Analysis in Geology*. New York: Wiley.

Delbaere, B. and H. Gulinck (1994) A review of landscape ecological research with specific interest to landscape ecological mapping. In B. Delbaere and H. Gulinck (eds), *Remote Sensing in Landscape Ecological Mapping*. Office for Official Publications of the European Communities, pp. 3–27.

deLepper, M.J.C., H.J. Scholten and R.M. Stern (eds.) (1995) *The Added Value of Geographical Information Systems in Public and Environmental Health*. Dordrecht: Kluwer Academic.

Densham, P.J. and G. Rushton (1992) Strategies for solving large location-allocation problems by heuristic methods. *Environment and Planning A*, **24**, 289–304.

Deveau, T.J. (1985) Reducing the number of points in a plane curve representation. *Proceedings of Auto-Carto VII*, American Congress on Surveying and Mapping, pp. 152–160.

Diament, A.D., W.G. Rees and J.A. Dowsdeswell (1993) Using GIS to study Arctic ice caps. *GIS Europe*, **2**(6), 22–25.

DiBase, D., A.M. MacEachren, J.B. Krygier and C. Reeves (1992) Animation and the role of map design in scientific visualisation. *Cartography and Geographic Information Systems*, **19**(4), 201–214, 265–266.

Ding, Y. and A.S. Fotheringham (1992) The integration of spatial analysis and GIS. *Computers, Environment and Urban Systems*, **16**, 3–19.

Doerschler, J.S. and H. Freeman (1989) An expert system for dense-map name placement. *Auto-Carto 9*, Baltimore, ACSM/ASPRS, pp. 215–224.

Dong, D. and Y. Bock (1989) Global positioning system network analysis with phase ambiguity resolution applied to crustal deformation studies in California. *Journal of Geophysical Research*, April.

Dorling, D. (1992) Stretching space and splicing time: from cartographic animation to interactive visualization. *Cartography and Geographic Information Systems*, **19**(4), 215–227, 267–270.

Dorling, D. (1993) Map design for census mapping. *The Cartographic Journal*, **30**, 167–183.

Douglas, D.H. and T.K. Peucker (1973) Algorithms for the reduction of the number of points required to represent a digitized line or its caricature. *The Canadian Cartographer*, **10**(2), 112–122.

Dunn, R., A.R. Harrison and J.C. White (1990) Positional accuracy and measurement error in digital databases of land use: an empirical study. *International Journal of Geographical Information Systems*, **4**, 385–398.

Dutton, G. (ed.) (1978) *Harvard Papers on Geographic Information Systems*. Reading, Massachusetts: Addison-Wesley.

Dutton, G.H. (1981) Fractal enhancement of cartographic line detail. *The American Cartographer*, **8**(1), 23–40.

Dutton, G.H. (1984) Geodesic modelling of planetary relief. *Cartographica*, **21**(2), 188–207.

Dutton, G. (1989) Modeling locational uncertainty via hierarchical tesselation. In M. Goodchild ad S. Gopal (ed.) *Accuracy of Spatial Databases*. London: Taylor and Francis, pp. 125–140.

Eastman, J.R., J. Toledano, W. Jin and P.A.K. Kyen (1993) Participatory multi-objective decision-making in GIS. *Auto-Carto 11*, Minnesota: ACSM/ASPRS, pp. 33–42.

Egenhofer, M.J. (1991) Extending SQL for cartographic displays. *Cartography and Geographic Information Systems*, **18**, 230–245.

Egenhofer, M.J. and A.U. Frank (1987) Object-oriented databases: database requirements for GIS. *International GIS Symposium: The Research Agenda*, US Government Printing Office, pp. 189–211.

Egenhofer, M.J. and A.U. Frank (1990) Lobster: combining AI and database techniques for GIS. *Photogrammetric Engineering and Remote Sensing*, **56**(6), 919–926.

Egenhofer, M.J. and R. Franzosa (1991) Point-set topological spatial relations. *International Journal of Geographical Information Systems*, **5**(2), 161–174.

Egenhofer, M.J. and J.R. Herring (eds) (1995) *Advances in Spatial Databases*. Lecture Notes in Computer Science 951. Berlin: Springer.

ESRI (1995) *Understanding GIS: The Arc/Info Method*. GeoInformation International.

Fegeas, R.G., J.L. Cascio and R.A. Lazar (1992) An overview of FIPS 173, the spatial data transfer standard. *Cartography and Geographic Information Systems*, **19**(5), 278–293.

Fischer, M.M. and P. Nijkamp (ed.) *Geographical Information Systems, Spatial Modelling and Policy Evaluation*. Berlin: Spring-Verlag.

Fisher, P.F. (1991) Modelling soil map inclusions by Monte Carlo simulation. *International Journal of Geographical Information Systems*, **5**(3), 193–208.

Fisher, P.F. (1993) Algorithm and implementation uncertainty in viewshed analysis. *International Journal of Geographical Information Systems*, **7**(4), 331–348.

Floriani, L.de (1989) A pyramidal data structure for triangle-based surface description. *IEEE Computer Graphics and Applications*, **9**(2), 67–78.

Floriani, L.de and P. Magillo (1994) Visibility algorithms on triangulated terrain models. *International Journal of Geographical Information Systems*, **8**(1), 13–41.

Floriani, L.de and E. Puppo (1988) Constrained Delaunay triangulation for multiresolution surface description. Ninth International Conference on Pattern Recognition. Rome: IEEE, pp. 566–569.

Flowerdew, R. and M. Green (1989) Statistical methods for inference between incompatible zonal system. In M. Goodchild and S. Gopal (eds), *Accuracy of Spatial Databases*. London: Taylor and Francis, pp. 239–247.

Foley, J.D., A.van Dam, S.K. Feiner and J.F. Hughes (1990) *Computer Graphics Principles and Practice*. Reading, MA: Addison-Wesley.

Foody, G.M. (1995) Land cover classification by an artificial neural network with ancillary information. *International Journal of Geographical Information Systems*, **9**(5), 527–542

Fotheringham, A.S. and M.E. O'Kelly (1989) *Spatial Interaction Models: Formulations and Applications*. Dordrecht: Kluwer Academic.

Fotheringham, A.S. and P.A. Rogerson (eds.) (1994) *Spatial Analysis and GIS*. Technical Issues in Geographic Information Systems. London: Taylor and Francis.

Fournier, A., D. Fussell and L. Carpenter (1982) Computer rendering of stochastic models. *Communications of the ACM*, **25**(6), 371–384.

Fowler, R.J. and J.J. Little (1979) Automatic extraction of irregular network digital terrain models. *Computer Graphics*, **13**(1), 199–207.

Frank, A.U. and I. Campari (eds.) (1993) *Spatial Information Theory*. Lecture Notes in Computer Science 716. Berlin: Springer.

Frank, A.U. and W. Kuhn (ed.) (1995) *Spatial Information Theory: A Theoretical Basis for GIS*. Lecture Notes in Computer Science 988. Berlin: Springer.

Frank, A.U., I. Campari and U. Formentini (ed.) (1992) *Theory and Methods of Spatio-Temporal Reasoning in Geographic Space*. Lecture Notes in Computer Science 639. Berlin: Springer-Verlag.

Franklin, W.R. (1984) Cartographic errors symptomatic of underlying algebra problems. *First International Symposium on Spatial Data Handling*, Zurich, International Geographical Union, pp. 190–208.

Franklin, W.M. (1989) Uniform grids: a technique for intersection detection on serial and parallel machines. *Auto-Carto 9*, Baltimore, ACSM/ASPRS, pp. 100–109.

Freeman, H. and J. Ahn (1984) AUTONAP – an expert system for automatic name placement. *First International Symposium on Spatial Data Handling*, Zurich, International Geographical Union, pp. 544–569.

Freeman, J. (1975) The modelling of spatial relations. *Computer Graphics and Image Processing*, **4**, 156–171.

Gargantini, I. (1982) An effective way to represent quadtrees. *Communications of the ACM*, **25**, 905–910.

Gargantini, I. (1989) Linear octrees for fast processing of three-dimensional objects. *Computer Graphics and Image Processing*, **20**, 365–374.

Gatrell, A.C. (1989) On the spatial representation and accuracy of address-based data in the United Kingdom. *International Journal of Geographical Information Systems*, **3**(4), 335–348.

Ghosht, A. and G. Rushton (eds.) (1987) *Spatial Analysis and Location-Allocation Models.* New York: Van Nostrand Reinold.

Gold, C.M. (1994) Three approaches to automated topology and how computational geometry helps. *Sixth International Symposium on Spatial Data Handling SDH 94*, Edinburgh, International Geographical Union, pp. 145–158.

Gold, C.M., T.D. Charters and J. Ramsden (1977) Automated contour mapping using triangular element data structures and an interpolant over each triangular domain. *Computer Graphics (ACM)*, **2**, 170–175.

Gonzalez, R.C. and P. Wintz (1987) *Digital Image Processing.* Reading, MA: Addison-Wesley.

Goodchild, M.F. (1986) *Spatial Autocorrelation.* Concepts and Techniques in Modern Geography. Norwich: Geo Books.

Goodchild, M.F. (1991) Issues of quality and uncertainty. In J.C. Muller (ed.), *Advances in Cartography.* London: Elsevier, pp. 113–139.

Goodchild, M.F. (1992) Geographical data modelling. *Computers and Geosciences*, **18**(4), 401–408.

Goodchild, M.F. and S. Gopal (eds.) (1989) *Accuracy of Spatial Databases.* London: Taylor and Francis.

Goodchild, M.F. and Y. Shiren (1992) A hierarchical spatial data structure for global geographic information systems. *Computer Vision, Graphics and Image Processing: Graphical Models and Image Processing*, **54**(1), 31–44.

Goodchild, M.F., S. Guoqing and Y. Shiren (1992) Development and test of an error model for categorical data. *International Journal of Geographical Information Systems*, **6**(2), 87–104.

Goodchild, M.F., B.O. Parks and L.T. Steyaert (eds.) (1993) *Environmental Modeling with GIS.* New York: Oxford University Press.

Gore, C.G. (1991) The spatial separatist theme and the problem of representation in location-allocation models. *Environment and Planning A*, **23**, 939–953.

Greene, D. and F. Yao (1986) Finite-resolution computational geometry. *Proc. 27th IEEE Symposium on Foundations of Computer Science*, pp. 143–512.

Greenlee, D.D. (1987) Raster and vector processing for scanned linework. *Photogrammetric Engineering and Remote Sensing*, **53**(10), 1383–1387.

Grimshaw, D.J. (1994) *Bringing Geographical Information System into Business.* Harlow: Longman.

Gunther, O. (1989) The design of the cell tree: an object-oriented index structure for geometric databases. *IEEE Fifth International Conference on Data Engineering*, Los Angeles, IEEE, pp. 598–605.

Gunther, O. and H.-J. Schek (ed.) (1991) *Advances in Spatial Databases.* Lecture Notes in Computer Science 525. Berlin: Springer-Verlag.

Guting, R.H. and M. Schneider (1993) Realms: a foundation for spatial data types in database systems. In D. Abel and B.C. Ooi (eds.) *Advances in Spatial Databases.* Lecture Notes in Computer Science, 692. Berlin: Springer-Verlag, pp. 14–35.

Guttman, A. (1984) R-trees: a dynamic index structure for spatial searching. *Proceedings ACM SIGMOD International Conference on the Management of Data*, ACM, pp. 47–57.

Haggett, P., A.D. Cliff and A. Frey (1977) *Locational Models.* Edward Arnold.

Haines-Young, R. and D. Green (1993) *Landscape Ecology and Geographical Information Systems.* London: Taylor and Francis.

Haining, R. (1990) *Spatial Data Analysis in the Social and Environmental Sciences.* Cambridge: Cambridge University Press.

Hake, G. and E. Hoffmeister (1978) Zem Begriffsystem der Generalisierung. Computer Gestutze Gebaudengeneralisierung. *Kartographische Nachrichten*, **2**, 47–55.

Hamilton, D.E. and T.A. Thomas (ed.) (1992) *Computer Modeling of Geologic Surfaces and Volumes.* Tulsa, Oklahoma: The American Association of Petroleum Geologists.

Hart, P.E., R.O. Duda and M.T. Einaudi (1978) PROSPECTOR – a computer-based consultation system for mineral exploration. *Mathematical Geology* **10**(5).

Hay, A. (1979) Sampling design to test land use map accuracy. *Photogrammetric Engineering and Remote Sensing*, **45**, 529–533.

Hearnshaw, H.M. and D.J. Unwin (ed.) (1994) *Visualisation in Geographical Information Systems.* Wiley.

Heller, M. (1990) Triangulation algorithms for adaptive terrain modeling. *Fourth International Symposium on Spatial Data Handling*, Zurich, International Geographical Union, pp. 163–174.

Herring, J. (1991) TIGRIS: a data model for an object-oriented geographic information system. *Computers and Geosciences*, **18**, 443–452.

Heuvelink, G.B.M. and P.A. Burrough (1993) Error propagation in cartographic modelling using Boolean logic and continuous classification. *International Journal of Geographical Information Systems*, **7**(3), 231–246.

Heuvelink, G.B.M., P.A. Burrough and A. Stein (1989) Propagation of error in spatial modelling with GIS. *International Journal of Geographical Information Systems*, **3**(4), 303–322.

Hirsch, S.A. (1982) An algorithm for automatic name placement around point data. *The American Cartographer*, **9**(1), 5–17.

Hobbs, M.H.W. (1994) Analysis of a retail brand network: a problem of catchment areas. In P. Fisher (ed.), *Innovations in GIS 2*. London: Taylor and Francis, pp. 151–158

Hoel, E.G. and H. Samet (1991) Efficient processing of spatial queries in line segment databases. *Advances in Spatial Databases (SSD '91)*, Lecture Notes in Computer Science, 525. Springer-Verlag, pp. 237–256.

Hogg, J., J.E. McCormack, S.A. Roberts, M.N. Gahegan and B.S. Hoyle (1993) Automated derivation of stream-channel networks and selected catchment characteristics from digital elevation models. *Geographical Information Handling – Research and Applications*. Chichester: Wiley, pp. 208–235.

Howe, D.R. (1989) *Data Analysis for Data Base Design*. London: Edward Arnold.

Hull, R. and R. King (1987) Semantic database modeling: survey, applications and research issues. *ACM Computing Surveys*, **19**(3), 201–260.

Hutchinson, M.F. (1995) Interpolating mean rainfall with thin plate smoothing splines. *International Journal of Geographical Information Systems*, **9**(4), 385–403.

Imhof, E. (1975) Positioning names on maps. *The American Cartographer*, **2**(2), 128–144.

Isaaks, E.H. and R.M. Srivastava (1989) *Applied Geostatistics*. Oxford: Oxford University Press.

Jackson, M.J. and P.A. Woodsford (1991) GIS data capture hardware and software. *Geographical Information Systems*, Vol. 1. Harlow: Longman, pp. 239–249.

Jancaitis, J.R. and J.L. Junkins (1973) Modelling irregular surfaces. *Photogrammetric Engineering and Remote Sensing*, **39**(4), 413–420.

Jenks, G.F. (1981) Lines, computers and human frailties. *Annals of the Association of American Geographers*, **71**(1), 1–10.

Jenson, S.K. (1985) Automated detection of hydrologic basin characteristics from digital elevation model data. *Auto-Carto 7*, Washington, DC: ASPRS/ACSM, pp. 301–310.

Jones, C.B. (1989a) Data structures for three dimmensional spatial information systems in geology. *International Journal of Geographical Information Systems*, **3**(1), 15–31.

Jones, C.B. (1989b) Cartographic name placement with Prolog. *IEEE Computer Graphics and Applications*, **9**(5), 36–47.

Jones, C.B. (1990) Conflict resolution in cartographic name placement. *Computer-Aided Design*, **22**(3), 173–183.

Jones, C.B. and I.M. Abraham (1986) Design considerations for a scale-independent database. *Second International Symposium on Spatial Data Handling*, Seattle, International Geographical Union, pp. 384–398.

Jones, C.B. and I.M. Abraham (1987) Line generalisation in a global cartographic database. *Cartographica*, **24**(3), 32–45.

Jones, C.B. and L.Q. Luo (1994) Hierarchies and objects in a deductive spatial database. *SDH'94 Sixth International Symposium on Spatial Data Handling*, Edinburgh, International Geographical Union, pp. 588–603.

Jones, C.B., A.C. Cook and J.E. McBride (1991) Rule-based control of automated name placement. *15th Conference of the International Cartographic Association*, Bournemouth, ICA, pp. 675–679.

Jones, C.B., D.B. Kidner and J.M. Ware (1994) The implicit triangulated irregular network and multiscale spatial databases. *The Computer Journal*, **37**(1), 43–57.

Jones, C.B., G.L. Bundy and J.M. Ware (1995) Map generalisation with a triangulated data structure. *Cartography and Geographic Information Systems*, **22**(4), 317–331.

Jones, C.B., D.B. Kidner, L.Q. Luo, G.L. Bundy and J.M. Ware (1996) Database design for a multi-scale spatial information system. *International Journal of Geographical Information Systems* **10**(8), 901–920.

Kasper, J.E. and D. Arns (1993) *Graphics Programming with PHIGS and PHIGS PLUS*. Reading, MA: Addison-Wesley.

Kennie, T.J.M. (1990) Field data collection for terrain modelling. *Terrain Modelling in Surveying and Civil Engineering*. Caithness: Whittles Publishing, pp. 4–16.

Khoshafian, S. (1993) *Object-Oriented Databases*. New York: Wiley.

Kidner, D.B. and D.H. Smith (1992) Compression of digital elevation models by Huffman coding. *Computers and Geosciences*, **18**(8), 1013–1034.

Kidner, D.B., C.B. Jones, D.G. Knight and D.H. Smith (1990) Digital terrain models for radio path profiles. *4th International Symposium on Spatial Handling*, Zurich, International Geographical Union, pp. 240–249.

Killen, J.E. (1979) *Linear Programming: The Simplex Method with Geographical Applications*. Concepts and Techniques in Modern Geography, 24. Norwich: Geo Books.

Kirby, G.H., M. Visvalingham and P. Wade (1989) Recognition and representation of a hierarchy of polygons with holes. *Computer Journal*, **32**(6), 554–562.

Knaap, W.G.M.van der (1992) The vector to raster conversion: (mis)use in geographical information systems. *International Journal of Geographical Information Systems*, **6**(2), 159–170.

Knuth, D. (1973) *The Art of Computer Programming, Volume 3: Sorting and Searching.* Addison-Wesley.

Koop, R.O. and F.J. Ormeling (1990) New horizons in thematic cartography in the Netherlands: The national atlas information system. In J. Harts, H.F.L. Ottens and H.J. Scholten (eds), *EGIS'90 First European Conference on Geographical Information Systems.* Amsterdam: EGIS Foundation, pp. 614–623.

Lam, S.-N. (1983) Spatial interpolation methods: a review. *The American Cartographer*, **10**(2), 129–149.

Lammi, J. and T. Sarajoski (1995) Image compression by the JPEG algorithm. *Photogrammetric Engineering and Remote Sensing*, **61**(10), 1261–1266.

Lang, T. (1969) Rules for robot draughtsmen. *Geographical Magazine*, **62**(1), 50–51.

Langford, M., D.J. Maguire and D.J. Unwin (1990) Cross area population estimation using remote sensing and GIS. *4th International Symposium on Spatial Data Handling*, Zurich, International Geographical Union. pp. 541–550.

Langran, G.E. and T.K. Poiker (1986) Integration of name selection and name placement. *Proceedings 2nd International Symposium on Spatial Data Handling.* International Geographical Union, pp. 50–64.

Lay, V.H.-G. and W. Weber (1983) Waldgeneralisierung durch digitale Rasterdatenverarbeitung. *Nachrichten as dem Karten-und Vermessengswessen*, **1**(92), 61–71.

Lee, J. (1991) Comparison of existing methods for building triangular irregular network models for terrain from grid digital elevation models. *International Journal of Geographical Information Systems*, **5**(3), 267–285.

Leifer, L.A. and D.M. Mark (1987) Recursive approximation of topographic data using quadtrees and orthogonal polynomials. *Auto-Carto 8*, Baltimore, MD, ASPRS/ACSM, pp. 650–659.

Leung, Y. and K.S. Leung (1993a) An intelligent expert system shell for knowledge-based Geographical Information Systems: 1. The tools. *International Journal of Geographical Information Systems*, **7**(3), 189–199.

Leung, Y. and K.S. Leung (1993b) An intelligent expert shell for knowledge-based Geographical Information Systems: 2. Some applications. *International Journal of Geographical Information Systems*, **7**(3), 201–213.

Ley, R. (1992) The Digital Geographic Information Exchange Standard – DIGEST. *Geographic Information 1992/3 The Yearbook of the Association for Geographic Information.* London: Taylor and Francis, pp. 392–397.

Lichtner, W. (1979) Computer-assisted processes of cartographic generalization in topographic maps. *Geo-Processing*, **1**, 183–199.

Lillesand, T.M. and R.W. Kiefer (1994) *Remote Sensing and Image Interpretation.* New York: Wiley.

Longley, P. and G. Clarke (ed.) (1995) *GIS for Business and Service Planning.* GeoInformation International.

MacDougall, E.B. (1992) Exploratory analysis, dynamic statistical visualisation, and geographical information systems. *Cartography and Geographic Information Systems*, **19**(4), 237–246.

MacEachren, A.M. (1994) Visualisation in modern cartography: setting the agenda. In A.M. MacEachren and D.R.F. Taylor (ed.) *Visualisation in Modern Cartography.* Oxford, Pergamon, pp. 1–12.

MacEachren, A.M. and D.R.F. Taylor (eds) (1994) *Visualisation in Modern Cartography.* Oxford: Pergamon.

Mackaness, W.A. (1994) An algorithm for conflict identification and feature displacement in automated map generalization. *Cartography and Geographic Information Systems*, **21**(4), 219–232.

Mackaness, W.A. and M.K. Beard (1993) Use of graph theory to support map generalization. *Cartography and Geographic Information Systems*, **20**(4), 210–222.

Mackaness, W.A. and P.F. Fisher (1987) Automatic recognition and resolution of spatial conflicts in cartographic symbolisation. *Auto-Carto 8*, Baltimore, Maryland, USA, ACSM/ASPRS, pp. 709–718.

Maguire, D.J., M.F. Goodchild and D.W. Rhind (ed.) (1991) *Geographical Information Systems: Principles and Applications.* Harlow: Longman.

Makarovic, B. (1973) Progressive sampling for digital terrain models. *ITC Journal*, **3**, 397–416.

Maling, D.H. (1989) *Measurements from Maps: Principles and Methods of Cartometry.* Oxford: Pergamon.

Maling, D.H. (1991) Coordinate systems and map projections for GIS. *Geographical Information Systems*, Vol. 1. Harlow: Longman, pp. 135–146.

Maling, D.H. (1992) *Coordinate Systems and Map Projections.* Oxford: Pergamon.

Mandelbrot, B. (1982) *The Fractal Geometry of Nature.* San Francisco: W.H. Freeman.

Marino, J. (1979) Identification of characteristic points along naturally occurring lines: an empirical study. *The Canadian Cartographer*, **16**(1), 70–80.

Mark, D.M. and F. Csillag (1989) The nature of boundaries on 'area-class' maps. *Cartographica*, **26**(1), 65–78.

Mark, D.M. and A.U. Frank (ed.) (1991) *Cognitive and Linguistic Aspects of Geographic Space.* Dordrecht/Boston/London: Kluwer.

Marr, D. and E. Hildreth (1980) Theory of edge detection. *Proceedings of the Royal Society of London*, **B204**, 301–328.

Martin, D. (1989) Mapping population data from zone centroid locations. *Transactions of the Institute of British Geographers*, **14**, 90–97.

Martin, D. (1996) An assessment of surface and zonal models of population. *International Journal of Geographical Information Systems*, **10**(8), 973–989.

Marx, R.W. (ed.) (1990) The Census Bureau's TIGER System. *Cartography and Geographic Information Systems*, **17**(1), special issue.

Mason, D.C., M.A. O'Conaill and S.M.B. Bell (1994) Handling four-dimensional geo-referenced data in environmental GIS. *International Journal of Geographical Information Systems*, **8**(2), 191–215.

Mather, P.M. (1987) *Computer Processing of Remotely-Sensed Images*. Chichester: Wiley.

McCullagh, M.J. and C.G. Ross (1980) Delaunay triangulation of a random data set for isarithmic mapping. *The Cartographic Journal*, **17**, 93–99.

McHarg, I.L. (1969) *Design With Nature*. New York: Doubleday/Natural History Press.

McKeown, D.M. (1990) Toward automatic cartographic feature extraction. In L.F. Pau (ed.), *Mapping and Spatial Modelling for Navigation*. NATO ASI Series, F65, Springer-Verlag. pp.

McMaster, R.B. (1987) Automated line generalisation. *Cartographica*, **24**(2), 74–111.

Medyckyj-Scott, D. and H.M. Hearnshaw (ed.) (1993) *Human Factors in Geographical Information Systems*. London: Belhaven Press.

Monmonier, M. (1987) Displacement in vector- and raster-mode graphics. *Cartographica*, **24**(4), 25–36.

Monmonier, M. (1992) Authoring graphic scripts: experiences and principles. *Cartography and Geographic Information Systems*, **19**(4), 237–246.

Montanari, U. (1969) Continuous skeletons from digitized images. *Journal of the ACM*, **16**(4), 534–549.

Morehouse, S. (1985) ARC/INFO: a geo-relational model for spatial information. *Auto-Carto 7*, Washington, DC, ASPRS/ACSM, pp. 388–397.

Morrison, J.L. and K. Wortmann (ed.) (1992) *Implementing the Spatial Data Transfer Standard*. Cartography and Geographic Information Systems **19**(5). American Congress on Surveying and Mapping.

Morton, G.M. (1966) *A Computer-oriented Geodetic Database and a New Technique in File Sequencing*. Ontario, Canada: IBM.

Muller, J.C. (1990) The removal of spatial conflicts in line generalization. *Cartography and Geographic Information Systems*, **17**(2), 141–149.

Muller, J.-C. and Z. Wang (1992) Area-patch generalization: a competitive approach. *The Cartographic Journal*, **29**, 137–144.

Muller, J.-C., J.-P. Lagrange and R. Weibel (ed.) (1995) *GIS and Generalization Methodology and Practice*. London: Taylor and Francis.

Nickerson, B.G. (1988) Automated cartographic generalisation for linear features. *Cartographica*, **25**(3), 15–66.

Nievergelt, J., H. Hinterbereger and K.C. Sevcik (1984) The grid file: an adaptable, symmetric, multikey file structure. *ACM Transactions on Database Systems*, **9**(1), 38–71.

Nunes, J. (1991) Geographic space as a set of concrete geographical entities. *Cognitive and Linguistic Aspects of Geographic Space*. Dordrecht: Kluwer Academic, pp. 9–33.

Nyerges, T.L. (1989) Schema integration analysis for the development of GIS databases. *International Journal of Geographical Information Systems*, **3**(2), 153–183.

Nyerges, T.L. (1991) Representing geographical meaning. In B.P. Buttenfield and R.B. McMaster (eds), *Map Generalization: Making Decisions for Knowledge Representation*. London: Longman, pp. 59–85.

Nyerges, T.L. and P. Jankowski (1989) A knowledge base for projection selection. *The American Cartographer*, **16**(1), 29–38.

Okabe, A., B. Boots and K. Sugihara (1992) *Spatial Tessellations – Concepts and Applications of Voronoi Diagrams*. Chichester: Wiley.

Okabe, A., B. Boots and K. Sugihara (1994) Nearest neighbourhood operations with generalized Voronoi diagrams: a review. *International Journal of Geographical Information Systems*, **8**(1), 43–71.

Oliver, M.A., R. Webster and J. Gerrard (1989a) Geostatistics in physical geography. Part 1. *Transactions of the Institute of British Geographers*, **14**, 259–269.

Oliver, M.A., R. Webster and J. Gerrard (1989b) Geostatistics in physical geography. Part 2. *Transactions of the Institute of British Geographers*, **14**, 270–286.

Openshaw, S. (1984) *The Modifiable Areal Unit Problem*. Concepts and Techniques in Modern Geography, 38. Norwich: Geo Books.

Openshaw, S. (ed.) (1995) *Census User's Handbook*. London: Longman.

Openshaw, S., A.E. Cross and C. Wymer (1987) A Mark 1 Geographical Analysis Machine for the automated analysis of point data. *International Journal of Geographical Information Systems*, **1**, 335–343.

Opheim, H. (1981) Fast data reduction of a digitised curve. *Geo-Processing*, **2**, 33–40.

Pannekoek, A.J. (1962) Generalization of coastlines and contours. *International Yearbook of Cartography*, **2**, 55–75.

Pavlidis, T. (1982) *Algorithms for Graphics and Image Processing*. Rockville, Maryland: Springer-Verlag.

Perkal, J. (1966) On the length of empirical curves. *Michigan Inter-University Community of Mathematical Geographers, Discussion paper 10*. Ann Arbor: University of Michigan, pp. 257–286.

Petrie, G. (1990a) Terrain data acquisition and modelling from existing maps. In G. Petric and T.O.M. Kennie (eds.), *Terrain Modelling in Surveying and Civil Engineering*. Caithness: Whittles Publishing, pp. 85–111.

Petrie, G. (1990b) Photogrammetric methods for data acquisition for terrain modelling. *Terrain Modelling in Surveying and Civil Engineering*. Caithness: Whittles Publishing, pp. 26–48.

Petrie, G. (1995) Photogrammetry and remote sensing. *The AGI Sourcebook for GIS 1995*. AGI, pp. 73–85.

Petrie, G. and T.J.M. Kennie (ed.) (1990) *Terrain Modelling in Surveying and Civil Engineering*. Caithness: Whittles Publishing.

Peucker, T.K. and D.H. Douglas (1975) Detection of surface-specific points by local parallel processing of discrete terrain elevation data. *Computer Graphics and Image Processing*, **4**, 375–387.

Peucker, T.K., R.J. Fowler, J.J. Little and D.M. Mark (1978) The triangulated irregular network. *ASP/ACSM Digital Terrain Models (DTM) Symposium*, ACSM, pp. 516–540.

Peuquet, D.J. (1981) An examination of techniques for reformatting digital cartographic data – Part 1: the raster-to-vector process. *Cartographica*, **18**(1), 34–48.

Peuquet, D. (1984) A conceptual framework and comparison of data models. *Cartographica*, **21**(4), 66–113.

Peuquet, D.J. (1988) Representations of geographic space: toward a conceptual synthesis. *Annals of the Association of American Geographers*, **78**, 375–394.

Pflug, R. and J.W. Harbaugh (ed.) (1992) *Computer Graphics in Geology: Three-Dimensional Computer Graphics in Modeling Geologic Structures and Simulating Geologic Processes*. Lecture Notes in Earth Sciences 12. Berlin: Springer-Verlag.

Pickles, J. (ed.) (1995) *Ground Truth: The Social Implications of Geographic Information Systems*. New York: The Guildford Press.

Plazanet, C., J.-G. Affholder and E. Fritsch (1995) The importance of geometric modeling in linear feature generalization. *Cartography and Geographic Information Systems*, **22**(4), 291–305.

Preparata, F.P. and M.I. Shamos (1988) *Computational Geometry*. Texts and Monographs in Computer Science. New York: Springer-Verlag.

Raper, J.F. (ed.) (1989) *Three Dimensional Applications in Geographical Information Systems*. London: Taylor and Francis.

Raper, J.F., D.W. Rhind and J.W. Shepherd (1992) *Postcodes: The New Geography*. London: Longman.

Reumann, K. and A.P.M. Witkam (1973) Optimizing curve segmentation in computer graphics. *International Computing Symposium*, Amsterdam: North-Holland Publishing Company, pp. 467–472.

Rhind, D.W. (1991) Counting the people: the role of GIS. *Geographical Information Systems*, Vol. 2, Harlow: Longman, pp. 127–137.

Richardson, L.F. (1961) The problem of contiguity: an appendix to statistics of deadly quarrels. *General Systems Yearbook*, **6**, 139–187.

Robinson, A.H., P.C. Muehrcke, A.J. Kimerling and S.C. Guptill (1995), J.L. Morrison, 6th edn. *Elements of Cartography* New York: John Wiley.

Robinson, J.T. (1981) The K–B–D-tree: a search structure for large multidimensional dynamic indexes. *ACM SIGMOD*, **10**, 10–18.

Rogers, D.F. (1985) *Procedural Elements for Computer Graphics*. New York: McGraw-Hill.

Rogers, D.F. and J.A. Adams (1990) *Mathematical Elements for Computer Graphics*. New York: McGraw Hill.

Romanowicz, R., K. Beven and R. Moore (1993) GIS and distributed hydrological models. *Geographical Information Handling – Research and Applications*. Chichester: Wiley, pp. 197–205.

Rosenfeld, A. and A.C. Kak (1982) *Digital Picture Processing*, Vol. 2, Academic Press.

Rosenfeld, A. and J.L. Pfaltz (1966) Sequential operators in digital picture processing. *Journal of the Association for Computing Machinery*, **13**, 417–494.

Rosenfield, G.H. and K. Fitzpatrick-Lins (1986) A co-efficient of agreement as a measure of thematic classification accuracy. *Photogrammetric Engineering and Remote Sensing*, **52**, 223–227.

Roussopoulos, N. and N. Liefker (1985) Direct spatial search on pictorial databases using packed R-trees. *ACM SIGMOD*, **14**, 17–31.

Saalfield, A. (1988) Conflation: automated map compilation. *International Journal of Geographical Information Systems*, **2**(3) 217–228.

Samet, H. (1990a) *The Design and Analysis of Spatial Data Structures*. Addison-Wesley.

Samet, H. (1990b) *Applications of Spatial Data Structures: Computer Graphics, Image Processing and GIS.* Addison-Wesley.

Samet, H. and R.E. Webber (1985) Storing a collection of polygons using quadtrees. *ACM Transactions on Graphics* **4**(3), 182–222.

Scholten, H.J. and J.C.H. Stillwell (ed.) (1990) *Geographical Information Systems for Urban and Regional Planning.* Dordrecht: Kluwer.

Schylberg, L. (1993) Computational methods for generalization of cartographic data in a raster environment. Royal Institute of Technology, Department of Geodesy and Photogrammetry.

Sedgewick, R. (1988) *Algorithms.* Reading, MA: Addison-Wesley.

Sellis, T., N. Roussopoulos and C. Faloutsos (1987) The R^+ tree: a dynamic index for multidimensional objects. *Thirteenth International Conference on Very Large Databases,* pp. 507–518.

Serra, J. (1982) *Image Analysis and Mathematical Morphology.* London and New York: Academic Press.

Shea, K.S. and R.B. McMaster (1989) Cartographic generalization in a digital environment: when and how to generalize. *Auto-Carto 9,* Baltimore, ACSM/ASPRS, pp. 56–67.

Sibson, R. (1978) Locally equiangular triangulations. *The Computer Journal,* **21**(3), 243–245.

Skidmore, A.K., P.J. Ryan, W. Davies, D. Short and E. O'Loughlin (1991) Use of an expert system to map forest soils for a geographical information system. *International Journal of Geographical Information Systems,* **5**(4), 431–445.

Smith, J.L., H.Q. Mackenzie and R.B. Stanton (1988) Knowledge based decision support for environmental planning. *Third International Symposium on Spatial Data Handling,* Sydney, International Geographical Union, pp. 307–320.

Smith, J.M. and D.C.P. Smith (1977) Database abstractions: aggregation and generalization. *ACM Transactions on Database Systems,* **2**(2), 105–133.

Snodgrass, R.T. (1992) Temporal databases. *Theories and Methods of Spatio-Temporal Reasoning in Geographic Space,* 639. Berlin: Springer, pp. 22–64.

Snyder, J.P. (1987) *Map Projections – A Working Manual.* United States Geological Survey Professional Paper 1395, United States Government Printing Office.

Spiegelhalter, D.J. (1986) Uncertainty in expert systems. In W.A. Gale (ed.), *Artificial Intelligence and Statistics.* Reading, MA: Addison-Wesley, pp. 17–55.

Starr, L.E. and K.E. Anderson (1991) a USGS perspective on GIS. *Geographical Information Systems,* Vol. 2. Harlow: Longman, pp. 11–22.

Staufenbiel, W. (1973) Zur Automation der Generalisierung tographischer Karten mit besonderer Berucksichtigung grossmassstabiger Gebsaudedarstellungen. Universtat Hannover.

Stone, M., W. Cowan and J. Beatty (1988) Color gamut mapping and the printing of digital color images. *ACM Transactions on Graphics,* **7**(3), 249–292.

Stonebraker, M. and G. Kemnitz (1991) The POSTGRES next generation database management system. *Communications of the ACM,* **34**(10) pp. 78–92.

Taylor, P.J. and R.J. Johnston (1995) Geographic information systems and geography. *Ground Truth: The Social Implications of Geographic Information Systems.* New York: The Guildford Press.

Teitz, M.B. and P. Bart (1968) Heuristic methods for estimating the generalised vertex median of a weighted graph. *Operations Research,* **16**, 955–961.

Tenenbaum, A. and M.J. Augenstein (1981) *Data Structures using Pascal.* Englewood Cliffs, New Jersey: Prentice Hall.

Teory, T.J., D. Yang and J.P. Fry (1986) A logical design methodology for relational databases using the extended entity-relationship model. *ACM Computing Surveys,* **18**(2), 197–222.

Thapa, K. (1988) Automatic line generalization using zero-crossings. *Photogrammetric Engineering and Remote Sensing,* **54**(4), 511–517.

Tobler, W.R. (1966) *Numerical Map Generalization.* Department of Geography, University of Michigan.

Tobler, W.R. (1979) Smooth pycnophylactic interpolation for geographic regions. *Journal of the American Statistical Association,* **74**, 519–536.

Tobler, W.R. (1986) Pseudo-cartograms. *The American Cartographer,* **13**(1), 43–50.

Tobler, W.R. (1987) Experiments in migration mapping by computer. *The American Cartographer,* **14**(2), 155–163.

Tomlin, C.D. (1990) *Geographic Information Systems and Cartographic Modelling.* Englewood Cliffs, New Jersey: Prentice Hall.

Tomlinson, R.F. (1985) Geographic information systems – a new frontier. *The Operational Geographer,* **5**, 31–36.

Topfer, F. and W. Pillewizer (1966) The principles of selection. *The Cartographic Journal,* **3**(1), 10–16.

Tsai, V.J.D. (1993) Delaunay triangulations in TIN creation: an overview and a linear-time algorithm. *International Journal of Geographical Information Systems,* **7**(6), 501–524.

Tufte, E.R. (1983) *The Visual Display of Quantitative Information*. Connecticut: Graphics Press.

Tufte, E.R. (1990) *Envisioning Information*. Cheshire, CT: Graphics Press.

Tukey, J.W. (1977) *Exploratory Data Analysis*. Reading, Massachusetts: Addison-Wesley.

Turner, A.K. (ed.) (1992) *Three-Dimensional Modeling with Geographic Information Systems*. Dordrecht: Kluwer Academic.

VanAken, J.R. (1985) An efficient ellipse-drawing algorithm. *ACM Transactions on Graphics*, **4**(2), 147–169.

van Oosterom, P.J.M. (1993) *Reactive Data Structures for Geographic Information Systems*. Oxford: Oxford University Press.

van Oosterom, P.G. and E. Claasen (1990) Orientation insensitive indexing methods for geometric objects. *4th International Symposium on Spatial Data Handling*, Zurich, International Geographical Union, pp. 1016–1029.

van Roessel, J.W. (1987) Design of a spatial data structure using the relational normal form. *International Journal of Geographical Information Systems*, **1**(1), 33–50.

Vincent, R.K. (1973) An ERTS multispectral scanner experiment for mapping from compounds. *8th International Symposium on Remote Sensing of Environment*, Ann Arbor, University of Michigan, pp. 1239–1247.

Visvalingam, M. and J.D. Whyatt (1990) The Douglas–Peucker algorithm for line simplification: re-evaluation through visualisation. *Computer Graphics Forum*, **9**(3), 213–218.

Voogd, H. (1983) *Multicriteria Evaluation for Urban and Regional Planning*. London: Pion

Wallace, G.K. (1991) The JPEG still picture compression standard. *Communications of the ACM*, **34**(4), 31–44.

Walsh, S.J., D.R. Lightfoot and D.R. Butler (1987) Recognition and assessment of error in geographic information systems. *Photogrammetric Engineering and Remote Sensing*, **53**(10), 1423–1430.

Wang, Z. and A. Brown (1991) A knowledge-based system for selection of map area colours from a colour chart. *ITC Journal*, **3**, 122–126.

Ware, J.M. and C.B. Jones (1992) A multiresolution topographic surface database. *International Journal of Geographical Information Systems*, **6**(6), 479–496.

Ware, J.M., C.B. Jones and G.L. Bundy (1995) A triangulated spatial model for cartographic generalisation of areal objects. *Spatial Information Theory: A Theoretical Basis for GIS*. Berlin: Springer-Verlag, pp. 173–192.

Watson, D.F. (1992) *Contouring: A Guide to the Analysis and Display of Spatial Data*. Oxford: Pergamon Press.

Waugh, T.C. and R.G. Healey (1987) The GEOVIEW design: a relational database approach to geographic data handling. *International Journal of Geographical Information Systems*, **1**(1), 101–112.

Weibel, R. (1991) Models and experiments for adaptive computer-assisted terrain generalization. *Cartography and Geographic Information Systems*, **19**(2), 133–153.

Welch, T.A. (1984) A technique for high performance data compression. *IEEE Computer*, **17**(6), 8–19.

White, E.R. (1985) Assessment of line generalization algorithms using characteristic points. *The American Cartographer*, **12**(1), 17–27.

White, M. (1991) Car navigation systems. *Geographical Information Systems*, Vol. 2. Harlow: Longman, pp. 115–125.

White, M.S. and P.O. Griffin (1985) Piecewise linear rubber-sheet transformation. *The American Cartographer*, **12**(2), 123–131.

Williams, R. (1987) Preserving the area of regions. *Computer Graphics Forum*, **6**(1), 43–48.

Wilson, A.G. and R.J. Bennet (1985) *Mathematical Models in Human Geography and Planning*. Chichester: Wiley.

Wilson, A.G., J.D. Coelho, S.M. MacGill and H.C.W.L. Williams (1981) *Optimisation in Location and Transport Analysis*. Chichester: Wiley.

Winston, P.H. (1992) *Artificial Intelligence*. Reading, MA: Addison-Wesley.

Winston, W.L. (1994) *Operations Research Applications and Algorithms*. Belmont, CA: Duxbury Press.

Worboys, M.F. (1994) Object-oriented approaches to geo-referenced information. *International Journal of Geographical Information Systems*, **8**(4), 385–399.

Worboys, M.F., H.M. Hearnshaw and D.J. Maguire (1990) Object-oriented data modelling for spatial databases. *International Journal of Geographical Information Systems*, **4**(4), 369–385.

Worrall, L. (ed.) (1990) *Geographic Information Systems: Developments and Applications*. London: Belhaven Press.

Worrall, L. (ed.) (1991) *Spatial Analysis and Spatial Policy using Geographical Information Systems*. London: Belhaven Press.

Wright, J., C. Revelle and J. Cohon (1983) A multiobjective integer programming model for the land acquisition problem. *Regional Science and Urban Economics*, **13**, 31–53.

Wyatt, B., D. Briggs and H. Mounsey (1988) CORINE: An information system on the state of the environment in the European Community. In H. Mounsey, *Building Databases for Global Science*. London: Taylor and Francis, pp. 378–396.

Yoeli, P. (1972) The logic of automated map lettering. *The Cartographic Journal*, **9**(2), 99–108.

Zimmermann, H.J. (1985) *Fuzzy Set Theory – and Its Applications*. Dordrecht: Kluwer.

Index